I0824133

THE IMPOSSIBLE FACTORY

ALSO BY JOSH DEAN

The Taking of K-129

Show Dog

THE IMPOSSIBLE FACTORY

THE REMARKABLE TRUE STORY OF
KELLY JOHNSON AND THE LOCKHEED SKUNK WORKS,
AMERICA'S INNOVATION MACHINE

JOSH DEAN

DUTTON

DUTTON
An imprint of Penguin Random House LLC
1745 Broadway, New York, NY 10019
penguinrandomhouse.com

Copyright © 2026 by Josh Dean
Penguin Random House values and supports copyright. Copyright fuels creativity, encourages diverse voices, promotes free speech, and creates a vibrant culture. Thank you for buying an authorized edition of this book and for complying with copyright laws by not reproducing, scanning, or distributing any part of it in any form without permission. You are supporting writers and allowing Penguin Random House to continue to publish books for every reader. Please note that no part of this book may be used or reproduced in any manner for the purpose of training artificial intelligence technologies or systems.

DUTTON and the D colophon are registered trademarks of Penguin Random House LLC.

Book design by Laura K. Corless

LIBRARY OF CONGRESS CATALOGING-IN-PUBLICATION DATA

Names: Dean, Josh author
Title: The impossible factory: the remarkable true story of Kelly Johnson and the Lockheed Skunk Works, America's innovation machine / Josh Dean.
Description: New York, NY: Dutton, [2026] | Includes bibliographical references and index.
Identifiers: LCCN 2025025606 (print) | LCCN 2025025607 (ebook) |
ISBN 9781524745516 hardcover | ISBN 9781524745523 ebook
Subjects: LCSH: Lockheed Aircraft Corporation.
Advanced Development Projects—History | Aircraft industry—United States—History |
Airplanes, Military—United States—Design and construction—History |
Johnson, Clarence L., 1910–1990 | Lockheed Aircraft Corporation—Employees
Classification: LCC TL568.L63 D43 2026 (print) | LCC TL568.L63 (ebook) |
DDC 629.1300973—dc23/eng/20260128
LC record available at https://lccn.loc.gov/2025025606
LC ebook record available at https://lccn.loc.gov/2025025607

Printed in the United States of America
2nd Printing

The authorized representative in the EU for product safety and compliance is Penguin Random House Ireland, Morrison Chambers, 32 Nassau Street, Dublin D02 YH68, Ireland, https://eu-contact.penguin.ie.

For three of the world's best humans:
Gill, Charlie, and Nicky

The Skunk Works is a concentration of a few good people solving problems far in advance—and at a fraction of the cost—of other groups . . . by applying the simplest, most straightforward methods possible to develop and produce new projects. All it really is is the application of common sense to some pretty tough problems.

—KELLY JOHNSON

Development of some of the country's most spectacular projects—the atom bomb, the Sidewinder missile, the nuclear-powered submarine—all were accomplished by methods other than the conventional way of doing business outside the system.

—KELLY JOHNSON

CONTENTS

PART III
A VERY SPECIAL NEED

PART IV
BLACKBIRD

CONTENTS

PART V
A LION IN WINTER

PROLOGUE

OUT OF THE SHADOWS

On October 18, 1982, *60 Minutes* correspondent Morley Safer introduced America to a burly man in a gray pin-striped suit, with rosy skin, a broad nose, and thinning white hair. Most of the tens of millions of people watching at home on their pleather couches had probably never heard of this man sitting in a chair on his sprawling California ranch, not far from where the president of the United States at that time, Ronald Reagan, also ranched. But this was one of the great engineering minds of the twentieth century—an innovator and machine builder on the level of Henry Ford.

And like Ford, this man hadn't just overseen evolutionary leaps in technology. His ideas and work would prove to be just as influential on management philosophy, and the ways in which smart people thrive inside an organization, as they were on innovation in his chosen field of aviation.

The man's name was Clarence Johnson, but pretty much everyone who knew him called him Kelly. "Kelly is a story about aviation and it's about a legendary character whose name would have been as well known as Charles Lindbergh, Amelia Earhart, even the Wright brothers if the nature of his work had not made it necessary for him to live a life in the shadows," is how Safer explained it, from a yellow chair in front of a large photo of the segment's subject. "Kelly is Kelly Johnson, designer of airplanes. He rarely talks to reporters, so the public knows little about him," Safer continued. "But for half a century he's been celebrated in the private world of aviation design, and this is why. His life is a virtual summary of modern aviation. Forty different airplanes began life on Kelly Johnson's drawing board."

Most of them were born inside the legendary institution he created—the Lockheed Skunk Works. What began under a literal circus tent in a Burbank parking lot (inside which America's first jet fighter was designed and built in a ridiculous 150 days) would grow into arguably the greatest innovation hub in aviation history: a thriving, rapidly iterating division that rewrote rules for design, for management, and for doing business; that trained many of the century's most audacious aerospace engineers; and that created both the fastest plane ever flown and the stealth fighter, which upended modern-day air defenses but looked like a flying paperweight.

The projects Kelly Johnson led at the Skunk Works, over all those decades, were largely top secret—classified at the highest possible levels. And so his life was spent mostly in the shadows. He did very few interviews over his half-century career, and despite possessing enormous self-confidence and being a large presence (literally and figuratively) in every room he entered, Kelly did not seek out fame or notoriety. "I learned a long time ago that you can't put your foot in your mouth if you keep your mouth shut," he told Safer.

By the time Kelly Johnson showed up on *60 Minutes*—easily the single most popular TV show in America at that time—he was seventy-two years old and in the final decade of a life that spanned nearly the entire history of airplanes. He'd fallen in love at his first glimpse of a plane flying over a rural Michigan airfield when he was a young boy, sketched his first plane design at twelve, and focused his life, singularly, around a very specific goal—of making beautiful machines to soar across the sky.

Over the course of a fourteen-minute segment, Safer ran through some of Johnson's greatest accomplishments—from the P-38 Lightning, an iconoclastic design that bedeviled Axis fighter pilots in World War II, to the legendary SR-71 Blackbird, an even more iconoclastic design that is arguably the greatest plane ever built—and asked Kelly what the world might expect to see next.

And there, way back in 1982, the creator of the world's first operational Mach 3 plane was already looking far past the horizon. He saw no need for a Mach 4 plane, because the future, Kelly said, was in unmanned aircraft. He envisioned technology disrupting his own industry in many new ways. There'd be little reason to fly around the world to Bahrain for a meeting when a person could just beam in via satellite "at the speed of light." Decades before either thing was widely adopted, Kelly was telling America that drones and videoconferences would eventually become commonplace.

Kelly Johnson was technically retired from his job running Lockheed's Skunk Works division by this point, Safer said, but he still consulted for his old shop, including on a new plane that was rumored to be in the works.

What was Kelly Johnson willing to say about this so-called stealth fighter, which was said to be invisible to radar?

It was a question Safer knew his subject wouldn't answer. The kind of question Kelly had been not answering for a half century.

Kelly just stared back at his interlocutor and smirked. "If I can talk about it," he said finally, "it's obsolete."

THE IMPOSSIBLE FACTORY

PART I

THE MAKING OF A MASTERMIND

1

THE AMERICAN DREAM

The story Kelly Johnson liked to tell about how his father got from rural Sweden to upstate Michigan is not 100 percent verifiable, but it goes something like this: In November 1888, twenty-four-year-old Per Jonsson—as he was known in Sweden—made his way south to Copenhagen, Denmark, and boarded a ship, bound for New York City, alone. His wife, Christine Anderson, stayed back to wait until her husband was settled, a particular kind of devotion required during difficult times.

His father's impetus to flee, Kelly would say, was not what you'd expect, given those facts. This wasn't the typical immigrant tale—a desperately poor or persecuted European man seeking a better life for his family in the land of opportunity. No, Per Jonsson was escaping Sweden's compulsory military service because he was a pacifist and didn't want to carry a gun. So he decided to pack up his life and start over in a place that offered basically unfettered freedom. Including the freedom *not* to bear arms.

On December 10, 1888, or thereabouts, Per Jonsson landed in New York, and that's probably where he—or, more likely, some immigration official—Americanized his name to Peter Johnson.

Peter had six hundred dollars, which was his life savings, and a dream of buying a ranch in the fertile promised land of Nebraska. But things went a little sideways for him in Chicago.

There, Peter fell for a scam, handing over a large chunk of his savings for a train ticket that he thought would take him to Nebraska, the land of plenty. Instead, Peter was put on a train heading north and ended up in the town of Marquette, on Michigan's Upper Peninsula, where the sight of Lake Superior—vast as an ocean, as viewed from the shore—surely would have revealed the harsh truth to him. That he was definitely not in Nebraska.

Nonetheless, Peter rolled with the punches. He parked himself there, in frigid, snowy far northern Michigan—still remote country, even today—and took the best work he could find. He laid railroad ties until he'd taught himself a more valuable trade: bricklayer.

As soon as he was stable—and had the funds to cover it—Peter sent for his wife, and once she too arrived from Sweden, he and Christine moved fifteen miles inland to Ishpeming, a small town settled because of its rich iron ore surface deposits and also known as the birthplace of "organized skiing" in America, thanks to the many Norwegians who settled there.

In 1893, Peter and Christine had their first child, a girl named Ida, and kept going from there. The couple had nine children in total.

Kelly, born on February 27, 1910, was the seventh of those kids, and the third son, but no one called him Kelly yet. He was then, and for the early years of his life, Clarence Leonard Johnson.

Peter Johnson was a good mason, and a tireless one; he could lay more than two thousand bricks a day, and that work ethic was one of many qualities that Clarence picked up from his father. Christine, meanwhile, took care of the kids and hand-washed laundry for other families as time permitted. Theirs was not a life of plenty. Money was always short.

Clarence was self-conscious about this; being poor bothered him.

He took back streets and alleys on his walk home from parades so that his friends wouldn't see where he lived, and young Clarence vowed to return one day with a chip on his shoulder—to walk proudly down Ishpeming's main streets as a man of real means.

The Johnsons pushed their children, and especially encouraged Clarence—who showed signs of being an intellectual standout from an early age—to take as much schooling as possible, education being the best way to escape poverty and make something in this vast and curious new country, where everyone seemed to have a hustle. Where you could show up looking for Nebraska and be sold train tickets to remote northern Michigan.

Anyway, the Johnsons didn't have to force Clarence to study. He would rush out the door to Ridge Street School, hoping to be first in line to enter and learn. This enthusiasm for education got him picked on, mostly by a rich kid named Cecil, who liked to push Clarence out of line and make fun of his name.

Young Clarence got sick of the abuse, especially the way that Cecil called him "Clara" in front of the other kids. And one day, Clarence just snapped.

The next time Cecil called him Clara, Clarence stuck his leg behind Cecil's—"kicked him behind the knee" is how he once described it—and then jumped on the bully. There was a loud crack, and Cecil's leg broke.

Cecil's parents were prominent folks, and they raised a fuss. They complained to school administrators and, despite the fact that he'd been bullied, Clarence was punished for fighting back with such overwhelming force. His teacher, Miss Hawes, smacked him so hard across the knuckles with a twelve-inch ruler that it snapped.

None of which deterred the boy. "I didn't care," Clarence later wrote. "I had accomplished my end." As in, he shut the bully up. And the fact that Clarence didn't cry when the teacher broke a ruler on his hand impressed his classmates even more. They told him that he needed a nickname—something stronger, and Irish (because he nearly always wore a green tie), to replace that awkward given name once and for all.

Someone suggested Kelly, from a song. Throughout his life, when Clarence Johnson told this story, he would say that the song was "Kelly with the Green Necktie," but that's wrong. The real title, it turns out, is "Has Anybody Here Seen Kelly?" an old English ditty that ended up in the 1910 Broadway musical *The Jolly Bachelors*.

Regardless, it stuck.

Kelly Johnson was an entrepreneurial kid and an adventurous one. Ishpeming was surrounded by wilderness, and he liked to build camps in the woods and sit on the bluffs outside town, watching iron ore trains travel to and from the mines. He did whatever work he could find, mostly odd jobs, and made thirty-one dollars one summer by picking fruit with his aunt at a nearby farm.

Kelly pocketed the cash, came home, and gave the entire sum to his mother. "She was so touched that I had not kept anything at all for myself," he wrote decades later. "No contribution I have ever made since has made me feel happier; none has been more important to me."

At ten, Kelly rode his horse, Mac, across town to his older brother Emil's house to learn how to lathe. Two years later, in 1922, he was earning ten dollars a week producing laths and paying seven dollars of that to his parents for room and board.

"From then on," he said, "I was self-supporting."

With eight siblings, there was always a kid, if not a few of them, around to play with. Once—the precise year is lost to history—Kelly nearly lost sight in one eye when his little sister Helen accidentally shot him in the face with an arrow during a game of cowboys and

Indians. The arrow barely missed his eye, but did some real damage. It took two weeks before Kelly's sight was normal again.

It was always a safe bet that if Kelly wasn't home, at school, or out camping, he was at the Carnegie Library—endowed by the steel mogul Andrew Carnegie, who owned Ishpeming's mine—with his dog Putzie, who liked to follow him into the building and inevitably got thrown out.

On one of those trips, Kelly picked up and devoured Francis Collins's *The Boys' Book of Model Aeroplanes*, which was written in 1910, just six years after the Wright brothers took flight. He also devoured the Rover Boys and Tom Swift book series. He especially loved the Swift stories, about planes and submarines and airships—"anything to do with mechanics"—and read entire volumes in a single sitting, becoming "convinced," he later said, "that I should be like Tom Swift."

Peter Johnson encouraged his son's curiosity. He built Kelly a workshop and let him use his tools, as long as he took care of them, which included making sure that every tool was back in its proper place. "If I dulled them or broke them, too bad, that was experience," Kelly later wrote. The only thing that would get him in trouble was losing a tool.

Working with his hands, turning ideas into objects—this thrilled Kelly. By the time he was twelve, Kelly had chosen his future. He wanted to build airplanes, and from that moment forward he prepared for it. This influenced every aspect of his life, from schooling to hobbies. He sketched planes, built models, and sometimes leaped from fence posts with a makeshift glider on his back—inevitably crashing right back down to earth.

Kelly's first design work was a book of collages and sketches titled *Aviation*. It contained mostly annotated clippings, but also his first-ever airplane design. He called it the Merlin 1 Battleplane—after the wizard Merlin from the King Arthur poems of Alfred, Lord

Tennyson—and though it bore a strong resemblance to an existing plane (the Curtiss JN-4 Jenny), Kelly referred to this sketch throughout his life, right up to the very end, as his first original design.

That same year, Kelly gave his first public speech, too—at a Lions Club luncheon. Kelly was still so small that he had to stand on a chair to be seen. He talked until they told him to stop and must have made quite an impression on the assembled, because the following Saturday, April 28, 1923, the local paper devoted two columns to the talk by this "bright little lad" who'd become "greatly interested in the navigation of the air."

It sounds, from the paper's report, as if Kelly did what kid scholars tend to do: He regurgitated the history of aviation that he had absorbed during his daylong stints at the library. Kelly told wild stories of early experiments in flight, such as the tale of a Frenchman who was the first human to go aloft in a balloon, in 1783.

The Frenchman, young Kelly told his audience, traveled about a mile and caused quite a stir when the balloon landed in a field. "The peasants who saw it, being ignorant of the experiment and naturally superstitious," wrote the reporter, "attacked it with scythes, pitchforks and other keen-edged and pointed agricultural implements that were at hand, believing it to be some monster with evil intent." (Kelly was nearly right: it was actually two Frenchmen, the Montgolfier brothers, who stood on a wooden platform attached to the bottom of a large silk-and-paper balloon, and they had traveled about five and a half miles. The terrified peasants part seems to be true, too, and it's rumored the pilots served them champagne, to ease nerves and keep the peasants from goring them to death with pitchforks.)

Kelly talked enthusiastically about all the progress in aviation's short history and predicted that the greatest strides were still to come, most likely in the United States. "The speaker thought that the US would finally outclass all other countries in the air because of its inventive genius and energy," the reporter wrote. "He predicted immense planes of several hundred feet spread, capable of carrying immense loads at great speeds. . . . Master Johnson's talk was heartily applauded."

This young boy, people thought, reflected very well on Ishpeming's schools.

When Kelly was thirteen, the Johnson family moved three hundred miles south, from the remote Upper Peninsula to the relatively urban environs of Flint, a thriving automotive town where there was better-paying work to be had.

Kelly attended Flint Central High, did lathing for money, and made model airplanes in his free time. At sixteen, he won second prize, and twenty-five dollars, in a Kiwanis Club model airplane contest for an updated, 3D version of his original sketch, now called the Merlin One.

He also went aloft himself for the first time—paying five dollars for an abbreviated ride in a four-passenger biplane. The plane took off and flew 700 feet before its engine cut out, forcing the pilot to make an emergency landing. Even so, Kelly was thrilled, summing the experience up like this: "It was fun! It was noisy, it was drafty, it was great!"

By the time he graduated from high school, Kelly had saved six hundred dollars. But he was temporarily burned out on education and planned to use that money to travel the world for six months. Until his French teacher, Miss Davis, talked him out of it. "What you should do instead," she said, "is enroll at Flint Junior College and keep building that big brain of yours."

Kelly took her advice and thought of a new use for those savings. He would learn to fly. It's what the industry's fathers—men like the Wright brothers, Glenn Curtiss, and Glenn Martin—had done. But

Jim Bishop, who ran the local flight school, refused to take Kelly's money for lessons.

"You don't want to start off on your career by giving me three hundred dollars to learn to fly," Bishop told him. "You have good grades, you will go a lot farther if you go on to the university. I won't take your money."

In 1929—on the eve of the stock market crash that would spin his country into the Great Depression—Kelly enrolled at Flint Junior College, with a focus on physics and math. He tutored other students in calculus and averaged ten to twelve dollars a weekend as a lather. That summer, he took a job at the local Buick factory, "swinging fenders on the production line or working on motor repair and block test." He also set out to read and understand the writings of Albert Einstein.

"Only twelve people in the whole world were supposed to be able to do so," he later wrote. "I wanted to be the thirteenth!"

Kelly developed his first ulcer in Flint, too. Ulcers, and gastro problems, would bedevil him for his entire life. Much later, doctors would periodically force him to take vacations, simply to allow his stress levels to drop, and this tendency to overwork was likely the trigger in Flint, too.

Kelly Johnson wasn't just smart. He was also a bear of a man, powerful and athletic. He excelled in all the sports he tried, especially football, and after starring at Flint Junior College, Kelly was offered a scholarship to play for a major college in the South whose name remains a mystery.

Wherever it was, Kelly went there during the summer break after he finished at Flint, to practice and get to know the team and school. He planned to enroll in aeronautical engineering classes, to embark upon the technical education that would underpin his career while playing a sport for fun. Then the football coach handed him a class schedule.

Your studies, the coach told him, *will be in physical education. Because football comes first.*

That was all Kelly needed to hear. He packed up his Ford Model T and headed north, to attend school in his home state, at the University of Michigan, which had also offered him a football scholarship.

Just one problem: Cars weren't allowed on campus. So Kelly set out to drive the Model T back to Flint. Along the way, a large Pontiac clipped his car's front hubcap while passing on a gravel road, causing Kelly to veer out of the way and into a ditch. He wasn't seriously injured, just a cut forehead. But that gash got infected, which made it impossible for Kelly to attend football tryouts.

This was painful but fortuitous. Fate, as it sometimes does, intervened. Without football, Kelly could turn his full attention to his studies. He would later say, in fact, that crashing into a ditch and cutting his forehead "was the best thing that ever happened to me."

Kelly thrived in Michigan's famous aeronautical engineering program—Michigan and MIT being the country's top two schools in the still-nascent field of aircraft design. Michigan had several star professors, including the program's founder, a Polish émigré and amateur philosopher named Felix Pawlowski, who had previously worked with the renowned inventors Igor Sikorsky (on the world's first four-engine plane) and Gustave Eiffel, designer of the Eiffel Tower.

Pawlowski had built the wind tunnel at Michigan and Kelly worked with him on the design of a Union Pacific train, a project to help clean smoke from the air over Chicago, and on a novel concept for generating energy using a wind machine. One of Pawlowski's most unconventional lessons was delivered to Kelly at a bank vault, where the professor kept a set of creepy wax "spirit" hands from a séance that were, as Kelly later wrote in his slim autobiography, "entwined in a manner that could not be explained."

The point of this field trip to a bank vault containing only some creepy wax hands? This esteemed scientist was teaching his star student a valuable lesson: "Don't automatically write anything off," Pawlowski told him. Or, in other words: *Consider all possibilities, and keep an open mind.*

Kelly also worked for another famous professor, Edward A. Stalker, in the university's wind tunnel, a paying job that allowed him to quit his less glamorous gig, washing dishes at a fraternity house. There Kelly saw opportunity. He asked Stalker if he could rent the wind tunnel when it wasn't in use and then, along with his close friend and fellow engineering student Don Palmer, began to take on freelance projects.

One of their first clients was Studebaker, a car company that was interested in streamlining its Pierce Silver Arrow sedan. Kelly and Don studied the car for aerodynamic weak points and calculated that the "big ugly headlamps . . . were eating up 16 percent of the power the engine developed at 65 miles-per-hour." To fix this, the students suggested that the lamps be "shaped into the fenders."

Despite being a work in progress, Kelly did not lack confidence in his own abilities. He had great respect for his professors but was not afraid to push back—or, as he later put it, "not so deferential that I would not argue back if I disagreed." For instance, when Pawlowski gave him a B on the results he calculated for wind tunnel tests on a "little biplane," Kelly simply refused to accept that subpar grade.

He argued that his numbers were correct, and it worked. Pawlowski changed his grade to an A. "He kept an open mind, as he advised," Kelly wrote.

Kelly's professors taught engineering from all angles. Despite being an aeronautical student, he was put in charge of a team of mechanical engineers and asked to study the energy balance of the university's steam power plant. He learned to apply scientific methods to the machining of metals from Professor O. W. Boston, an expert in the field and, according to Kelly, the first person to "imbed thermocouples on a lathe or milling machine to measure tool temperatures." And the Timoshenko brothers, from Russia, taught him about structure and vibration, which would later prove critical in designing airplane wings and tails.

Kelly devoured it all. He finished three years of study in two years, impressing his professors and, according to him, earning extra cash by beating them at poker. Kelly also tutored students in calculus—for, he later claimed, a shocking $7.50 per hour.

The one downside of obsessive schooling: Kelly had little if any social life—a total of two dates over four years, some card games, and sports, as time allowed, especially tennis. He often played Willis Hawkins, another top engineering student, who became a good friend (and, later, a close co-worker).

According to Hawkins, Kelly played tennis the same way he attacked problems—unconventionally (or, as Hawkins put it, "different from the rest of the world"). His serve, in particular, was confounding. Kelly served backhand and, Hawkins later said, "If you have ever tried to receive one of those, you'll find that it is very difficult." So difficult for Hawkins that he just gave up. He decided to be Kelly's doubles partner instead.

These were the days of Prohibition, but students are students, and they made up for the lack of retail alcohol by making their own. Kelly didn't approve, or drink, because he'd heard the "stories of my hard-drinking Swedish ancestry."

Also: Stress, and poor diet, exacerbated the ulcers he'd first gotten in high school. At Michigan, the problem was incessant and Kelly self-treated by trying to keep his stomach filled. His go-to meal was cheap and unhealthy: two doughnuts and a glass of milk, which cost him twenty cents. One semester, he ate 647 doughnuts, which we know because he kept track. This helped the ulcers but inflated his belly.

Another of Kelly's future engineers, Carl Haddon, had been an electrical engineering major at Michigan State when Charles Lindbergh flew across the Atlantic in 1927. He decided to shift his focus to airplanes and transferred to Michigan.

Haddon was a year ahead of Kelly, but the younger student was already casting a long shadow. The two first met in an engine design class. Kelly was, Haddon later recalled, already "recognized as a genius." That term's project was to design an airplane engine. "Most of us just copied engines that were already on the market or that we found a blue print of," but not Kelly. He designed his from scratch, while working covertly. According to Haddon, "through a little square on a piece of paper that he covered his drafting board with because he didn't want any of us to see. He was secretive even then."

The job market was a mess when Kelly Johnson graduated. It was 1932, and America was in the throes of the Great Depression. But Kelly tried to be proactive. He and Don Palmer drove to the East Coast in a Chevrolet they borrowed from a professor and cruised around, knocking on doors at aircraft manufacturers like Sikorsky, Martin, and Curtiss in the hopes that one of them might be hiring.

A bold effort, but not a productive one. The two engineers got no job offers, so Kelly adjusted his plans. Rather than fumble around looking for work in a terrible market, he'd learn to fly, by enlisting in the Army Air Corps. That plan lasted as long as the physical exam. Poor vision in Kelly's left eye—lingering damage from when his sister accidentally shot him in the face with the arrow—disqualified him from flying.

This, Kelly would later admit, was some of the best luck of his life, on par with the car crash that caused him to skip playing football.

Instead, Kelly and Don set off again in the borrowed Chevrolet, this time heading west. The two young men were broke. They survived on cheap homemade sandwiches and camped out in fields, in schoolyards, and alongside streams. To lower their gasoline costs, they modified the professor's car—drilling a small hole in the intake manifold so that they could insert a valve and open it on demand while cruising, which allowed more air to enter and increased fuel economy by three to four miles per gallon.

Their destination, ultimately, was Burbank, home base of a new and promising aviation company known as the Lockheed Corporation.

2

THE RISE OF LOCKHEED

Allan Loughead was fourteen on December 17, 1903, when news of what the Wright brothers had just done on a sand dune in North Carolina reached him in California's Bay Area. The idea that humans had harnessed flight blew Loughead's mind and dictated the course of his life. This magical thing—flying!—was what he needed to do.

Nearly seven years later, in 1910, Allan finally got airborne himself, in a motorized box kite. And just two years after that, the company that would become Lockheed was born when Allan and his brother Malcolm built a V-8-powered two-wing seaplane on a ramp along San Francisco Bay.

The brothers called their plane, which had seats for a pilot and two passengers, the Loughead Seaplane Model G. And on June 15, 1913, Allan disregarded "the jeers of the skeptics" who'd gathered to watch the Lougheads try—and likely fail—to fly what was then the largest seaplane ever built. Allan hopped into the cockpit, taxied out into San Francisco Bay on the plane's giant pontoons, and took off.

Next, the Lougheads took their flying machine on the road, charging five dollars a head to take people aloft. They raised three thousand dollars and used that as seed capital to found the Loughead

Aircraft Manufacturing Company in a Santa Barbara garage. The brothers began work on a new design—a twin-engine flying boat that could carry ten people—with design assistance from a brilliant teenager named Jack Northrop.

Northrop—who would grow up to become a legend in aviation himself—did the design's three-view and detail drawings, as well as the stress analysis, which was then practically an unknown art for airplanes.

The Lougheads' new craft had twin booms, a triple tail, and a 150-horsepower engine. They called it the F-1.

The future seemed bright. In 1918, the U.S. government put in an order for fifty scout seaplanes for the war in Europe, but peace was reached before a single one was built. So the Lougheads pivoted back to tourism; they took clients on flights over the ocean—ten dollars for twelve miles, or twenty dollars for thirty miles. The brothers' planes were an attraction. In 1919, they flew King Albert and Queen Elizabeth of Belgium during a visit.

Meanwhile, in the shop, Jack Northrop and the company engineers pushed innovation. Their two-cylinder light sport plane, the S-1, had folding wings and air brakes, which were inspired by . . . seagulls. Engineers plopped some dead fish on a beach and observed how gulls slowed down during dives, in preparation for landing. So, credit seagulls for airplane flaps.

The S-1 was a bust, unfortunately. There just wasn't a market for it. And with the company struggling, Malcolm left to start a different outfit to manufacture another of his inventions, in a different nascent industry—making hydraulic brakes for cars.

But Allan kept at it, and by 1926 he was tired of people butchering his last name. He rebranded his company as the Lockheed Aircraft Corporation and moved to a new shop in Hollywood. A year later, he unveiled a remarkable new design—a high-wing monoplane with a 41-foot span and a five-cylinder engine. He called it the Lockheed Vega.

And on July 4, 1927, Edward Bellande—an accomplished barnstormer, crop duster, and test pilot—flew the Vega for the first time, over an oceanside farm that is now the Los Angeles International Airport. "Boy, she's a dandy," he cooed upon landing.

Airplanes were no longer some novelty; they were now a cultural phenomenon. Charles Lindbergh crossed the Atlantic. Admiral Byrd had his sights set on flying to the South Pole. And pineapple mogul James D. Dole put up twenty-five thousand dollars for the first person to reach Honolulu by plane from the continental United States. (Second place would get ten thousand dollars.)

A total of fourteen planes were chosen for the field, out of thirty-three entries, but just eight made it to the starting line in Oakland. Several crashed en route to the race, including a triplane whose crew included Hoot Gibson, star of cowboy movies, which missed the runway when approaching the field and plunged into the ocean. Hoot—and the rest of crew—survived but had to swim back to shore in the frigid bay waters.

On August 16, thousands of fans standing atop buildings and crowded onto boats in the harbor watched as the first entry—a blue-and-yellow monoplane named the *Oklahoma*—took off and crashed. The second plane also took off and crashed. So did the third. The fourth plane, *Golden Eagle*, was a Lockheed Vega owned by the son of newspaperman William Randolph Hearst. Jack Frost (yes, his real name) was at the controls, and he, fortunately, did not crash on take-off. He left the field, flew over the horizon, and, unfortunately, vanished. The *Golden Eagle* didn't reach Hawaii and was never seen again.

The winner of this cursed race? Colonel Arthur Goebel, a World War I fighter ace turned Hollywood stunt flyer who arrived in Honolulu in his Travel Air monoplane 27 hours, 17 minutes, and 33 seconds after leaving Oakland. Goebel loved flying, had a big personality, and was one of aviation's greatest evangelists. "I am an aviation preacher," he declared, "and I am going to broadcast the possibilities of the air to everyone I meet."

In August 1928, Goebel was back in the news, this time in a Lockheed Vega 5 monoplane named the *Yankee Doodle*, as he set a cross-country speed record, flying west to east across America from Mines Field in Los Angeles to Curtiss Field, on Long Island, in 18 hours, 51 minutes.

Goebel's feat was an excellent advertisement for Lockheed planes, and Allan's business boomed in the wake of the publicity. Soon he needed more space, so he moved again, to Burbank, in the San Fernando Valley—specifically, to the corner of Empire Avenue and Victory Place, where Lockheed would establish its base and build airplanes for more than fifty years.

Lockheed went public in 1928, and in 1929, the Detroit Aircraft Corporation (DAC) assumed control in a hostile takeover. Numerous players were vying to build the Ford Motors of the air, including DAC, which almost immediately ran into trouble in the form of the Wall Street crash and the ensuing Great Depression.

By 1931, Lockheed was the only part of DAC still functioning. On October 27, 1931, it was placed into receivership. And four months later, in February 1932, the receiver shut it down.

That May, an ad was placed in the *Los Angeles Times* advertising that Lockheed's assets were up for sale in a bankruptcy, valued at $130,000. Interested buyers were to submit sealed bids that would be opened by a U.S. District Court judge on June 6. Highest bidder wins.

A man named Robert Gross, co-founder of the Viking Flying Boat Company, showed up with a design for a plane with a metal fuselage, as well as a kind of miracle—a consortium of investors willing to invest a large sum in a new and speculative industry, even in the midst of the Great Depression.

Gross adored planes, and he found partners by appealing to emotion, by seeking out other people who, like him, loved the idea of flight: Walter T. Varney, whose Varney Speed Lines airline flew Lockheed planes; Carl B. Squier, a World War I pilot and a Lockheed plant manager for Detroit Aircraft; the investor Randolph C. Walker; Walker's cousin, the ex–Army pilot Cyril Chappellet; and Lloyd Stearman, an airplane designer who had previously worked with Gross and Varney at another aircraft manufacturer, Stearman-Varney, Inc.

So on June 21, 1932, the Lockheed Aircraft Corporation was reborn, from the ashes of Allan Loughead's original company and on the site of its original Burbank plant—inside a factory that was originally built to manufacture pottery. According to a 1939 story in *Time* magazine, the grass around the factory was overgrown "and the factory had one employee—a watchman who had started working" for the Loughead brothers "and saw no reason to quit because he was not paid."

Lloyd Stearman was named president and CEO, with Carl B. Squier as vice president, Gross as treasurer, and Cyril Chappellet as secretary.

Balance sheets in those early days were not encouraging. The company survived mostly on spare parts and service. The first month, Lockheed reported just $1,622.94 in sales, including two checks from a loyal customer by the name of Amelia Earhart. One was for $150, the other for $6.

Money was so tight at times that there wasn't enough in the treasury to buy a single "calculating machine" for the engineering department, which also explains why chairs were also at a premium, and some workers sat instead on wooden boxes acquired from the local grocery store.

Everyone multitasked. One of quality-control specialist Harvey Christen's jobs, for instance, was to make lumberyard runs in a Model T truck so light that a load of lumber would cause the front end to come off the ground. To compensate, Christen brought the heaviest worker he could find to ride shotgun and provide a counterweight for the truck's nose.

Robert Gross's company inherited a few key employees who would play a huge role in the company's development, including Christen and lead test pilot—the only test pilot—Marshall Headle, who helped ensure quality production on the line by reminding the assembly crew of what they were building: an airplane that he himself would be flying.

Lockheed's main plane at the time was the Vega, with a fuselage made from Sitka spruce, and Headle took each new Vega off the line for a check flight to make sure it was flight-worthy.

Prior to doing this, however, he'd stroll through the assembly hangar and choose one lucky worker at random—a fitter, welder, or woodworker—and tap him on the shoulder.

You, Headle would tell the man, *have been selected to fly with me. Put on a parachute and let's go.*

This almost always terrified the craftsman or line worker, but he couldn't exactly tell the test pilot that he was too scared to fly in the plane he'd helped build.

"You can be sure," Christen later said, "that an aircraft worker would be mighty careful in his craft after an incident like that."

The factory mascot was a bulldog named Contact, who slept wherever he felt like and was almost impossible to move once he'd fallen asleep. If Contact was in your space, an electrician told the company's internal newspaper, *The Lockheed Star*, in 1932, "business was conducted elsewhere."

The drooly dog was also a good luck charm. Flight line mechanics referred to Contact as the boss and would insist that he be on hand for the first flight of every plane. Contact was also a free-range animal, welcome to wander across Victory Boulevard to the local sandwich shop, where the owners would feed him whatever was on hand and put it on his tab, paid by Lockheed.

Lockheed's design department was led by a young engineer named Hall Hibbard, who was born in 1903, the same year the Wright brothers launched themselves off a dune in North Carolina and proved that human flight was possible. Hibbard said that he knew he would fly, and be around planes, from the first time he saw one flying overhead himself as the son of a Presbyterian minister who'd moved his family to the Philippines to spread the word.

He ended up at MIT—Michigan's rival for aeronautical engineering supremacy—and was hired after graduation by the Stearman Aircraft Corporation for a job in Wichita, which was then calling itself the "air capital of the world." Hibbard's job was to do stress analysis on Stearman's biplane, built of stick and wire, with fabric for wings.

When Lloyd Stearman joined Bob Gross to start the Stearman-Varney Company, in Oakland, he brought his young engineer along, and Hibbard was immediately put to work on a twin-engine plane with a metal fuselage that would become one of the most important airplanes in Lockheed's early history: the Model 10 Electra.

Hibbard, then, was part of the core team that took over Lockheed, when Gross shut down Stearman-Varney in Oakland and moved the gang to Los Angeles. "We bundled up everything we had and came to Burbank," Hibbard later said, to pick up right where they'd left off.

This was the Lockheed that Kelly Johnson and Don Palmer found in the summer of 1932, in an old brick factory along Victory Avenue in Burbank, which was then mostly farmland in the foothills of the San Gabriel Mountains. It was a scrappy start-up with a small staff, and the company's production manager, Richard von Hake, told the two young men that it wasn't *quite* time to hire any new engineers.

This was only temporary, he told them. There would be a need, soon enough, for fresh blood. In the meantime, von Hake advised, Kelly and Don should go back to Michigan, further their technical educations, and come back in a year.

"I think we'll have something for you," he told them.

Kelly took that advice to heart. He went back to Ann Arbor and picked up right where he'd left off, compiling A's, tackling ambitious jobs, and occasionally ruffling feathers. He pursued a master's with a focus on supercharging engines—"to get high power at high altitude"—and boundary layer control.

He and Palmer also picked up some more design work for Studebaker. This time, they tackled a streamlined version of the Pierce-Arrow called the Silver Arrow, which was eventually put into production. They also modified five semi-stock Studebakers for the 1933 Indianapolis 500 field. All five qualified for the race, at speeds ranging from 110 to 116 miles per hour, and had dramatically better fuel economy than the other cars in the field. The Silver Arrows' efficiency jumped from 7 to 11.6 miles per gallon at 113 mph, which meant fewer pit stops for the racers.

But the bigger job, the one that arguably changed the course of Kelly's life—or at least steepened the trajectory of his rise—came from none other than Lockheed.

Robert Gross had a bold vision for the company he'd rescued. He was also in fairly desperate straits, trying to build an airplane company during the Great Depression. On March 3, 1932, Robert wrote to his brother and business partner, Courtlandt, that the situation was dire: "[I have] a few cents over five dollars in my pocket and we have a twenty dollar piece in the vault." And yet Robert was still seeing a half-full glass. "Strangely enough, our business here is looking up," he wrote. Depression be damned, customers were still buying Lockheed planes.

What Gross really needed was a hit, something new and innovative to pry open doors that were currently closed to him. He needed a plane so good that the airlines wouldn't be able to ignore it. And that plane—with two engines and a metal fuselage—was coming.

He called it the Electra.

The idea came to Gross in the fall of 1932 as he sat in the Union Air Terminal at Burbank Airport, dunking a doughnut in his coffee and watching passengers board their planes. Some went to a Lockheed Orion, powered by a single engine, while others climbed aboard a twin-engine Boeing.

Gross asked himself: Which of those planes would he rather fly, if given a choice? The answer was obvious. The one with two engines, because that was just safer.

Lockheed was already experimenting with a two-engine design, along with a new single-engine plane, and when Gross returned to his office, he told Hall Hibbard—who had just replaced Lloyd Stearman as chief engineer—to drop the latter project and focus entirely on the twin-engine concept.

The Electra was, in almost every way, Gross's baby. The bossman was at Hibbard's side as he did the initial sketches at a drawing board in the little farmhouse that housed the engineering office, telling his engineer where the lines ought to go. (Hibbard claimed to not mind the wingman. Gross just had "a wonderful intuitive sense," he later said.) Or at least that's what Gross was doing when he wasn't running around L.A. selling Lockheed shares to pay for the company's overhead.

The company's coffers were constantly at risk of running empty. In March 1933, Robert wrote to Courtlandt that the company had "roughly $28,000 in cash," plus "another $6,000 due on Doolittle's plane"—as in legendary aviator Jimmy Doolittle—which would be paid upon delivery in a few weeks. "This of course is no sum with which to build a business in the face of panic, but we are a lot better off than we might be."

Gross worried that he might require a fifty-thousand-dollar loan to build the all-important Electra. More troubling, the Electra had problems. The plane, on which the company's future depended, was unstable under certain conditions. The design, without changes, just wouldn't be safe to fly.

So Gross and Lockheed president Lloyd Stearman called the best person they could think of for the job: Professor Stalker at the University of Michigan. Stalker, in turn, asked his trusted assistant, Kelly Johnson, to help.

Wind tunnel testing of scale models was not the norm in airplane design at this point. Other manufacturers didn't use it at all. But it was important to Hall Hibbard, who'd been trained on model testing at MIT. And critical here, if they were going to solve the Electra's stability issues.

Stalker and Stearman ran some wind tests and signed a report stating that the plane was okay as is. It wasn't that they failed to notice the plane's bad stability curve; it was that nearly all airplanes on the market had this same issue.

Kelly Johnson, the graduate assistant, wasn't so sure. He had been studying stability—specifically, elevator effectiveness for pitch control—as part of his graduate studies. He had, as luck would have it, studied the Electra while doing that work. Kelly imagined bad things happening to this airplane. He worried about instability on both axes. But he was just a student assistant.

He wasn't in a place to disagree. Yet.

In May 1933, Kelly Johnson finished his master's and used some savings to buy a used Chevrolet sedan. Then he and Don Palmer set off for California—and Lockheed—for the second time. As promised, Kelly got a job, as an eighty-three-dollars-per-month tool designer, and was placed under the wing of Hall Hibbard. Palmer landed at the Vultee Aircraft Company, in nearby Glendale.

Lockheed's engineering department had just five people at this point, and Hibbard, its thirty-year-old leader, was a phenom himself. The company was in search of new ideas, which often came from young hires who weren't yet colored by tradition. Kelly Johnson fit right in.

"He looked so young, I was almost afraid he couldn't read or write," Hibbard later said. But he liked that Kelly was also an athlete, perfect for the company's after-hours sporting events.

Kelly's job was technically tool designer, but that didn't even begin to cover the actuals of his day-to-day. Lockheed was a scrappy place that worked only because its hires were Swiss Army knives. "In those days, you couldn't specialize," Kelly later said. "You had to learn everything. . . . It was the best thing that ever happened to me." Kelly designed tools, sure, but also worked on aerodynamics, flight testing, stress and weight analysis, and wind tunnel testing.

The company's focus at that point was entirely on the Electra, and pre-sales of the new plane continued to drive growth. By November, the company had 308 men, and a prototype Electra "in a three-quarter state of completion," according to Gross.

Gross predicted a first flight as soon as December.

Kelly, remember, had doubted that the Electra was safe to fly back when Professor Stalker had run the first set of wind tunnel tests. And one of the first things Kelly did upon arriving in Burbank was to tell his bosses as much.

"When I announced that the new airplane, the first designed by the reorganized company and the one on which its hopes for the future were based, was not a good design, actually was unstable, Chappellet and Hibbard were somewhat shaken," Kelly later recounted.

"It's not the conventional way for a young engineer to begin employment."

Unconventional, but a sign of things to come.

Specifically, Kelly noted, the Electra had a "kink in the stability curve that won't quit—it's going to pitch up and have real troubles." The plane, in his opinion, was unstable in all directions.

Hibbard was taken aback by the young man's gall. But something in his new engineer's tone convinced him to heed the warning. He told Kelly to go back to Michigan and work out a solution. Kelly drove east again, with a wooden wind tunnel model of the Electra prototype in his backseat. And using Stalker's *Principles of Flight* textbook as one of his guides, he went to work.

Day after day, Kelly ran tests on the model, making various modifications to stabilize the plane. Finally, on the seventy-second test, he had a breakthrough. Kelly added "controllable plates on the horizontal tail" to increase stability, removed the wing fillets, and then decided that the Electra needed twin tails because a single rudder wasn't enough to control the plane if one engine were to fail.

That design solution—twin tails—would become Lockheed's signature.

Shortly after Kelly's twin-tail design was tested and proven, Hall Hibbard fired off a letter to his protégé, written late one night on a factory typewriter. "You may be sure that there was a big celebration around these parts when we got your wires telling about the new find and how simple the solution really was," Hibbard wrote. "It is apparently a rather important discovery and I think it is a fine thing that I would never have guessed at . . . and it is very surprising to find out how much good is actually done by these tip sections."

Kelly returned to Burbank a company hero and was immediately promoted to engineer, just the sixth in Hibbard's department, then housed in a rickety shack with a leaky roof.

By the time he was twenty-five, Kelly had spent as much time in the wind tunnel as anyone at Lockheed—probably as much as anyone in the industry. He also insisted on participating in the flight tests, as engineer. Doing this helped him to intimately, intuitively understand the way that air flows over the wings, around the fuselage, through the engine, and out the exhaust. This internal airflow, he liked to say, is "just as exciting and important as the flow over the outside."

Hall Hibbard would famously say that Kelly Johnson could "literally see air," and this is what he meant. The young engineer could foresee many problems just by looking at a design, before testing, and spotting potential issues.

Kelly also considered it his duty to truly stand behind his work. He believed that those who design planes, who choose their shapes and materials, should absolutely fly in them. "I figured I needed to have the hell scared out of me once a year in order to keep a proper balance and viewpoint on designing new aircraft," he later wrote.

From the outset, Kelly flew alongside the company's test pilots. He did that on the Electra and also on experimental versions of the many planes he modified for star aviators like Wiley Post, Laura Ingalls, Amelia Earhart, and Sir Charles Kingsford Smith. He was on the first flights of the Model 12, 14, Lodestar, Hudson, XC-35, and the Constellation. "If a plane held anybody besides the pilot, you could bet that the man would be Kelly," one test pilot said.

On February 23, 1934, test pilot Eddie Allen was ready to take the Model 10 Electra on its first-ever flight. He did that one solo, but Kelly Johnson was riding shotgun on every subsequent flight in the test program. He was there, in the right seat, for dive tests, stall tests, and spin tests.

During one particularly hairy test, Kelly asked Allen to push the plane to its maximum dive speed, around 320 mph, to make sure it was free of flutter and control problems. The plane had been stuffed with lead bars to simulate the weight of a full passenger load, and Allen had it flying at about 12,000 feet over Burbank when—at Kelly's orders—he initiated a "steep, screaming, power-on dive," which is about as terrifying a maneuver as you might imagine: full throttle with the nose pointed at the ground.

Halfway to the ground, at around 6,000 feet, Kelly heard a horrendous bang and looked over to see Allen scrambling to clear debris from his face as he yanked back on the stick to pull out of the high-speed dive. The windshield had blown in, and scraps of insulation covered Allen's face. He landed safely and Kelly immediately redesigned the windshield.

When the Civil Aeronautics Authority (a precursor of today's FAA) finally gave the Electra its airworthiness certificate after a successful test at Mines Field (now Los Angeles International Airport), Kelly and Hall Hibbard drove home in the latter's bright red Buick LaSalle cabriolet smoking cigars, and test pilot Marshall Headle flew the plane back to Burbank with a CAA inspector onboard.

Hibbard and Kelly got to Burbank just in time to see Headle's landing, and it did not go as planned. As the pilot descended toward the field, his left-side landing gear got stuck. There was no radio communication with the plane at that time, so the ground personnel began frantically waving at Headle, who apparently got the message, because he aborted the landing, climbed, and circled around.

Disaster averted, temporarily. But they still needed to get the message to Headle, who was only aware that he had a problem. He would have had no way to know what the problem was or what to do about it.

Inside the factory, all hands scrambled for a solution as Jimmy Doolittle—a factory regular—suggested a Hail Mary of sorts. Doolittle volunteered to take off and fly past Headle with a message written on his plane. Hall Hibbard told some workers to scrawl the words "Try landing at United—good luck" on the side of Doolittle's plane's black fuselage, United being the name of a nearby airport with a much longer runway, as well as firefighters on standby.

Headle did just that and was fine.

But the failure of the plane's retractable landing gear was poorly timed, right after certification and just before it was scheduled for delivery to Northwest Airlines. The company was still young and far less stable than the competitors. Lockheed's future—its survival—essentially depended on winning this contract.

Engineers scrambled to fix and redesign the mechanism that retracted the gear, but the setback was huge. Gross had no choice but to lay off nearly everyone, and even those who stayed weren't paid for several weeks while Kelly and the engineering staff worked without pay to fix the problem and save the company.

Failure at this point just wasn't an option. Lockheed didn't have the war chest to survive disaster. So the company's future, again, was in doubt.

Gross's boldness, and his team's comfort with risk, saved the company.

By July, Lockheed delivered the first Electra to Northwest Airlines, for fifty thousand dollars. To celebrate, Lockheed threw a party at Neil's Drugstore and, as Robert Gross played piano, Kelly had the first drink of his life, a Snug Harbor bourbon. It tasted, he later said, "more like dregs from the bay."

Within a year, Lockheed had introduced a second, smaller version of the plane, known as the Model 12 Electra. This six-passenger plane cost $40,000 and was also a hit. Sales of the two planes took off, and Lockheed turned a net loss of $190,891.39 in 1934 into a net profit of $217,986.37 in 1935. That September, the Polish airline LOT purchased four Electra 10s, making Lockheed's first international sale.

The Lockheed Aircraft Corporation was really flying.

3

YOU GOTTA START SOMEWHERE

In those fast and loose early days of Lockheed, everyone multi-tasked. Hall Hibbard counted nuts and bolts and measured tubing. Test pilot Marshall Headle worked in the employment office and hired with an eye on which newbies would also be valuable to the company's baseball team, which played heated practice games during the lunch hour, and for which Kelly Johnson excelled, as a utility man who played every position—but especially left field—and hit for average and power. Legend has it, he struck out only one time.

Employees rented rooms nearby, almost without fail. The newest hires tended to live, at first, in a boardinghouse and just walked to work, which was helpful for late nights and for saving money on transportation.

The confident, precocious Kelly Johnson flourished at this small, nimble operation, where everyone contributed to design. His natural inclination was to drive extremely hard to get a job done, to push himself and others to solve problems at all costs. Kelly was friendly until someone disappointed him, which didn't always go over well in the shop. Hall Hibbard helped his protégé temper this trait, without squelching the fire. He taught Kelly that brute force had a place but didn't always work. And as useful as sticks can be, it was also important to know when to use a carrot—to learn to persuade people, and lead them to the outcome you desire.

Kelly even found himself a personal life, finally, in the form of a "tall, good-looking young woman"—his own words—from Wisconsin named Althea Young, who worked as assistant treasurer in the corporate office, which occupied a two-story brick, ranch-style house. The company's telephone operator, one of the three other women who worked at the company, got Kelly thinking about Althea, who had apparently referred to the brash, broad-shouldered engineer as a "snippy young kid."

Kelly took this as a challenge and asked Althea out.

"Vera told me you were a brain," Young apparently said, and accepted. They went for a steak dinner and split the dollar-fifty bill.

The romance bloomed from there. Lockheed's star engineer and the assistant treasurer would rent horses and ride into the canyons above Burbank once or twice a week, and also golfed together. Kelly even got his new girlfriend onto a test flight of the new Electra, which wasn't technically allowed. So he snuck Althea onboard the prototype, which had no passenger seats, so she sat on bare metal, at least pretended to enjoy herself, and made it back to base in one piece.

Althea was smart, athletic, and independent. She often beat Kelly at golf and insisted on going Dutch at dinner until he was promoted and making more money. Only then did she let him start picking up the checks.

Althea was up to the challenge of dating a man whose personality—even in his twenties—occupied most of the space in a room. She was comfortable with him, confident enough to quell his ego, and up for a good time, as evidenced by Robert Gross's memory of the first time he met the couple in person, at a Lockheed picnic—shortly after they'd begun dating and around the time Kelly was impressing his bosses by fixing the Electra.

The young couple came "tearing down the road in a stripped-down Ford and bounced right into the middle of the picnic grounds in a cloud of dust, a swirl of hard-boiled eggs and sandwiches, and a barrage of bathing suits."

Kelly and Althea were fully and madly in love but didn't rush things, which wasn't exactly a typical pattern in the 1930s. The two dated for *four years* before marrying, finally, in 1937, in a church on Wilshire Boulevard, even though neither attended services regularly.

They honeymooned at Yosemite National Park and settled into a "fine old" rental house in the foothills above Burbank, which at that time was the edge of a wilderness.

Althea quit her job and became a housewife, never an easy thing with a husband like Kelly, whose first love was always going to be his career. Weekends, though, were for leisure, and every Sunday Kelly and Althea rode horses at a local stable.

When that place went under, they bought their favorite horses, put them at a boarding stable, and rode as much as possible in the canyons of Malibu. "Althea shared my love for the outdoors," Kelly wrote. She also shared his dream—to have a ranch of their own.

There was never a dull day at Lockheed in that era. Even before Robert Gross and his partners bought Lockheed out of bankruptcy and remade it into one of this nascent industry's most innovative companies, this was a brand beloved by flyers. Nearly every famous aviator flew a wooden Lockheed plane, and the company's modest Burbank headquarters was a kind of home base for the era's great, often eccentric flyers.

Perhaps the most colorful of them all was Roscoe Turner, a dapper World War I pilot who became a barnstormer, stunt pilot, and chaser of records. On chilly days, Turner would show up to the factory in Burbank wearing a bowler hat and full-length fur coat over his tailored suit and jodhpurs, but his most whimsical flourish, by a long shot, was Gilmore, the adopted lion that flew as his copilot.

Turner got Gilmore as a cub, and the young lion became a regular in the Lockheed hangar where Turner kept his plane. When Gilmore grew too big to roam free—when such freedom became a safety risk to humans in his vicinity—Turner walked Gilmore around on a chain, but when he annoyed the mechanics or threatened to eat the company mascot, Gilmore would be banished to a cage in the engineering building.

Still, the lion sometimes escaped. Kelly was eating lunch with Roscoe one day when Gilmore strolled through the room, causing both men to scramble for the rafters. On another occasion, Althea got the scare of her life when the pet lion wandered into the main office and chased her up the stairs.

Such quirks came with the territory. Pilots were daredevils and rogues—often rich men and war veterans who lived for the thrills. But Kelly absolutely loved these characters, who shared his obsession with flight and adored the challenge of building their dream machines. He worked personally on planes for all the star aviators, often in his spare time, above and beyond his regular duties as engineer on corporate projects.

Charles Lindbergh pretty much lived in the Burbank plant while Lockheed built his custom Sirius, running down parts and even doing errands for his mechanic. He was, Kelly thought, an "absolutely dandy pilot," talented enough to be a test pilot.

One of the first famous flyers to befriend Kelly, not long after his arrival in Burbank, was Wiley Post, the test pilot turned adventurer who'd flown around the world in 1931 and then again in 1933—both times in a Lockheed Vega nicknamed the *Winnie Mae*.

Kelly marveled at Post's flying skill, especially considering his physical limitations. Post lost an eye in his earlier years as an Oklahoma oil field roustabout and, with it, his depth perception. But Post turned tragedy into triumph. He used the eighteen hundred dollars in insurance money he got from that injury to buy his first plane, covered the lost eye with a patch, and took to the air, depth perception be damned.

This was a peculiar niche in the burgeoning field of aviation. These eccentric adventurers were thinking far in advance of the current trends. They had big dreams, and often the money—or at least the panache—to bet on miracles. It was, in that way, the perfect environment for Kelly Johnson.

Wiley Post had a very specific project in mind. His goal was to reach the stratosphere, where the thinner air would provide less resistance and enable him to reach higher speeds. Post wanted to fly to 40,000 feet or higher and cross America at 400 mph. This may as well have been supersonic; it was nearly that ambitious.

To achieve this audacious goal, Post needed to optimize his plane for high-altitude flight; specifically, he wanted a custom skid installed on the belly of his plane so that he could eject his landing gear after takeoff to reduce weight, while leaving himself something to land on, since taking off is only half the problem, and you can't just land on a fuselage. He also commissioned a supercharger, to push air into the engine, and a pressure suit he developed with the B. F. Goodrich tire company that, to Kelly, "looked like a deep sea diver's outfit."

Kelly worked with an engineer named Jimmy Gerschler to make the landing gear that detached after takeoff, as well as the skid.

And the wildly experimental *Winnie Mae*, with its supercharger, detachable gear, and personalized pressurization system, worked. On September 4, 1934, Wiley Post reached 40,000 feet over Chicago, and pushed even higher from there. On a subsequent flight, Post reached 50,000 feet and discovered the jet stream in the process.

But setting height records wasn't Wiley's goal. Speed was what he was really after. Remember that Kelly was a veteran user of the still relatively new wind tunnels; he'd modified race cars to increase speed and win the Indy 500 as a side project. And he streamlined the *Winnie Mae* to optimize speed, too. But no matter how many tweaks and tricks Kelly tried, the top speed he could achieve in models of Wiley's plane was 260 mph. And yet Wiley wanted more, so Kelly kept at it.

Most days after work, he and Wiley would reconvene at Neil's Drugstore, where Kelly would sip on a sherry as Wiley drank beer and urged Kelly to keep looking for ways to add speed, no matter what the math said.

Wiley was determined to be the first person to exceed 400 mph in flight. And he made a bet with Kelly, who had serious doubts that this was possible. This bet was extremely reasonable. No one here was gambling beyond his means. If the new plane could reach 400 mph, Kelly would have to buy Wiley a book on gardening. But if Kelly was right, Wiley owed him a twenty-one-inch slide rule.

The *Winnie Mae* never did get there; in fact, it never got past 340 mph. But Wiley never paid the bet, either, and Kelly was still griping about it decades later.

Wiley also commissioned a second experimental plane, this one a combination of two wrecks. He took a Lockheed Orion, which had a plywood fuselage, and added wings from a Lockheed Sirius, a giant engine, and the biggest propeller he could fit. Finally, Post asked for floats so he could land on water.

The plan was for him and his friend Will Rogers—one of the most beloved actors in Hollywood—to fly around Alaska and northern Canada hunting and visiting wilderness areas. Rogers named Post's new plane the *Aurora Borealis* in honor of the atmospheric phenomenon they hoped to see, but the engineers called it "Wiley's Bastard" or "Wiley's Orphan," which tells you all you need to know about their confidence in its Frankenstein design.

Kelly warned Wiley in advance that the giant engine and propeller made the plane's nose dangerously heavy. He worried that it lacked the elevator power to get off the ground, but Wiley figured out how to wrestle his awkward experimental plane into the air and got it CAA certified.

And the experimental setup worked, for a while. Wiley and Rogers took off from Seattle on August 7, 1935, and headed north. On August 15, the two adventurers bid farewell to a group of onlookers in Anchorage and stopped later that day at an Eskimo seal hunters' camp three hundred miles south of the Arctic Circle, to deal with engine problems.

Wiley popped the hood and worked on the engine and then hopped back into the cockpit, alongside his pal Rogers. They taxied across the lagoon, took off, and then the engine failed again, causing the plane to crash into the water at 8:18 p.m.—which we know because that's when the watch found on Post's wrist had frozen. Both men perished.

Then there was Amelia Earhart, a kind, gentle woman, according to Kelly, who "paid attention to what people told her" and flew, he would often say, as well as any of her male rivals. Earhart had also fallen in love with aviation at her first sight of a plane, in her case at the Iowa State Fair when she was ten, and she, too, took her first flight with a barnstormer.

She moved to California after World War I specifically so that she could learn to fly—pawning a fur coat and jewelry to pay for lessons—and flew her first solo flight after just ten hours of instruction.

Earhart became famous as the first woman to cross the Atlantic Ocean by air on June 17, 1928, but that was as a passenger. This bothered her, so Earhart poured everything she had into becoming an accomplished pilot in her own right. She founded the Ninety-Nines, a group of female aviators, set an altitude record in a Pitcairn autogyro, and then, on May 20, 1932, cemented her place in history after flying her bright red Lockheed Vega (modified by Kelly Johnson) across the Atlantic, enduring fog, wind, ice, a leaky reserve tank, a spot fire from a bad weld on her exhaust pipe, and a broken altimeter and tachometer.

She'd hoped to reach Paris but settled for Ireland.

More record flights followed. Earhart flew from Los Angeles to Newark in 1932, Honolulu to Oakland in 1935, and Mexico City to Newark in 1935.

Then she settled on the boldest mission of all—flying around the world.

Women were a rarity on the Burbank lot, at least outside the secretarial pool, but Earhart made an impression on the men around the factory, where even those large male egos respected this bold woman's talent and guts. She was "serious, silent as a shipmaster, blond and tousled-headed, with something of Lindbergh in her face," according to engineer Harvey Christen, who, like Kelly, was a big fan. "No flyer has exerted a better or stronger influence in the promotion of interest in flying."

When Earhart began preparations for her around-the-world attempt in 1937, she chose a Model 10 Electra and worked closely with Kelly to maximize the range. Again and again, Kelly rode as a passenger as Earhart took off from Burbank to test her plane with a variety of weights, balance conditions, and engine settings.

In preparation for her record attempt, Kelly put together a detailed memo for Earhart that read like a pilot's manual, complete with meticulously calculated estimates for fuel load and range, as well as tips for saving fuel, such as when to deploy flaps during take-off and when it was ill advised.

Amelia Earhart departed the United States, from Miami, on June 2, 1937, heading east, and got as far as the central Pacific Ocean a month later when she, and her plane, simply vanished.

Kelly was shaken by the news. This was a personal friend—someone he cared about as well as the kind of pilot he rooted for. And the fact that there was no explanation for her disappearance haunted him, as it did people around the world.

So Kelly set about trying to solve the mystery of what happened. He did some calculations and found that the plane was far off course when it crashed into the ocean well north of Howland Island, an uninhabited hunk of coral just north of the equator and quite far from any human populations.

Earhart's infamous end, Kelly believed, was not her fault. There are many theories about Earhart's demise—entire books have been written about it—but to Kelly, it was clear: The plane simply ran out of fuel while in flight, because Earhart was off course and lost, due to her "incapacitated" navigator, Fred Noonan, a celebrated navigator who had worked for Pan Am and charted many air routes over the Pacific.

"The preponderance of evidence both at the time of her crash and at the present indicates that Amelia was lost without the help of a competent navigator at the time of the flight," Kelly told one inquirer who wrote to him in 1982. He was less diplomatic during the *60 Minutes* interview, in 1980, when Morley Safer asked what he thought about Earhart's disappearance. Kelly's answer: Fred Noonan was "either sick or drunk."

Anyone who worked at Lockheed in those days would have seen a lot of one other very famous face: Howard Hughes, the film tycoon and aviation mogul who had founded his own company, Hughes Aircraft, in 1932, when he was just twenty-seven years old.

Hughes, who would go on to be one of the most famous and eccentric businessmen of the century, was a good customer for Lockheed, and a loyal one. As such, he got special treatment—or, at least, he operated in such peculiar ways that Hall Hibbard and Kelly Johnson learned to work around him. "Some of us had to devise ways and means to be with this guy," Hibbard said, and often the unenviable task of babysitting the mogul fell to him.

Hughes preferred to work at night, which meant that Hibbard would often need to work late, too. Many a night, Hibbard would meet Hughes in Burbank at 10:00 or 11:00 p.m. and work with him until 4:00 a.m., at which point Hibbard would run home, shower and eat, then race back to Lockheed so that he'd be at his desk by 8:00 a.m. to work his regular hours. Hughes, of course, went home and slept.

The strangest favor Hughes ever asked of Hibbard was when he petitioned Culver City to allow him to extend a runway at his own nearby factory to accommodate bigger, faster planes. Doing so would mean permanently blocking three or four city streets. So Hughes needed Hibbard, an engineer, to explain to the local government why this bold ask was necessary.

Hughes instructed Hibbard to meet him on a corner in Hollywood and showed up, according to Hibbard's recollection, in a "terrible old Chevrolet."

Hibbard sized up the beater of a car and said that they should take his Buick to the meeting instead. Hughes's clothing was even worse. He was in no way dressed to testify at an important meeting—wearing an open-collared shirt, partly unbuttoned, and some pants that may as well have been plucked off the rack at a thrift shop.

Fortunately, Hibbard had an extra suit in the trunk that he'd been planning to take to the cleaners, so he told Hughes to change into it and stopped en route to pick up a one-dollar tie that he paid for, because Hughes—a multimillionaire—rarely carried cash.

Permission to extend the runway was granted, and Hibbard never saw his suit again.

Around the factory, Hughes was a loud presence, and engineers never knew what to expect when he was on the premises. A young engineer named Jack Duffendack met the eccentric mogul for the first time when Hibbard asked him to go to the runway and check out a Lodestar that had been modified for an unnamed client. This plane had, Hibbard told him, "a fancy fuel system" and he wanted Duffendack to "check it good," because this client was important, and picky.

Duffendack was on board giving the plane a thorough inspection when Hughes climbed in. He overheard Duffendack telling the chief mechanic that he was worried about a fuel trap in the system. The mechanic said that to fix this was a "hard change," but Hughes told the mechanic to go ahead and do it. In the meantime, Hughes said, he was taking "Duffy" to lunch. The two got in Hughes's Ford sedan, drove to a nearby hamburger joint for burgers and milk, and ate lunch in the car.

Howard Hughes wasn't just an extremely rich and sometimes annoying client. He founded his own company, Hughes Aircraft, in a section of a hangar he rented from Lockheed. But his greatest impact on Kelly, and the company, was to spur Lockheed's first major commercial project, in the form of the Constellation—a passenger plane project that truly transformed the company.

The Constellation began as a request from Hughes, who approached Robert Gross in 1938 with a very specific need. Hughes had become the majority shareholder of Transcontinental & Western Airlines (later just TWA) and, along with TWA president Jack Frye, wanted to commission an aircraft that would shake up the industry. Specifically, Hughes and Frye wanted a plane that could fly coast to coast without stopping—at 300 miles an hour while carrying twenty sleeping passengers, thanks to a pressurized cabin, for a total payload of six thousand pounds.

This was a big ask and a giant undertaking for a company that was still on relatively unsteady legs. But giant undertakings were exactly what Hall Hibbard and his star pupil, Kelly Johnson, wanted to tackle. And they weren't starting from scratch.

It was clear by 1938 that aviation was on the verge of a boom. The future was in larger planes that could carry more people, and Lockheed—like its primary competitors, Douglas and Boeing—had begun to explore a concept that had four engines instead of two. Lockheed called its version the Model 44, or Excalibur.

So, there was an existing design concept, and model, for a pressurized four-engine plane, but Hughes's request was a step beyond this. Hibbard, with Kelly, went to work.

They took their initial ideas to the mogul's house in the Hancock Park section of L.A. and urged Hughes to think even bigger: to imagine a plane that could pull aviation out of its niche as a fringe pursuit—one that "would carry more people farther and faster than ever before, and economically enough to broaden the acceptance of flying as an alternative to train, ship and automobile."

In other words, to completely democratize the air.

The biggest difference that Kelly and Hibbard recommended was to increase the passenger load from twenty to one hundred, which Lockheed could do by putting those people in seats instead of beds. Doing this made far more sense if TWA wanted to make money on these flights. What's more, a design like the one they had in mind could fly across the ocean, thus creating an entirely new travel niche—transatlantic commercial flights. Lockheed planned to use a new engine that was already in development for the B-29 bomber—the largest air-cooled engine in the world—as its power source.

Lockheed's proposal was fine with Hughes. He was primarily looking to give TWA a competitive advantage, a plane offering service that no other airline could match, and this one fit the bill. But he asked Kelly and Hibbard for discretion. He wanted to keep development of this plane secret. Hughes treated the project "with all the subterfuge that secret weapons require," according to a company history of the plane, and insisted that Lockheed could not sell the plane to any other commercial airlines until it had delivered thirty-five of them to him.

The Constellation project completely transformed Lockheed, at least in terms of image. "Up to that time we were sort of 'small-time guys,' but when we got to the Constellation"—aka the Connie—"we had to be 'big time guys,'" Hall Hibbard said.

The engineering group was still small when the Constellation concept spun up, and every project went through Hibbard, but Kelly was his right hand. He was, Hibbard later said, a "powerhouse on this airplane." It's impossible on a plane as big as the Constellation to declare any one person the designer, but when Hibbard was later asked who was most responsible for the Connie's creation, he was unequivocal in response: "The answer is Kelly Johnson."

Lockheed's wunderkind engineer did the early sketches, conducted the wind tunnel tests, and oversaw all development up through flight-testing the plane that put Lockheed in the position to create and win all the radical work that was to follow. Kelly, Hibbard said, "could spot trouble" better than any person he'd ever met. Whenever a problem occurred, he just had an intuitive ability to identify the cause almost instantly.

Lockheed's design had numerous industry firsts. Most notably, the Constellation was the first plane with hydraulically assisted power controls (roughly, the equivalent of power steering in cars). This was an idea that Kelly had been kicking around for a while, especially as he envisioned a future with much larger planes that flew farther and carried more passengers. But Robert Gross, who signed off on budgets, had been skeptical; this seemed like added complexity. The customers weren't demanding power controls.

This chafed Kelly. Customers don't drive innovation, and they probably aren't going to ask for something that doesn't exist yet. He was struggling to convince his boss when, one day, he saw a chance to make his point in the company parking lot.

Kelly spotted Gross walking to his car, a late-model Chevrolet that he knew had power steering. You don't need power steering to drive this nice car, he told Gross, but it sure makes it a hell of a lot easier. And, according to Kelly, that's all it took. "I never heard another word of dissent about power steering in aircraft."

The Constellation was fast (350 mph) and had a cabin pressurized to fly at 20,000 feet, which allowed it to get above 90 percent of bad weather. The Connie was so big, and required so much aerodynamic testing, that Hibbard and Kelly had to use six different wind tunnels. There was so much development work on this plane, in fact, that Kelly finally had one of his wishes granted; Gross approved the construction of an on-site research facility, right next to his wind tunnel, large enough to house a full-size mock-up.

Then, on December 7, 1941, Japan attacked Pearl Harbor. Overnight, Lockheed shifted its operations—including the Connie project—to preparations for war.

When the Air Force heard about a pressurized plane with power controls and a top speed of 350 miles per hour—faster than many fighters of that era—it took notice. Such a craft was perfect for carrying troops and supplies to Europe. So in 1942, "the Connie was drafted," as Kelly later put it.

On October 5, 1942, Lockheed filed a patent for the plane, with Clarence L. Johnson and Hall L. Hibbard as co-inventors. Either for alphabetical or seniority reasons, Hibbard was listed first.

The Air Force took over production of the plane from Lockheed and began work on a modified version of the Constellation with olive-green paint that it called the C-69.

After two days of delays caused by unusually high winds, the Connie took off from Burbank for its maiden test flight on January 9, 1943, with a test pilot named Eddie Allen at the helm. Allen was borrowed from Boeing, a competitor, because of his experience on large planes and because he was, in Hibbard's estimation, the best test pilot in America at the time. Lockheed's own Milo Burcham flew as Allen's co-pilot and Kelly, as he liked to do, was onboard, too, as the flight-test engineer.

The flight itself was just a short hop, from Burbank to Muroc Dry Lake, on the other side of the San Gabriel Mountains, and took just fifty minutes, but it was such a success that Eddie Allen signed off on the spot and returned to Seattle.

The sight of this giant plane—Lockheed's biggest, boldest gamble yet—soaring into the sky brought both Hibbard and Gross to tears. "It was the only time I ever saw Bob Gross cry," Hibbard recalled. "But of course it was from joy."

Joy that his enormous gamble on a giant plane—commissioned at a time when Lockheed's financial health was far from certain—actually paid off, transforming his company overnight. Or at least, over the course of a year. Lockheed's production tripled from 1941 to 1942, and C-69s for the war effort were primary drivers of that business.

After the war, the Air Force had no need for all these giant troop carriers, so TWA—and Howard Hughes—bought back all the C-69s it could, converted them for civilian passengers, and put the Constellation into commercial service in February 1946.

In preparation for the plane's introduction, Hughes had a clever idea for a publicity stunt. He wanted to fly a Constellation across the country himself and set a new transcontinental record in the process.

Hughes was an accomplished pilot, and a bold one, unafraid to take risks. "He scared me," Kelly later said. "I didn't like his sense of judgement in certain instances," like when Kelly forced Hughes to undergo some flight training on the Connie before he could make his record attempt.

Lockheed had a rule: Anyone who wanted to fly a new plane had to first be checked out by the company's test pilots. In this case, the job fell to Milo Burcham.

Burcham took Hughes up, along with a few other people, including TWA's Jack Frye (who planned to accompany Hughes on his cross-country mission) and Kelly Johnson. The plane was barely in the air, just a few thousand feet over the foothills behind the plant, when Hughes made an unusual demand. He wanted Burcham to put the plane into a stall, which is not at all like stalling a car, where the engine cuts out. Basically, a stall in a plane means that some action by the pilot—nearly always a mistake—causes the plane to lose lift and fall from the sky.

This wasn't technically the proper or safe altitude for such a test, but Burcham knew the Constellation was up to it, so he said okay. He eased the plane into a gentle stall, then quickly and easily recovered.

Hughes scoffed. "That's no way to stall!" he barked, and asked for the controls. Burcham acceded, and with Kelly standing in the cockpit between the two men, Hughes yanked back as hard as he could on the control, to induce a stall.

"That's one of the few times I've seen zero air speed on an airplane in the air," Kelly later said, because that's where they were. Dead but aloft over the mountains. The plane dropped from sheer inertia, and Kelly, who wasn't belted in, was thrown up against the ceiling of the fuselage.

"Up flaps!" he screamed. "Up flaps!"

Burcham grabbed the controls, pulled the flaps up, pushed the nose down, and stabilized the plane, with 2,000 feet of altitude to spare. The group then flew on to Palmdale, where they were to perform a series of takeoff and landing tests.

Burcham demonstrated a few times, then handed the controls to Hughes, who had difficulty handling the engine torque during takeoff. Each takeoff was slightly hairier than the last, with Hughes letting the plane pull into a tighter angle, coming closer and closer to crashing into the airport tower.

Finally, Kelly ducked into the back of the plane and asked Jack Frye what the hell he should do. Frye basically shrugged.

So Kelly marched back into the cockpit and ordered Burcham to take over.

"Milo," he said, "take this home."

Hughes was not happy. He looked at Kelly as though he had stabbed him in the back and said nothing for the rest of the flight. He just sat in a jump seat, livid with rage.

This was what the party eagerly awaiting the return of Hughes's maiden flight in Lockheed's critically important new commercial plane saw when the Connie landed back in Burbank: an extremely valuable customer who was extremely pissed off because he'd been disallowed from putting his new plane through a series of dangerous maneuvers that made that plane's designer and his chief test pilot uncomfortable.

When word got back to Robert Gross, he was furious, too—with Kelly. Hughes was the company's most important buyer, and he'd been embarrassed.

Kelly didn't care one bit. He simply went home and poured himself a whiskey and soda to celebrate having survived the day.

Hughes got over it and agreed to come back for a series of orientation flights that would help him get control of the plane. Kelly offered his test pilots bonus pay for this job and handed his role as flight-test engineer over to Rudy Thoren, for his sake, and Hughes's. The two men never flew together again, he says. "It was mutually agreeable."

The second time out, Hughes listened to instructions and worked hard. He was determined to get it right. By Kelly's own estimation, Hughes made nearly seventy takeoffs and landings in the Lockheed Constellation over a single weekend.

And then he was ready.

On April 17, 1944, Howard Hughes—with Jack Frye onboard, and Rudy Thoren riding shotgun—flew the Model 049 from Burbank to Washington, D.C., in 6 hours, 57 minutes, 51 seconds, setting a new speed record for transcontinental flight.

As Hughes approached Washington's National Airport—which sits right on the Potomac, across from the city—the tower told him to slow his approach and wait for instructions to land when a space opened. That wouldn't do. Hughes was out to set a record. So he just flew right on in and landed, scaring Thoren so much that he refused to make the return flight west if Hughes was flying.

Robert Gross—the Hughes whisperer—had to fly to Washington to broker a peace between the two men so that Thoren would actually make the return flight.

That same Connie made history again not long after, when it was returned to the Air Force and taken to Wright Field, outside Dayton, Ohio, and put in the hands of seventy-two-year-old Orville Wright.

Then the man who made history's first powered flight flew for the last time in his life on a Lockheed Constellation.

Over time, the idea that Howard Hughes was the architect of the Constellation percolated through the rumor mill. It was, to some, a secret truth of this famous plane, and Kelly shot it down numerous times later in his life, including in a 1963 letter that clarified the record. Hughes "did not even have a representative in the Lockheed factory," Kelly wrote. "So I would say Mr. Hughes's main contribution to the Constellation program was in buying some."

Hall Hibbard was a little more generous. According to Hibbard, Hughes "helped us a great deal in the design," especially on the cockpit and landing gear. He "looked at our drawings of the landing gear" and "made some very creditable statements which actually changed our landing gear design for the better."

The difference in opinions probably had a lot to do with where the two men sat. Kelly was a young engineer, only beginning to build his reputation. So he probably cared most about credit. Hibbard was an executive, worried more about business, and especially about keeping a wealthy customer happy.

And Hughes meant a lot to Lockheed. The infamous mogul also imposed his will upon the place. The size of his fortune, and his mercurial personality, meant that Hughes enjoyed his own set of rules around Lockheed. He kept a small fleet of five planes, mostly Constellations, parked around the field in Burbank and he liked to come and visit those planes, even when he wasn't flying them. Though he often did that, too. He'd just stroll out to the runway, fire up the engines, and take off.

He assigned college students to guard those planes from, well, no one really knew what they were guarding against. The students maintained a twenty-four-hour watch, working in four- or five-hour shifts in the backseat of cars that had been fitted with portable desks so that they could work on their studies. According to engineer Jack

Duffendack, Hughes "put 40 students through school that way." At a minimum, then, this particular manifestation of Hughes's legendary paranoia did result in a pretty helpful charity.

Whether or not Hughes deserved a little credit or a lot for the Connie's design, there's no question that he was the driving force in the creation of a plane that was more revolution than evolution.

Compared with its closest competitor at the time, the Connie was 25 percent more powerful and 24 percent faster, with 24 percent more range and 14 percent more payload capacity—and yet carried passengers at 23 percent lower cost. It advanced technology while helping to democratize an industry, making air travel achievable for tourists. There's no question it pushed aviation into a new era.

When Kelly sat down with Hibbard to write a detailed history of the Constellation, he pointed out that the Connie was a great leap forward precisely because they had aimed so high. Designs that endure—the ones that define companies—are those that require the biggest risks. Incremental leaps might win in the short term, but both Kelly and Hibbard strongly believed that you had to look far ahead and take giant leaps to endure. "Failure through timidity" is how Kelly phrased it.

Because no matter how advanced or triumphant a certain plane may be, the truth, Hibbard said, is that "it is necessary to start the design of its successor long before the sale of the first design has shown a profit."

4

LOCKHEED GOES TO WAR

Here's how little regard the U.S. government had for airplanes as a national defense priority in the first few decades of the twentieth century. In 1928, the U.S. Army asked President Calvin Coolidge for twenty-five thousand dollars to buy a squadron of fighter planes. The president's reply: Why can't we just buy one plane and have all the pilots share it?

Coolidge just wasn't exactly a fan of aviation—in fact, he was the last U.S. president to never fly in an airplane. Nonetheless, Coolidge did make a substantial contribution to the industry by signing the 1926 Civil Aeronautics Act, which established federal oversight of airplanes. He subsequently created the Army Air Corps—precursor to the Air Force—and named an assistant secretary for aviation in the War Department.

From that moment on, America would fight wars in three ways—on land, at sea, and in the air.

Robert Gross had long coveted this new, potentially enormous business sector, but well into the 1930s, he struggled to crack it. That wasn't for lack of trying. And in the meantime, he set his sights on a foreign military that had more of an appetite for warplanes—the U.K. Air Ministry.

Throughout 1936, Robert and his brother Courtlandt, who worked on the East Coast, maintained a dialogue with British attachés about the possibility of selling planes to the United Kingdom. With a war looming, the British military recognized that it would very likely have to buy American planes.

On September 1, 1939, Adolf Hitler's tanks rolled over the border into Poland, and World War II began. Things escalated quickly, in Europe and Asia, and as global peace unraveled, the U.S. government tried to stay out of the fray. Congress passed the Neutrality Act in 1939, to stake its official position as not interested in fighting, despite the fact that the American public was strongly in support of Allied forces.

That year, 1939, the U.S. aircraft industry totaled $225 million in sales. The next May, President Franklin D. Roosevelt recognized the inevitable—that the United States would be drawn into the war—and called for the urgent manufacture of fifty thousand planes.

How unprepared were America's air forces for war prior to that point? General George C. Marshall later testified that the "few partially equipped squadrons" in the United States "could hardly have survived a single day of modern aerial combat."

America wasn't officially in the war, but it wasn't totally out, either. Because of the Lend-Lease program, U.S. companies were making and shipping equipment and ammunition to England. One such project was the Hudson bomber.

A year earlier, at the start of 1938, Hall Hibbard had fallen ill and was out of work for more than a month. Gross noticed, in his absence, that the engineering department lacked organization and was far too reliant on Hibbard, especially with such dramatic growth on the horizon. So Gross made some changes. Among them was the promotion of Kelly Johnson, at just twenty-eight, to chief research engineer.

And it was around this time that England went hunting for a solution to a problem that had plagued its forces in World War I: German U-boat attacks on its ships. The British needed a plane to patrol its coasts and hunt submarines, to help secure supply lines.

A British contingent traveled to America in search of a company that could build them this sub-hunting bomber, and there was no plan—at first, anyway—to visit this scrappy upstart, Lockheed. Until someone changed their mind.

A salesman named Otto Graf caught wind that the British delegation was planning to visit Douglas, based nearby, and recognized the opportunity. Graf studied articles and obtained as much information as he could about the contingent, then went to the train station, recognized the Brits from photos, and introduced himself upon their arrival.

Graf told the group that his company, Lockheed, wished to present a design, too, and invited them to come to Burbank. A few days later, the Englishmen did just that.

Kelly and Hall Hibbard were ready to make their case. They saw a perfect solution for the problem in Lockheed's Model 14 Super Electra. The plane was fast, approximately the right size, and versatile enough to carry all sorts of gear. So they hurriedly built a full-scale wooden mock-up of an anti-submarine version.

That mock-up was just a guess at the Brits' requirements. And it wasn't exactly right. The British had more specific needs and offered Lockheed a do-over. The visitors assumed that this total redesign would take a week, if not longer.

This was a Friday. By Monday, Kelly and his engineers had modified the mock-up to better fit those needs. The Brits were so impressed by how quickly Lockheed had addressed their concerns that they invited the company to England to meet with the Air Ministry's technical staff.

Courtlandt Gross, the company math whiz, led a small entourage from Burbank that included Lockheed's lawyer, vice president of sales Carl Squier, and Kelly Johnson.

On May 25, 1938, the men departed for England on the *Queen Mary* and, according to a story Gross liked to tell later in his life, Kelly was so frustrated by the speed and efficiency of the giant cruise ship that he sketched a redesign for it in his notebook while en route.

Upon arrival, the Air Ministry staff ripped Kelly's latest plane design to shreds in a meeting that lasted only a half hour. The needs of the British couldn't be met with a few tweaks. It would require a total redesign, but rather than admit defeat and lose the job, Kelly asked for the chance to try again. His new design, he said, would be ready by the end of the three-day holiday weekend.

Lockheed's group bought a drafting table and some T-squares and set up an impromptu design studio in a Hotel Mayfair Court suite. There, Kelly dug in and worked for seventy-two straight hours. As he later explained it in his memoir: "I had to fit in all this new equipment, rearrange copilot and radio operator positions, make weight and structural analysis, figure contract pricing, and guarantee that the design would meet certain performance requirements."

Officers at the Air Ministry could not believe that a twenty-eight-year-old engineer had just led a wholesale redesign of their plane, in a hotel, over the weekend. (He'd also done it, according to Gross, without sleeping, or changing his suit, which by the end looked like "an unmade upper berth.")

"The combination of Kelly's youth on the one hand, and competence on the other, puzzled our British friends," is how Gross later described it.

And over subsequent days, additional questions were answered until the Brits were sold on Kelly's design, and on Lockheed.

Before giving Lockheed the green light, however, the Air Ministry's chief of the Air Staff pulled Courtlandt Gross aside and asked him one last question.

The ministry was pleased with the work but a little uncomfortable with its author. "We are not used to dealing with the technical say-so of a man so young as Mr. Johnson," were the minister's precise words.

If the project was to go forward, the Brits would need Gross's personal assurance that the plane Lockheed delivered would really meet all the specs set out by this young man in his design.

Could Jolly Old England truly count on the math of this child?

"We have great confidence in Mr. Johnson," Gross replied. "I promise that my brother and I will see to it that your confidence in this company is not misplaced."

The UK sub-hunter project was beyond top secret. The British forbade anyone in the U.S. group to discuss this purchase by phone, so it was a complete surprise to Robert Gross and Hall Hibbard when Kelly came home with a giant contract—and a totally different plane design.

The Air Ministry ordered two hundred planes for $25 million. It was the largest order, in dollars, ever placed in the United States during peacetime to date, and the largest contract ever given to an American company by a foreign buyer—not to mention a critical lifeline for a company that had limited credit and only $334,000 in cash on hand.

The order also came with an enticing bonus. The British would buy as many planes as Lockheed could build by December 1939, up to two hundred fifty.

"This was the order that put Lockheed on the map," Hall Hibbard later said. "It was one of the most thrilling times I can ever remember."

The Hudson contract was a lifesaver, but it put enormous strain on a company with fewer than two thousand employees. Robert Gross had to initiate a massive expansion to meet the need. He raised $1.25 million in short-term financing, and a hiring frenzy ensued.

Because unemployment in the United States was still so high—17 percent of the total workforce was unemployed—lines to apply for open jobs at the Burbank plant stretched for blocks.

The prototype Hudson flew on December 10, 1938, and production aircraft began to roll off the line the following February.

Kelly followed the first three Hudsons shipped to England, in order to create a flight-test and pilot-training program. Again, he sailed aboard the *Queen Mary*, but this time he didn't obsess about the pace because he'd brought Althea along. They danced every night and made the most of the pace that had frustrated him before. It was, he said, "a real vacation." One that could only have come because he had a plane to deliver to the war.

Kelly's understanding of aerodynamics, learned over hundreds of hours in wind tunnels, grew with every plane he designed and flight-tested. He was in the right-hand seat of the Hudson bomber for 551 intentional stalls. His test pilot, often Marshall Headle, would let the plane fall out of the sky, as Kelly visualized where the airflow was separated so he could tweak his design later to fix it.

He was on board in England, too, especially when test pilot Milo Burcham was demonstrating the plane's abilities to British pilots. The RAF wouldn't allow American pilots to be at the controls for familiarization flights, but they did put Kelly in a royal-blue RAF flight suit and make him flight engineer, so he could demonstrate certain key features of the Hudson.

When war broke out and the Hudson was put into active service,

Britons nicknamed the plane "Old Boomerang," because it seemed almost invincible. It wasn't uncommon to see one limping back to base, basically shot to pieces, but still flying, somehow.

Hudsons played a major role in the Battle of Dunkirk; a Hudson was the first American-built plane to shoot down an enemy craft, off the coast of Jutland, on October 8, 1939; and a Hudson performed the first—and perhaps still only—capture of a submarine by a plane.

How exactly does a plane capture a sub? By catching it on the surface and doing figure eights overhead until a destroyer arrived on scene to complete the takeover. This worked because German sub captains knew the plane and its armaments. A Hudson had forward and aft firing guns, as well as bombs and torpedoes. If the sub attempted a dive, the plane would have simply destroyed it. Later, the captured boat was converted and put into service as a British submarine.

Jack Duffendack, from Bartlesville, Oklahoma, was an "airplane rat" who applied to all four schools that offered degrees in aeronautics, and three of the four—Cal Tech, MIT, and Georgia Tech—were too hard to get into. So he went to the fourth, the University of Michigan. Duffendack worked at the wind tunnel and had been there when the newly hired Kelly Johnson brought the Electra models from Burbank for testing.

In his second year, he met a new student named Willis Hawkins at registration. The two of them founded the Aeronautical Sciences Club and, upon graduation, took Hawkins's car and drove west, for California, just as Kelly Johnson and Don Palmer had done.

Both men were hired as junior detail draftsmen at Lockheed. Hawkins was assigned to the wing group; Duffy—as he was known at the plant—went to engines.

Mostly, Duffy installed engines on the Hudson bomber. But his boss, Lon Story, had begun to work on a secret project and was frequently absent.

One Sunday morning, Duffendack's phone rang. It was Lon Story. "Duffy, come on over!"

Duffendack went to the plant and found Story, at 10:00 a.m., sipping a brandy.

He offered one to Duffendack, who felt that he couldn't say no, and so he, too, was sitting there sipping liquor on a Sunday morning when Kelly Johnson arrived.

Duffendack didn't know what was going on and things only got weirder when Hall Hibbard showed up.

"Have you told you him yet?" Hibbard asked.

The Australians had just ordered fifty Hudson bombers with a whole bunch of revisions, including a new engine. Because Story was busy on the secret project, he needed Duffy to design the installation for this new engine.

Kelly fixed him with a stare. "Damn you, Duffendack," he barked. "I'm going to fly this new power plant in three and a half months!"

Duffy was twenty-five. He had been out of college for only a year and a half. And now he was being asked to lead design the engine installation for a major order to support a country's war effort.

Kelly said that sketches had already been done, based on an existing Pratt and Whitney design. Duffy could use the five people in his group, plus another eight whom Kelly would round up by Monday night. By Friday, he'd have forty—and he was to have plans ready for them.

Duffy was, in many ways, over his head, but he trusted subject matter experts to do the detail work—exactly as Kelly wanted.

Many years later, when a historian interviewed Duffy for an oral history project, he took credit for something that would become a key rule of Kelly Johnson's Skunk Works, but which did not yet exist in 1938—that individual draftsmen should be responsible for the design.

It was a matter of necessity, Duffy said. If he was going to do this project in one hundred days, he had to save time, and one way was to remove a step—there would be no checkers to check drawings. The lead engineers were "fully, solely responsible" and designs would jump straight from the desk of the draftsman to production.

Kelly liked the idea. He was already pursuing some basic rules for any project he was given, such as: Any layer of a process that can be cut, should be, so long as it doesn't damage the end result. This certainly fit the bill.

Duffy worked almost nonstop, sleeping on his drafting board, but he made Kelly's deadline. The first test plane flew in three and a half months.

5

THE FORK-TAILED DEVIL

Even before they were pulled across the pond to deliver the Hudson for the Brits, Kelly Johnson and Hall Hibbard had been secretly at work on something far faster and more audacious. This was the secret project that Lon Story had been summoned to quietly work on.

Back in March 1936, Kelly and Hibbard began sketching a new concept, known as the Model 22. By now it was obvious to both men, and to Robert Gross, that the U.S. military would eventually come calling, and they wanted to be prepared. They tried to predict what the war machine might need.

And the following January, the Army Air Corps (the precursor to the U.S. Air Force, which wasn't officially formed until 1947) did indeed come calling. It sent word to the industry that it was in need of a new plane, specifically, a high-performance twin-engine "interceptor," to fly out and engage enemy bombers before they could reach their targets.

This interceptor should be able to reach 360 mph at 20,000 feet with a high-speed climb out to that altitude, to get out of the range of ground fire as quickly as possible.

Inside Lockheed's design shop—still located in the old ranch house—plans centered around a concept that could be based on existing parts and layouts. The idea, basically, was an extension of what Kelly and Hibbard, with help from new hire Willis Hawkins, were already doing on the Model 22 project.

Lockheed predicted that its plane would exceed all the Air Corps's onerous specifications—said to be the most ambitious specs ever requested—and when the company was awarded the contract for a single prototype, on June 23, 1937, the new design became known as the XP-38. (The X stands for "experimental"; the P for "pursuit.")

Best of all, Kelly Johnson's first major U.S. military project came with basically his dream conditions. The Air Corps wanted innovation, as fast as possible, with only minimal oversight. Music to his ears.

The project manager assigned by the Air Force to oversee this program was a young first lieutenant named Ben Kelsey.

Kelsey was head of fighter projects at Wright Field, the base named for Orville and Wilbur because it was on the site of their former facility, outside Dayton, Ohio.

Kelsey loved Lockheed's design and wanted to do whatever possible to bring it to fruition. He was there to help, not hinder. "If we asked Ben for a decision," Kelly later wrote, "we got it—on the spot."

In retrospect, the plane's shape is a classic. But when Lockheed's first fighter was conceived, the XP-38's design was radical, practically heretical. People, Kelly would later say, thought it was "funny looking." But they were reacting only to form, with no consideration of function. "There was a reason for everything that went into it, a logical evolution," he explained. "In design, you are forced to develop unusual solutions to unusual problems."

By the time he and Hall Hibbard had put a propeller on the giant liquid-cooled Allison engine, with a turbo supercharger and retractable landing gear, each of its two nacelles—a nacelle being the smooth, rounded shell that protects each engine and enables air flow around it—stretched nearly to the tail. Kelly decided to extend each of them another four feet, so that they did reach the tail, and called them booms.

Before the XP-38, all planes had propellers that turned clockwise, as viewed from the cockpit. In the event of an engine failure, especially at low speeds, the plane would turn sharply toward the bad engine. To help reduce this torque effect—the tendency of a plane to pull to one side after one engine fails—Kelly decided to make the propellers counter-rotating. They both turned inward, at least on the prototype. On production models, they turned outward, which didn't help with engine failure stability but did continue the other benefit of counter-rotating propellers.

This opposite rotation of the propellers—the left side turning left, while the right turned right—made the plane equally good at maneuvering left or right, which was an immediate asset in combat. Single-engine planes tend to experience engine torque when turning to the side opposite the rotation. The best fighter pilots knew, then, which way a bogey was likely to turn under pressure; he'd almost always go the way of his propeller rotation. But a P-38 was unpredictable. It could turn either way. This is one reason the plane would end up being so exceptional at air-to-air combat.

Some of the elements that would become tenets of Kelly Johnson's management ethos were born during the P-38 development program.

For starters, development was breakneck.

Kelly obsessed over every aspect of his planes, right up until they were handed off to the customer. It's not that he stopped caring after; he most certainly did not. But his primary responsibility was over.

He also insisted that his engineers run the flight-test program. Prior to the P-38, the industry standard for military planes was that engineers would design a prototype, then hand it off to the Army for testing. This was wrong, Kelly believed. It was the designer's *responsibility* to prove a plane did what it was supposed to do. Therefore, the engineers and pilots "must live together," and on the P-38, he dug in.

Until a plane was proven, its kinks worked out, he wanted his test pilots—masters of aviation, who can think fast and improvise—on the stick. "The production pilots do not need to be as skilled as the experimental pilots," he explained. "You can have a different type of person doing production flying. We should all be smart enough, and I hope I was, to take people out of flying the dangerous airplanes at a reasonable time."

On January 27, 1939, just fourteen months after Lockheed got the contract to build the XP-38, the prototype was disassembled, packed into a truck, and driven in secrecy from Burbank to its test site, at Riverside's March Field, one of America's oldest air bases.

The stakes here were extremely high, and a very special guest showed up for that first test: General H. H. "Hap" Arnold, chief of the Army Air Corps. Arnold—one of the first-ever military aviators and a man who was taught to fly *by the Wright brothers*—was responsible for the new plane's creation. He had wanted a 400-mile-per-hour plane, and this was it.

The man at the controls for this first flight was its benefactor, Lieutenant Ben Kelsey, who had handed Kelly the job and then stayed out of his hair.

And Kelsey nearly crashed the plane before it flew a single minute. During a speed test on the runway, he accelerated and then slammed the brakes—one of the last components installed, just a day prior—but they barely worked. It took every pound of effort Kelsey had, plus every inch of the runway, to bring the plane to a stop. It turned out there was residual grease on the shoes from the oily rags they'd been packed in.

The next day, with those brakes wiped clean, Kelsey took off and the plane flew perfectly. Shortly thereafter, he was dispatched on a speed run, to try to beat Howard Hughes's coast-to-coast record.

Kelsey flew the XP-38 prototype, which had fewer than five hours of flight time on it, across America—with stops in Texas and Dayton—to Marshall Field on Long Island, cruising at 420 mph and finishing the trip with a literal bang.

The engines failed on approach and Kelsey steered the plane away from a busy highway, plowed through a treetop, and crash-landed on a nearby golf course. The plane was totaled. And Kelsey's elapsed time of 7 hours, 43 minutes, missed Hughes's record by 15 minutes.

The crash was a setback for Lockheed but not for the Air Corps. Hap Arnold ordered Kelsey to travel immediately to Washington, where he was taken on a frantic tour of important office visits, assuring generals and politicians that this exciting new prototype—which had a total of twelve flight-test hours and had just crashed onto a golf course at the tail end of a very public flight—was ready for full production.

By most traditional measures, the plane *wasn't* ready. But there just wasn't time to waste. The U.S. government needed this fighter, so Robert Gross put it into production. He bought an old distillery on San Fernando Road and that's where the first P-38s were built.

Lockheed employed 17,000 workers by the end of 1940 and could only barely keep up with the growth driven by the war in Europe. Skilled workers, in particular, were in limited supply, so Lockheed decided to help its own cause by creating the country's first federally approved apprenticeship program. The company also partnered with Cal Tech to create an intense sixteen-week retraining program to convert engineers and other technical workers into aviation specialists.

Despite the slow drip of alarming news from Europe, there was very little stomach on Capitol Hill, or among the U.S. public, for war. Until December 7, 1941, when the Japanese attacked Pearl Harbor. "That fateful Sunday began typically for 53,000 Lockheed and Vega employees," reported *Of Men and Stars*, a corporate history. "Not until late morning did radio newscasts report the fantastic attack."

The effect on Lockheed, and its community, was immediate. The mood around the Valley was somber and everyone feared that the Japanese could attack California next. At night, the town lights were blacked out, and giant searchlights swept the sky. People covered their windows with plywood and were warned to drive only when necessary.

Lockheed's key executives and engineers met in Burbank on the night of the attack and began planning for a total shift to war production. Rumors circulated that Japanese ships were steaming for the coast, that submarines lay in wait outside harbors. And Lockheed's fleet of planes awaiting delivery on the lot at Burbank was a huge target. By the end of the day on December 7, the U.S. military had ordered "every ship into the air that can possibly fly."

Lockheed pilots took off for destinations in every direction, bound for any landing strip that could take them. Tony LeVier helped fly Hudson bombers forty miles inland, to the desert, making several trips.

Within months, Lockheed had more than 50,000 employees, making it the largest aircraft manufacturer in the country, and America's entry into the war forced the company to go to a seven day-a-week schedule. In the scramble to find space and add capacity, warehouses, stores, and even bowling alleys were converted over for manufacturing.

At its peak, in mid-1943, Lockheed employed 91,000 people in

Burbank and the company's campus was basically a city, with its own defenses. Bomb shelters were built around the grounds, and the Army stationed antiaircraft guns on the roofs of the plants.

Another innovation was born of necessity during World War II. In 1942, Lockheed created America's first airplane maintenance and training division. The idea was to teach others how to maintain the planes, to keep them flying. More than 40,000 men finished the twenty-eight-day crash course and then went off into the world, to help keep Lockheed planes in the war.

Urgency can force innovation. But that's rarely a perfect process. And even the greatest innovations often come into use before all the kinks are worked out. Which was certainly the case with the P-38. And Kelly was not at all surprised.

Kelly had predicted problems with the arrival of newer, faster planes. Back in 1937, he had written a report titled "Aerodynamics and Performance Study, Lockheed Model 22 Pursuit Airplane," which outlined his concerns, as envisioned on his own design (then just a prototype).

Kelly worried that a plane flying in excess of 400 mph, as his exciting new P-38 now did, might encounter something called compressibility effects—and his report devoted six pages to the problem.

In short, air passing over an airplane's wing or fuselage has to move out of the way to get around the plane's curves. When the P-38 was flying at high altitudes, at high speeds, that air had to move more than 40 percent faster, approaching Mach 1, or the speed of sound. During flights at lower speeds, the air just flows over and past the wing, like water past the hull of a boat. But air is a gas, so it is compressible and elastic. At higher speeds, it begins to act very differently from water. It can't move fast enough to get past the wing's leading edge and a kind of invisible wave forms—a shock wave—and in the wake of this shock wave, a "compressibility burble" happens. That makes the wing less effective, and when that burble hits the tail, it causes aerodynamic buffeting.

This could take down a plane.

Kelly consulted with two of the world's foremost experts, Drs. Theodore von Karman and Clark B. Millikan, both at Cal Tech, and found that neither of these aerodynamicists could provide a formula for calculating how and when compressibility might occur. So Kelly just pushed on with his design, quite aware that he'd have to confront and overcome this new problem when it cropped up for his pilots. As Warren Bodie later wrote in his book *The Lockheed P-38 Lightning*, "Compressibility was waiting—like the dark at the top of the stairs—for those brash enough and fast enough to venture into the realm."

The P-38, as the production version was known, was the first fighter to have high-lift maneuvering flaps; the first plane with a tricycle landing gear (meaning it had a wheel at the nose, too, and didn't sit on its tail when at rest); the first plane with a full arsenal of firepower—a 20mm cannon and four .50-caliber machine guns—in the nose; and the first plane on Earth to "get high enough and go fast enough to reach Mach numbers approaching the speed of sound."

It wasn't until the P-38 reached these speeds that a pilot finally encountered the problem that Kelly had been worrying about. And it was as bad as he feared.

The initial squadron of Lockheed P-38s was taken to Selfridge Field in Michigan for pilot training, and its first days as an active military fighter were disastrous. Somewhere between Mach .67 and Mach .80, the problem Kelly predicted started to occur, and once it did, things would spiral rapidly out of control.

An Air Corps major checking out in one of those first production planes encountered trouble as his air speed hit 350 mph in a dive. His P-38 began to "vibrate violently" and the major could not pull the plane out of the dive. Only some frantic cranking of the nose-up trim on his elevator trim control saved the plane as it plummeted toward the ground inside of 12,000 feet. When he got back to base, the major wrote a letter to Lockheed.

What the hell was wrong with this otherwise fantastic plane?

It was, as Kelly assessed, a tendency for the plane to "steepen up in a dive, and make it very difficult to recover." He surmised that this was what he'd predicted—compressibility—and not the "tail flutter" suggested by others. What he suspected was happening was the shock wave, which caused the air to start tumbling rather than flowing smoothly over the wing, resulting in a stall of the wing. Basically, drag increases, lift decreases. This also resulted in increased lift on the tail, making it harder to slow the plane and stop the dive. If the pilot had enough margin for error, he could slow down and recover in lower, denser air. But in the stress of battle, when evasive action is the norm, he might not have this cushion.

The Air Force demanded an urgent fix. But the P-38 was already in production, and war was raging. So Kelly and his team began a frantic search for a solution, trying all kinds of things—the direction of prop rotation, the tail's weight balance, increased cable tension on the elevator, even a higher tail. Nothing worked.

One test pilot barely avoided disaster, in a modified P-38 that had a heavy rope tied to the elevator's trailing edge. But on November 4, 1941, a different pilot was less fortunate.

A huge crowd of 25,000 employees was on hand that day, gathered to hear General Hap Arnold, the P-38's foremost champion, speak about the war effort.

Many at Lockheed believed that this problem of "flutter" was overblown and had mostly been fixed anyway. "We had difficulty convincing people it wasn't the funny-looking aircraft itself," Kelly later said, "but a fundamental physical problem."

Kelly was in Robert Gross's office that morning, explaining to his boss that he needed more data in order to identify the precise cause of the problem. He said that test pilot Ralph Virden was up over Burbank right that moment in a specifically modified plane, laden with instrumentation, with a spring-loaded tab on the elevator. The tab, he said, had so far produced promising results in dives at high altitude. Right about now, Kelly told Gross, "he's going to try it further down."

As in, closer to the ground.

Then both men went silent. Even through the walls of the building, they could hear the "banshee howl" of a P-38 in a dive.

Nearby, a real estate agent heard a terrifying roar and looked skyward as a piece of a plane's tail snapped off as its pilot attempted to pull up. "I could see the aluminum tailpiece glistening in the sun as it floated away from the rest of the airplane," he told reporters. "The plane itself fell off on its side and started to zigzag to the earth. I just knew it was going to crash."

Another woman was ironing at home when she also heard the roar of the plane's engines. She ran outside and saw the plane descending fast, clearly out of control, until it disappeared into some trees.

"It looked just like a toy airplane," she told the *Los Angeles Times*. "I knew the pilot didn't have a chance."

Virden crashed through the roof of a Glendale home, and the owner tried but failed to free the pilot from the burning wreckage. He did not survive.

"Ralph Virden was a great pilot but an even greater man," Robert Gross said in a statement issued in the aftermath. "If anyone ever had national defense at heart it was he, who every day was carrying the science of aviation into new and higher fields."

The crash was the result of two distinct problems. First, Virden had put the plane into a dive too close to the ground, allowing himself no margin for error. But also, Kelly's new fix was too strong, overstressing other parts of the tail until they broke.

Kelly had suffered from ulcers since college, and they often flared up under stress. That happened again in the midst of the compressibility crisis. He was eating dinner at the house he and Althea had purchased in Encino—on the same hill as Clark Gable—when ulcer pain flared so severely that he fell out of his chair. Rather than take him to the hospital, Althea offered him a double shot of brandy.

Kelly went to work, as always, the next day. He self-medicated and sucked it up for a few days and then finally told the company doctor, Lowell Ford, but only after the worst of the pain had subsided.

The booze, Ford told him, had probably relaxed the stress that inflamed the ulcers, and so Kelly convinced his doctor to allow further consumption in moderation. As he told it, this was critical. Because with a war raging, and his fighter killing pilots, this certainly wasn't a time to quit.

Kelly was summoned to a meeting at Wright Field, where he begged the National Advisory Committee for Aeronautics (NACA), the precursor to NASA, to let him use its high-speed wind tunnel at Langley Air Force Base, outside Washington, D.C., for model testing, because the one he had at Lockheed couldn't approximate speeds over 300 mph.

That December—the same month the Japanese attacked Pearl Harbor—a one-sixth scale model of the P-38 was placed into NACA's Langley tunnel, with the group's specialist in high-speed aerodynamics, John Stack, overseeing a series of tests. Stack's tests revealed the root cause—shock waves forming on the wing, causing a loss of lift and an increase in drag.

It was, in fact, compressibility. Exactly as Kelly suspected. And the fix was something that both Kelly and John Stack had landed on: so-called dive flaps.

Lockheed's chief of research, Ward Beman, had first suggested to Kelly that external dive flaps on the front wing spar could pull the nose out of a dive and stop the buffeting. And these flaps were a relatively simple, elegant solution. They could be deployed with a button in the cockpit and were very easily installed; mechanics could do it in the field without recalling planes.

Kelly called these newfangled flaps dive brakes and ordered them installed on all new planes coming off the line in Burbank. For the hundreds of P-38s already overseas, mostly in England, Lockheed designed a modification kit that could be installed in situ, and packed 487 of these kits into a C-54 for delivery to the U.S. Eighth Air Force. Unfortunately, RAF fighters mistook the cargo plane for a German condor and shot the C-54 down off the coast of Ireland.

Instead, the P-38s in England went on to fight and win dogfights in the European theater without the flaps.

Shortly after the Pearl Harbor attack, Air Force general George Kenney came calling to Burbank, in need of as many P-38s as Lockheed could turn out, in order to stop Japan's rapid advance across the Pacific. Lockheed commenced a "Snowman Project," to winterize Lightnings—over "a few frantic days of around-the-clock work"—so they could be sent north, to Alaska, to defend the Aleutian Islands. Their arrival is a big reason why the Japanese march across the Pacific, capturing islands, never reached Dutch Harbor in the Aleutians.

The P-38 was not conceived as a high-volume plane. When Kelly designed the peculiar-looking fighter, Lockheed executives guessed that the Army might buy fifty of them at most. It was difficult to build, but workers on the line iterated as the fighters went by. They invented new tools and processes to speed up production, including a patented magnetic riveter that cut riveting time on the P-38 in half. “Standardization and interchangeability became ‘musts’ wherever possible,” according to an internal Lockheed history. And Kelly was right there, watching and helping.

The company hired as frantically as it could. “Violinists and paperhangers, morticians and midgets, schoolboys and housewives, Hollywood extras and shoe clerks worked side by side,” a company history reported. Men left the line in huge numbers to serve—24,000 in total fought in World War II, and 394 were killed. And to fill those slots, every stone was turned over; literally nobody was overlooked. Lockheed hired the blind, the elderly, and disabled war vets. “One man with no hands was a tool dispatcher—and a capable one. . . . Numerous seeing eye dogs guided their masters through the maze of corridors to their work benches.”

More than 4,000 high school students worked in shifts, alternating work and school. And women, especially, answered the call, filling every possible job—riveting, sure, but also stress analysis, expediting, engineering, inspection, and tooling. “Big airplanes are made up of small parts,” Courtlandt Gross said. “And women build small parts to perfection.” By June 1943, 40 percent of all workers—35,000 of the 94,000—were women.

Lockheed wasn’t yet ten years old, but since Kelly first sketched the Hudson bomber in the summer of 1938, the company had built more than 2,000 military planes, including 250 P-38 Lightnings. Production of both models increased, and the floor also prepared to start assembly of the new B-17 Flying Fortress, a Boeing design that would be built in a partnership.

“The air industry was called upon to build thousands of something it had built only dozens of before,” Robert Gross later said. “It was like a youth who is suddenly expected to go to college before he graduated from primary school.”

By the middle of 1943, the P-38 assembly line snaked out the door into the parking lots, and Lockheed had streamlined processes so much that it now made fifteen a day. The first P-38 built took 360,000 man-hours. The five hundredth plane took 17,000. And the last one, 10,000 fighters later, took only 3,800.

With its two engines and tails and its shiny silver skin, the P-38 looked radically different from anything else in the air at that time. It was the crown jewel of the Air Force, and the public loved it, too. And that love continued for decades. Kelly would say near the end of his life, in 1982, that he was asked about the P-38 more often, still, than any other plane, including some extremely famous designs that would follow. "Few aircraft have captured the public imagination more swiftly or held it longer," Walter Boyne wrote in *Beyond the Horizons*.

Making a beautiful plane, Kelly thought, was the result of a successful design. It was a beautiful plane because it worked so well. Aesthetics and performance were closely related.

The Fork-Tailed Devil racked up wins and accolades in World War II, but Kelly and his engineering staff did not rest on the plane's laurels. The P-38's design was incredibly versatile and Lockheed's engineers modified it over and over, most importantly by adding new fuel tanks that extended the plane's range to three thousand miles.

By the end of the war, there would be eighteen different versions of the P-38, including a photo reconnaissance plane, a rocket-carrier, and a bomber fighter capable of carrying a bomb load greater than the original B-17 Flying Fortress.

This chameleon of a fighter was a favorite of pilots and it was especially effective against Japan's Zero fighter.

On January 1, 1943, *The Lockheed Star* proudly boasted of the P-38's first action in the Pacific, in a story titled "Lightnings Beat Zeros in First Pacific Clash." The newsletter reprinted a report from the *Los Angeles Times* that "in a series of screaming, scrambling dogfights which ranged from 5,000 to well beyond 20,000 feet," twelve American Lightnings "routed a force of 27 Jap Zeros," destroying thirteen Zeros and two of fifteen dive bombers that fought in the battle.

Meanwhile, the U.S. side lost only a single plane, which fought on and chased down a dive bomber even after having one of its two engines shot out. But the rear gunner on the dive bomber took out the plane's horizontal stabilizer, forcing the pilot to crash-land—and then catch a transport back to base.

At first, pilots were wary of the P-38, thinking it was too big to outmaneuver the smaller Zeros, but once aloft it was surprisingly agile, and a much better fighter all around, with more speed and a tremendous advantage at higher altitudes.

A pilot in a P-38, Kelly said, "could slow down to near-nothing speed, pull back on one engine, cartwheel without stalling, and reverse direction to face his adversary," and pilots flying Lightnings registered 1,358 kills in the Pacific.

Lightnings could fly faster, lower, and longer than the enemy planes, and the P-38 was beloved by pilots for its ability to fly even with significant damage.

In one case, an unarmed surveillance plane ran into a swarm of Japanese Zero fighters near the Coral Sea but escaped—after one engine was shot out—by flying straight up to high altitude, where the fighters couldn't reach it. And thank God it did. The plane's photos would later prove critical in the U.S. Navy's pivotal victory in the Battle of the Coral Sea.

Major Richard Bong, who flew with a picture of his girlfriend Marge painted on the nose of his plane, shot down forty Japanese fighters, and aces like Bong cherished the plane for its durability. One returned to base on a single engine, with its tails split and one boom shredded following an airborne collision. Another lost an engine, took five 20mm cannon shots and one hundred regular machine gun bullets, but got home just fine.

Arguably the Lightning's single greatest moment came in 1943, when U.S. intelligence cracked a Japanese code and learned that Admiral Isoroku Yamamoto, commander of the Japanese naval forces and architect of the Pearl Harbor attack, was due to visit the island of Bougainville at a specific time on April 18 as part of an inspection tour of naval bases.

Yamamoto was famously punctual, so there was great certainty in the U.S. command that a mission to intercept and shoot down his plane could work, provided U.S. pilots were up to the challenge of flying one thousand miles round trip, just over the ocean surface, with no air-conditioning.

Major John Mitchell, commander of the 339th Fighter Squadron on Guadalcanal, was put in charge of a mission code-named Operation Vengeance.

Obviously, Mitchell would use P-38s, because no other plane in the U.S. arsenal could fly under radar, just above the surface, over 435 miles of open ocean from Guadalcanal to Bougainville.

Mitchell and fifteen other pilots, all flying P-38s equipped with drop tanks, embarked on a two-hour, forty-five-minute mission loaded with challenges.

The pilots couldn't even take a direct route because that would require flying over islands with Japanese installations, including anti-aircraft guns and observation posts that could alert and scramble Zeros.

But exactly as expected, the four P-38s flying lead in the so-called killer group spotted Yamamoto's G4M Betty bomber, with six Zero escorts, at the precise time and location it was predicted—at 9:34 a.m., flying thirty-five miles off the coast of Ballale Island.

The P-38s routed the Japanese formation. They destroyed the escorts and then shot down Yamamoto's plane, which crashed into the thick jungle on Ballale, as well as another Betty carrying a second high-ranking Japanese general, while losing only a single P-38. The rest of the Lightnings returned to base, one of them flying well despite 104 bullet holes.

Lightnings were less effective in Europe, especially against German aces who forced the planes to fight at low altitudes, where the P-38's size made it less nimble. But P-38s more than made up for it off the coast of North Africa, terrorizing the German transport plane routes.

It was stalking these transports that earned the P-38 its German nickname "der Gableschwanz Teufel," or "the fork-tailed devil." Japanese pilots had a different nickname for the plane. It was the "two airplanes with one pilot." And American pilots called it the "round-trip ticket," for its ability to take a beating and stay in the air, covering huge distances even after losing an engine.

Its Air Force overseer, Ben Kelsey, had this to say: "[That] comfortable old cluck," he said, "would fly like hell, fight like a wasp upstairs, and land like a butterfly."

When the war ended, and German technical papers were found and distributed, American aircraft designers learned that the Germans had also battled with compressibility and ultimately solved the problem by using swept wings, which first appeared on planes at the Battle of Poland.

This, along with the incremental gains in speed that accompanied huge leaps in horsepower, convinced Kelly—and all his rivals—that the next generation of planes needed not just better wings and tails but also a whole new method of propulsion.

In the six years since Lockheed, then a scrappy start-up, began work on the XP-38, the aircraft industry had boomed; it was completely transformed. By the war's end, it was the single largest industry in America. And Lockheed—thanks in no small part to Kelly Johnson—was at the forefront.

PART II

THE SKUNK WORKS IS BORN

6

BORN OF NECESSITY

In early June 1943, Kelly Johnson arrived at Burbank Airport from a trip to Florida, raced to headquarters, and burst into Robert Gross's office, where he found his immediate boss, Hall Hibbard, also sitting. This was fortuitous, because what Kelly had to say he wanted both of them to hear.

The Army Air Corps wants a proposal for a plane designed around an English jet engine, Kelly told them. He had already run some math. *And they want it as fast as possible*, Kelly said. He was willing to promise delivery of this jet prototype in 180 days.

His two bosses stared at their young engineer as if he'd just spoken to them in German.

This bold promise wasn't quite as audacious as it sounded, either. Lockheed had been talking about making a jet since 1940, when Kelly and Hibbard recognized that propeller-powered fighters just couldn't get much faster than the P-38 and began to lobby the Air Corps to let them make a jet-powered plane.

A jet was a giant leap from the P-38, which itself had been a giant leap over its predecessors. But there are limits to how much power a propeller (or two) can generate, whereas a jet engine gets more power the faster the plane goes—high-velocity air sucked into the engine boosts its power, creating additional thrust. That, combined with thinner, swept-back delta wings, should allow such a plane to fly supersonic.

Still, what Kelly was proposing to his two bosses on this day seemed a little nuts: a radically new plane, from scratch, in six months. No airplane manufacturer had ever built a prototype in a full year, and their cocky young engineer was predicting that he could build a plane around an entirely new means of propulsion in half that time.

"Okay, Kelly, it's your baby," Gross replied. "I don't think anything will come of it, but set up shop."

In reality, Kelly had been pestering Gross and Hibbard for a few years to let him create an "experimental department" where engineers and shopworkers "could work together closely in development of airplanes without the delays and complications of intermediate departments to handle administration, purchasing, and all the other support functions." He wanted, he explained, a "direct relationship between the design engineer and mechanics and manufacturing"—as in, they should all work together, in the same space.

It was a new twist on a relatively old-fashioned style of urgent manufacture—the way airplanes were built in the early days of aviation, when every company was a scrappy start-up working out of a garage.

Inside a sprawling organization like Lockheed, though, a lean, flat special projects shop would allow the company to take technical risks cheaply. A failure doesn't cripple the larger company because it hasn't allocated too many resources, and a hit provides a massive leap in a short time.

This fire drill to build a fighter jet was the perfect opportunity to put that dream into action, while also tackling an aviation challenge that Kelly and Hibbard had been chewing over since 1940. That's when an engineer named Nathan Price—who'd been working in turbine power plants—was tasked with designing a turbojet engine.

Kelly and Willis Hawkins, in turn, had cooked up a stainless steel airframe to fit the engine and named it the L-133—a concept that, Kelly claimed, could approach Mach 1, the speed of sound, and fly at 50,000 feet.

Unfortunately, the Air Corps wasn't interested at that point, and Hap Arnold told Lockheed to keep doing what it was already doing well—churning out sixteen P-38s, four B-17s, and a handful of PV-2s and Lodestars a day for the war effort.

But it wasn't wasted effort, and Kelly's prescience about jet engines would prove to be critical for the future of the Skunk Works. This was clearly the future of propulsion and he, among a very small number of people, saw it coming.

Nazi scientists had been working on jets as far back as 1936, when Ernst Heinkel began tinkering with a new engine. And on August 27, 1939, a German test pilot flew a jet-powered airplane for the first time ever. But instead of deploying this new jet as a fighter, Hitler opted to deploy it as a low-altitude bomber. This decision, Kelly would later say, was one of Hitler's greatest mistakes. "If Hitler hadn't been so stupid to say they must be bombers," Kelly said, "they could have taken our B-17s and B-24s out of the sky."

What Kelly didn't know at that time was that the Air Corps, under the direction of Hap Arnold, had also been pursuing a jet fighter in the shadows. In the spring of 1941, a "Special Committee on Jet Propulsion" had been formed, and an effort was launched to build a prototype jet engine.

Then Arnold learned, on a visit to England, that the Royal Air Force was way ahead. The Brits had put Frank Whittle's W.1 jet engine into a small fighter known as the Gloster Meteor, and the prototype was already flying.

The Air Corps made some inquiries. It arranged for a Whittle engine to be sent to the United States, and on October 1, 1942, a Bell XP-59A with that engine flew at Muroc Dry Lake, making the first-ever flight by an American jet.

It took off and soared . . . all of twenty-five feet over the runway.

The Bell was a dog. The combination of an engine that lacked thrust with high drag from a poor airframe design meant that the Bell's performance against piston-engine planes was pathetic. Its maximum speed in level flight was just 404 mph. Lockheed's P-38, in particular, vastly outperformed it. So this was an easy call. The Army Air Forces Board declared that it was "not believed that the P-59 is operationally or tactically suited for combat."

But word of this secret test made its way to Kelly Johnson. On June 17, 1943, Kelly was in the Florida Panhandle, to see a new version of his P-38 fly at Eglin Field, when an Air Corps colonel shared some important news: An American jet—the Bell—was flying. But it wasn't flying very fast.

The colonel asked if Kelly still wanted to build a jet. If so, he had a different British engine to design around—the de Havilland H-1 Goblin, which was now ready for flight.

Kelly didn't need any encouragement, only the specs.

He got them on the spot and began sketching a design for a jet airplane on whatever paper he could find during the flight home to Burbank.

It was upon landing from that trip that he raced back, marched immediately in to see Gross and Hibbard, and was given the green light.

Kelly picked five of his favorite engineers and went to work on a more formal design to present. A week later, he flew to Wright Field to pitch that design to General Hap Arnold along with the base's commanding officer, Major General Frank Carroll.

The generals loved it and approved the project immediately. In fact, the Air Corps was going to give him a contract for both a plane *and* an engine.

Engines are complicated; they take time, and time wasn't a luxury the Air Corps had. So, for starters, Kelly was to design a plane that could fly the British engine that already existed. Once that was flying and deployed, he could take on the total package and build a wholly American jet.

The proposal Johnson delivered was short, not even a dozen pages.

"And we'll build it in one hundred and eighty days," he said, starting as soon as he was given a thumbs-up. "When can you give me that?"

It was 10:00 a.m. when he asked this question.

"There's a two p.m. flight to L.A.," General Carroll replied, and Kelly should be on it. "Your time starts then."

One day after he got the green light, on June 22, Kelly had already solved his first big problem: the lack of space for his new project. He found a temporary home for his engineers—a 28-by-40-foot space at the top of some stairs in the model shop for Lockheed's wind tunnel, in Building B-1. This would do for preliminary design until he found a more suitable base of operations.

The top secret jet project also had an org chart with Kelly at the top, plus four direct reports, and a list of important principles for the program.

And on June 23, Kelly officially started the clock. "The 150 days starts today," he wrote in the project log for the new jet, which he named the XP-80. "Can't take 'no' for an answer."

Kelly laid out nine key points in the log, including the size of the staff he'd need (twenty engineers, about eighty shop men), the schedule (ten hours a day, six days a week), and what would become one of his tentpole principles: "Secrecy of paramount importance."

A day later, he assembled twenty-five of his favorite engineers in a conference room in the B-1 plant and told them what he had promised the Air Corps. The group around this table was going to design America's first jet fighter in 150 days. That's what he told his team—that they had *150* days, not the 180 he'd promised Hap Arnold. And the engineering had to be finished in half that time.

Kelly's idea was to flatten the process; to shorten the flow of information and ideas by cutting out steps, literally and otherwise. Collapsing the walls increased competency, too. Later, Kelly began to grade his engineers on a scale of 1 to 5, with 1 being best. He could "consistently" take a 3 and "inside of several months" improve him to a 2 or 1. The key, he said, was that his engineers worked so closely with their final project, and that they were empowered to act decisively and take ownership.

And though Kelly himself wasn't deeply trained in every specific variant of engineering for flight, he was an autodidact, and that loomed large in the room. "He is unique," his assistant, Richard Boehme, later said. "You have to be careful or he knows more about your subject than you do."

The best way to increase speed was to cut fat. Kelly identified inefficiencies and removed them, often ruthlessly.

Take drawings. In a typical program, an engineer makes a drawing, then hands it off to a checker, who makes sure all the parts are labeled and spelled right, and that what's drawn can actually be produced. In Kelly's outfit, there were no checkers. The guy who draws the part delivers the plans directly to the one who makes it, and if the maker can't make it, the drawer has to fix his drawing.

Kelly later conducted a survey that showed a set of drawings within a typical aircraft company traveled 6,000 feet, more than a mile, while moving from the designer who drew them to the planner, and on to the tool designer, who made the part that had been drawn. He estimated that drawings in his shop traveled only 600 feet, at most.

The top engineers and shop men were handpicked by Kelly, and those men, in turn, picked their favorites to fill out the ranks. A few others heard whisper of a secret project and pulled strings to get into the program.

Like Warren King, an engine guy who was close with the test pilots, and asked Milo Burcham for help getting in. King showed up for an interview, got hired in, and was told to tell his wife that his new job might require long absences that couldn't be explained.

When King arrived for work, he still didn't know what it was he'd be working on. He just knew that whatever it was, he wanted to be a part of it.

Throughout his early projects, Kelly had also been studying where the gaps were in traditional engineering hierarchies. Delays tended to occur most often because of poor or slow communication and a lack of timely decision-making—either because no one accepted responsibility or too many people had it.

Probably the biggest rule, which is admittedly a little self-serving, was that there should be a single, strong, and knowledgeable leader. That was Kelly. He knew a lot and had the ability to learn and understand in great detail most technical aspects of airplane design. When he didn't know something, he knew what question to ask and who should answer it. The only thing that truly matters is the end result. Does this part serve it? What about this person? Or that meeting you want to have?

Also, don't tell me it's impossible. Always assume it can be done.

This scrappy new outfit was Kelly's fiefdom. In it, Lockheed's corporate policies did not apply. "The idea was to get the idea to the shop," engineer Joe Szep said. "No matter how you got it. We did not have to follow the company practices of drawing up things where you have a title block and certain things you had to include. They said, if it's a scrap of paper and it does the job, do it."

If Kelly's rules weren't iconoclastic enough, the facilities that housed his team spoke volumes for the scrappiness of the whole venture. The place may as well have had a pirate flag flying over it.

With no Lockheed space available for the larger tooling and production work, he just grabbed what he could, annexing a parking lot next to the wind tunnel, and across a set of railroad tracks from a plastics factory that provided an awful stench when the wind was blowing toward Lockheed. And that unlikely plot is where Kelly set about building a temporary structure.

He grabbed old crates (which had once held Wright engines for Hudson bombers) to tear apart for walls and rented the largest canvas he could find for a roof—a circus tent. Joe Szep, one of the twenty-five chosen engineers, called the 11,000-square-foot temporary factory a "lean-to," and had inspectors stopped by at any point, there is no way it would have passed code.

That lean-to, with its pungent air, was enough of a structure to house two stories. The twenty-five engineers worked upstairs, above the shop, and a locked door separated them from the eighty-three who worked on the shop floor. Every engineer was given a key and they called themselves, jokingly, "the Key Men." Shopworkers understood the order and stayed out.

Lockheed couldn't even spare tools, so Kelly bought an entire local machine shop to supply his shack. He also "stole" twenty-two engineers he admired "from around the factory," and named W. P. Ralston and his old college buddy Don Palmer as project engineers. Art Viereck, head of the experimental engineering department, was put in charge of the shop. The operation didn't look like much. But,

Kelly said, "it had every function we needed to operate independently of the main plant."

In the program's very early days, Kelly directed Don Palmer to write down a set of ten detailed principles for the jet project that would evolve, over time, into Kelly's "fourteen basic operating rules."

The initial rules, as Palmer wrote them, were:

1. Centralization of authority and responsibility. Project engineer is fully responsible for all decisions subject only to next higher authority. This includes design, materials, strength, weight, production design, costs, procurement of materials.

2. Everything possible will be done to save time in accomplishing the ultimate result. Project engineer will devise ways and means as required to avoid delays. These ways and means will be implemented by management without decentralization of control or authority.

3. After decisions have been made, there will be no changes without the consent of higher authority than the Project engineer.

4. Information will be passed on to the shop in the most direct and simplest manner. Advance information will be released at the earliest possible time for material procurement, tooling, and shop make-ready. Layouts on paper or metal will be used extensively. Sketches shall be used wherever possible. The airplane shall serve as the basic mock-up and parts may be designed on the spot and made to fit.

5. A purchasing setup working directly for the project and no other shall be made. This setup shall cooperate with the

established purchasing department but shall be allowed to operate without the red tape and policy encumbrances thereof. Keynote—get the stuff.

6. A separate stockroom shall be established and means shall be set up to get project materials into stock directly upon receipt.

7. Receiving, fabrication, assembly, and final inspection shall be one ground of men (one man).

8. Special parts, materials, shapes, fastenings shall be avoided whenever possible. Parts and materials from stock shall be used even at the expense of weight within reason. The chance of delay in delivery is too great.

9. Records of all information going into the manufacture of the airplane are to be kept by the men who develop and transmit the information. This includes changes as necessitated by the fabrication and assembly of parts.

10. Each engineer on this project shall be designer, shop contact, parts chaser, and mechanic as the occasion demands. Engineering shall always be within a stone's throw of the airplane. There shall be but one object—to get a good airplane built on time. Any cause for delay shall immediately be reported to C. L. Johnson in writing by the person anticipating the delay.

Palmer would get a lot of credit for this founding document, but he wasn't remembered fondly by all the key players. Like Irv Culver, who once said that Palmer "didn't do anything but kick cigarette butts under the table when Kelly came in"—butts and other trash being a constant because the custodial crew wasn't cleared to enter this secret facility.

As Culver tells it, Palmer lacked authority among the team because he wasn't a good engineer—"technically incompetent" in Culver's words—and was in his position only because he'd gone to Michigan with the boss. There was also this: "Kelly liked him because he could bawl him out to his heart's content, and there wasn't anything Kelly liked better than bawling somebody out."

The hierarchy was, as Kelly wanted, truly flat. Engineers worked alongside the production guys and supervisors didn't just supervise; they designed. They also determined how something would get made and what tooling it would require so that no further meetings were required. Individuals were not policed. They were trusted to do the jobs that Kelly had selected them for, because he knew they were experts at it. And they tended to get irritable when someone offered insight without being asked.

Kelly wasn't a constant presence, but he liked to show up and walk the line, asking production guys what they needed and then taking those needs straight to the engineers.

Speed and efficiency drove everything. "The trick in designing a plane is to do it as quick and dirty as you can, and then cover your mistakes before anyone sees them," Irv Culver explained. "The longer you take to design something, the worse it gets, because people invent more and more reasons to make it complicated."

That confidence broke down barriers. The group was open and democratic. Everyone was free to criticize everyone else.

"He would give you the job, expect you to do it, and if you didn't do it there'd be somebody in your place to do it," said one of Kelly's early lieutenants, Charles Van Der Zee. "He gave you the authority to do it and he backed you up on top of it. . . . He trusted us, and on top of it he expected us to do our job."

One night, Van Der Zee was still in the shop around 9:00, nearly four hours after the typical end of his workday, when Kelly strolled in.

"Hey, Van Der Zee," Kelly said. "If you can't do with brains, don't do with hours."

"Well, Kelly," he replied, "I can do more by accident when I don't have to worry about the shift than I can on purpose during the daytime."

"Then I'll get you an assistant."

That was unlikely to help, Van Der Zee replied. "If I've got to tell him everything I want done, I might as well do it myself."

Kelly considered this and walked away.

One of the keys, according to Irv Culver, was the lack of draftsmen. The industry norm, he said, is that an engineering shop is laden with them, and "the fancier drawings you make, the further they push you up the line until all the heads of the projects are draftsmen." Technical experts, then, don't tend to rise, and administrators make decisions. Not under Kelly.

Most of all, Kelly despised oversight, and he dreaded the stifling menace of bureaucracy. A founding rule of his nascent operation was that the military wasn't allowed inside.

Just nine days after commencing work, Kelly led a mock-up meeting to present the team's design to six representatives of the Air Corps, led by the project liaison officer, Major Ralph Swofford.

Lockheed's design wasn't radical. It was actually quite conventional-looking. Really, the only substantially novel aspect of the plane was its wing, which was not a typical airfoil shape. Rather, it had a low aspect ratio and a laminar flow surface. Kelly called it a "wind tunnel wing."

The Air Force approved Lockheed's proposal that night, and Commanding General Hap Arnold ordered his men to make available every resource that Kelly would need to build a working fighter plane. Within a week, he had guns, radios, wheels, and tires.

Kelly filled the lean-to with engineers of his choosing, some of them Michigan alums like Carl Haddon. Haddon had been a year ahead of Kelly, but the younger student cast a long shadow. He was, Haddon recalled, already "recognized as a genius."

Kelly encouraged his men to break rules, as he looked out for weak points in traditional design and manufacturing. Decision-making, in his opinion, was often a choke point. So this group would not let decisions fester. All decisions were made at the start of the day, in a 7:00 a.m. staff meeting that department heads dared not be late for. (And if Kelly needed an answer from the customer, he had a special line on his desk that rang through to the Air Force Special Projects Office, staffed by a small group empowered to give him answers.)

At seven on the dot, Kelly asked whoever was standing closest to the door to lock it so that it was literally impossible to sneak in. Staff engineers arrived at around 7:30, and definitely by 8:00, because that's when Kelly would stroll into the design area to review the work, especially any problems that had arisen.

Joe Szep, who led design of everything aft of the wings, explains how it typically worked: "When two engineers come together and have problems trying to decide, in a normal operation they say, 'Gee, we better form a committee.' And the committee says, 'Oh, we have to form a task force and study this.' Pretty soon a task force gives a report. Then they have to study that and come up with an answer. By that time a month has gone by. Kelly came in and says, 'We'll do it this way.' Every day he made a decision that allowed us to go ahead. And that was the secret of the [operation]."

At 9:00 a.m., Kelly would move on to the shop group. He conducted these two meetings separately so that each group could feel free to complain openly about the other. He also wanted the production crew—builders, blue-collar types—to know that they were just as important as the brainy, college-educated engineers. Morale—

what he called "esprit de corps"—was as critical to the project as timely decision-making.

Lunch was thirty minutes, which was not enough time to go out, so everyone packed food from home. Entertainment for this precious half hour was often electrical engineer Henry Rempt, who'd toured America with his family before the war, visiting national parks, and he had a very thorough record of those trips, complete with color slides. Rempt was the group's cruise director, a spirited social butterfly who twice threw Saturday-night parties for engineers and wives at his hacienda-style house in nearby Van Nuys.

Don Palmer and Bill Ralston policed progress and kept a seamless flow of information between the two departments. On day one, they posted a schedule on the wall with the engineering goals that needed to be met every day and week to make the deadline.

There were no offices, and not really even furniture, at least in the traditional sense—just eight-foot boards that served as desks, with a mechanical T-square on each one. For math, engineers used slide rules. To determine curves, they used one of about a hundred different French curves.

Designers would approximate the strength of a particular part, then give it to a structural engineer for stress analysis. They'd go to a desk with an adding machine and check the math, to be sure the parts were strong enough to do the job.

Space was limited, and so was equipment. And the new outfit went about requisition the same way their boss had approached recruitment. They skipped the paperwork and just took what they needed. At night, Kelly's men would go on raids. They found extra Beverly shears, for cutting sheet metal, in Building 7, and extra workbenches out on the P-38 production line.

Van Der Zee led that particular raid. He took six guys, divided into two crews, and gave strict orders: Only take the benches that don't have locked drawers, because locked drawers are likely to contain tools and the owner of those tools is going to miss them and stir up trouble with "Plant Protection," as the in-house security department was known.

Competition between individuals and crews over who hammered, twisted, and bent metal best was fierce. The noise in the shop was deafening as Lockheed's best metalworkers turned shiny rectangular sheets of difficult-to-handle aluminum into elegant silver wings.

"It was all handwork, every damn bit of it," one metalworker later recalled. "We had some crackerjack metal men, people who were just spectacular. You just cannot believe what can be done if you know how to do it."

Great highs were often followed by crushing lows. Within the first two weeks, Irv Culver and Dick Boehme were thrilled to find a way to squeeze 151 gallons of fuel into a fuselage tank, allowing them to eliminate two outboard tanks on the wing. One day later, however, that bubble burst: They'd mistaken imperial gallons for U.S. gallons. Back went the tanks.

Lockheed security stood outside the doors but had no idea what they were guarding. They were simply told to stop anyone who didn't have the proper credentials. Once, an arrogant security man—a top officer of some kind—marched toward the building and defied orders to halt and identify himself. Employees were warned that anyone who defied such orders might cause a warning shot and that's exactly what the guards did; they fired a shot well over the top of his head.

That problem never happened again.

The official name for America's first jet prototype was the XP-80, but to the men who built it, she was known as Lulu Belle, named for a character from the cartoon series *Betty Boop*.

Kelly obsessed over every detail, and as Lulu Belle came to life he would spend hours sitting in the cockpit, trying to determine the ideal places to put instruments and controls. A dramatically faster plane would require quicker reactions, which was likely to result in more pilot errors. Controls had to be simple and in the most logical place. Even a second lost hunting for a button could spell doom. Sitting there, inside the cockpit, Kelly tried to visualize the pilot's experience so that he could design around it.

The pressure, Kelly later admitted, was "intense," especially as days and then weeks went by. He set a schedule of six days a week, ten hours a day, with Sundays off, because he knew that even the hardest, most dedicated workers needed to rest. He hung a calendar titled "Our Days Are Numbered" and every day he would write in giant numerals the number of days remaining until the plane needed to fly.

There were times when the team would work twenty-four hours a day for three or four days in a row—until a co-worker noticed a guy's work slipping and sent him home to sleep it off. Some needed more rest than others, and the group policed itself.

Conditions in the tent were less than ideal. It was dark, cramped, and hot. On July 26, a log entry states, "No work started yet to relieve heat in lean-to. Men are complaining (100 F)." Fortunately, this misery lasted only a day. On July 27, an air cooler was installed.

Another problem was illness. The frantic schedule, and lack of time off, took its toll. The September 14 log reported that three men were out sick and that twenty man-days in total had been lost due to illness.

Of course, in winter, the place was frigid. And that first winter was colder than normal. "It was the most air conditioned place I ever worked in," Charles Van Der Zee recalled, "because the condition of the air on the inside was the same as it was on the outside."

The only heat source was a small potbellied stove that was barely suited for a cabin. When an engineer got cold, he could hover around this stove, which caused whatever side of the body was facing it to be warm, while the parts facing away continued to freeze. The only other warm spot in the entire factory was the cockpit of the prototype, which was insulated.

As the days ticked by and workers grew more frantic, they began to sneak in on Sundays. Kelly heard about this and put an end to it. "By coming back in here on Sunday you're hurting the project," he told them. "You don't get enough rest and you get sick. The next man I catch in here on Sunday goes back to the B-17s."

To make the deadline, Kelly urged simplicity. If something can be done in a simpler, cheaper way, it should be, even—maybe especially—if there's some more elegant solution. For instance, he wanted to avoid using hydraulics for the plane's nose landing gear doors, but the team working that problem couldn't find an alternate solution. So Irv Culver asked for one night to take a crack at it.

Around midnight, Culver grabbed Van Der Zee and went across the road for a coffee, then—feeling refreshed—came back and designed a solution that came to be known as "Little Goon Face."

When it was time to mount the front gear into the well, it didn't fit. Every time the gear closed, it smacked into the fuselage. Charles Van Der Zee called up to engineering for help and Culver, who'd done the drawing, came down to take a look.

"Anyone got a two-by-four?" Culver asked, and a board was produced. He told whoever was there to put the board inside the well, against the gear, and then kicked the board, hard.

"Okay," he said. "Check it."

Someone removed the board and closed the gear. It still hit the fuselage, but the fit was closer. Culver said to put the board back, and he kicked it again. This time, it fit.

A nearby supervisor couldn't believe it. Here they were, working with precision equipment, with great care and meticulous measurements, and how did Culver fix it? Brute force. But he wasn't mad. Quite the opposite. What good is it if it doesn't actually work?

From the outside, activity inside the tent probably looked like chaos, but Kelly's wildly unconventional methods worked. The prototype was progressing on time, and even ahead of schedule, so fast that on August 16, the project log noted a visit from design boss Hall Hibbard "to make sure the progress pictures were not faked!" Kelly was so satisfied by the pace of progress, in fact, that he extended the weekend and gave everyone Saturdays off, too.

That Christmas, most of Kelly's workers took a few extra days' break, leaving a skeleton crew. That's when Hall Hibbard made his rounds of the various departments around Lockheed to wish everyone a merry Christmas. When he got to the lean-to, it was nearly empty and so freezing cold that Benny Benson, the project's on-duty electrical engineer, was huddled over the stove.

He didn't look up when a voice wished him a merry Christmas. He just snapped back, "Merry Christmas my ass!" And then got the shock of his life when he finally saw his boss standing there. That's how Benson—one of the quieter, more reserved workers on the Lulu Belle project—got his nickname, Blabbermouth.

For the first few months, there was no engine for Kelly's team to design around. They had only specs and a blueprint. As late as October 11—nearly four months in—the project log revealed just how much the team was fretting. There was "no word on engine as yet," and only four of the twenty-six workers whose jobs centered around the XP-80's power plant were able to do any work with the mock-up.

The engine was flown, in five crates, from England to Muroc Dry Lake—future home of Edwards Air Force Base—with a stop in Canada, on an Air Force C-54, accompanied by a single British technician named Guy Bristow.

He and the engine arrived on November 3, just seven days before the airframe was finished.

Bristow didn't especially want to *stay* in the godforsaken desert on this brief escape to sunny California, so Lockheed got him a rental car and a hotel room in Hollywood. The first night, Bristow went out driving and slipped into old habits. He was cruising along on the wrong side of the road—which was, to be fair, the right side of the road for a British man—when a cop pulled him over and, hearing an accent, asked for his papers.

Bristow had no driver's license, and his passport bore no stamp of entry, because he'd arrived on a military transport, so the cop arrested this mysterious foreigner and put him in jail, presumably under suspicion that he could be a spy of some kind.

With his one phone call, Bristow called Lockheed and set off a minor panic. Finally, the right person was alerted, officials were called, and someone sorted the mess out.

That Monday, Guy Bristow arrived back to work at Muroc with a bruised face and one hell of a story.

Warren King, one of Kelly's original eight hires for the project, had been put in charge of the British engine, despite having never seen a jet engine in his life. At the time, he didn't even understand how such a thing worked.

So when the Goblin arrived, King asked for "the paperwork"—the manual—and learned that there wasn't any. Paper was a security risk, something that could be lost or stolen.

"They just kind of laughed at me and said do the best you can . . . and left to go back to England," King recalled.

So, King did the next best thing. He closely studied the machine, to identify its many parts, and then reverse-engineered a manual, without ever taking the thing apart.

On November 14, on the 139th day of work, the XP-80 was rolled out of the Burbank plant, under heavy guard, after midnight, when

Kelly knew the grounds—and local highways—would be deserted. It was time for engine tests at Lockheed's secret desert test site, at Muroc.

The plane, which had been painted a dark green, was lifted by crane and loaded onto a trailer, complete with the airplane version of a fake mustache—a propeller on the front to disguise its true identity.

Lulu Belle was 34½ feet long with a rounded nose, a small bubble cockpit, and a 39-foot wingspan, with blended wings mounted just about halfway down the fuselage from the nose.

Warren King rode in a car with "the head FBI man" at the front of a convoy that included two Army Jeeps and cleared the route so that the truck could drive straight up the middle of the two-lane road as it headed north through the Angeles National Forest.

The trip took most of the night, and upon arrival, the truck was directed into a hangar at Muroc's north base. It wasn't unloaded until the building's doors were shut and locked.

Once the engine had been mounted into the fuselage, the finished prototype was wheeled out of its hangar into the open and fired up. Or at least that was the idea. In reality, the engine wouldn't start.

Kelly's foreman, a burly bear of a man named Dorsey Kammerer, stared angrily at the stubborn plane, then stomped off and returned with a sixteen-pound sledgehammer that he set down in front of the plane.

"Now, start, you Lulu Belle," he bellowed. "Start or I'll smash your nose in."

And, if you believe Warren King's account, she started right up.

This prototype's fuselage had a set of ducts on either side that led into the engine air inlet, the opening where air was sucked in to feed the engine. "It looked like a set of pants," Charles Van Der Zee later said. Guy Bristow, the Brit, worried aloud that these ducts were going to collapse if the airplane was stressed.

The first tests, of a tied-down plane running its engine at limited rpm, went fine. But on November 18, the engine was pushed further during static tests, this time all the way to 8,800 rpm, and as soon as the engine shut down, the ducts collapsed, just as the British guy predicted.

Kelly was right there, standing between the two engine ducts, when it happened. "I almost lost my pants down the intake," he recalled.

So Kelly broke his Sunday rule. Art Viereck, Dorsey Kammerer, Kelly, and Bristow tore into the engine, using the Brit's tools, because the engine had been assembled with British nuts and bolts, which rendered American tools useless.

By very early the next morning, the engine was almost fully disassembled—ready to be repaired and rebuilt. But Bristow saw something the other three did not. According to *The Lockheed Star*'s version of events, Bristow "looked up into the grease-smeared, haggard faces" of Viereck, Kammerer, and Kelly. "He was one of the few men in the world who knew a jet engine from end to end."

"I'm frightfully sorry," he told the men. He pointed to a crack in the compressor housing; the engine wasn't fixable. They'd have to ask the British for another engine.

It was gutting news. This was the only Goblin jet engine in America. And it would take at least six weeks for a replacement to be built and shipped over.

The replacement Goblin didn't take quite that long. It was rushed from England and reached Muroc on December 28. Two days later, it was mounted in the plane and ready for a new set of static tests.

This time, the tests went fine. And on New Year's Eve, Lulu Belle's engine revved all the way up to her max, 9,600 rpm, with no complications.

She was ready to fly.

Kelly Johnson set Saturday, January 8, 1944, as the date.

Muroc is one of the driest places in America, but when a schedule has no margin for error, errors inevitably occur. And it rained very hard on the eve of the test flight, so the XP-80 team awoke the next morning to find the entire *dry lake*—chosen as a site for testing specifically because it is flat and hard and almost always *dry*—covered in a few inches of water.

Kelly identified one portion on the east side that the wind had blown dry and asked chief engineering test pilot Milo Burcham to test the firmness of the ground by landing a P-38. The strip held up. It would have to do.

Burcham was Kelly's pick to make the maiden flight in America's first jet fighter. Every test pilot was different. Some, like Tony LeVier, were collaborative. LeVier liked to camp out in the design studio, peering over shoulders, and wanted to know the backstory of every screw and bolt. Burcham was different. He just jumped in and went.

Prior to joining Lockheed, Burcham ran his own air service out of a hangar in Long Beach and was an acclaimed acrobatic pilot. In particular, he chased inverted records (as in, flying upside down), and it was not uncommon for clients of his air service to arrive at the hangar and find Burcham himself inverted—strapped into a chair that he'd attached, upside down, to the rafters.

This is how he trained his body to set inverted records, such as the one he set in 1934 when he flew upside down for 1 hour and 47 minutes between San Diego and Mines Field, in Los Angeles.

Burcham pushed limits with Lockheed, too. As Kelly's P-38 raised the flight ceiling up over 30,000 feet, Burcham took it upon himself to become Lockheed's guinea pig for testing the effects of high-altitude flying. He worked alongside doctors from Minnesota's renowned Mayo Clinic, alternating long periods in a high-altitude chamber and in the P-38, soaring above the clouds. There was no one better, in Kelly's opinion, to make the first true jet flight in U.S. history.

Kelly always built a short final lag into the schedule. He liked to let a new plane—especially an evolutionary leap like the XP-80—sit in the hangar for two days so that he could stare at it and think. He'd walk around it with the key engineers, asking pointed questions about specific components. *Did we do this? How about that? Have you thought of this possibility?*

After that, he'd make one last personal inspection of the plane.

Finally, the next day, it was time to fly.

He had promised his entire secret project team that if they finished the plane on time, everyone could attend the first flight and that there would be what he called a "beer bust" afterward. This was a Lockheed tradition. Everyone who worked on a new plane got to watch its first flight and then had a party on the company's dime.

Before dawn on a frigid January 8, Kelly's engineers and shopworkers, plus Robert Gross and Cy Chappellet—about 140 people in total—piled into three school buses and headed north, toward the North Base at Muroc.

The group unloaded at about 8:00 a.m. and gathered on some berms on the north side of the field, wearing overcoats and wool hats, and watched Milo Burcham prepare to make history. Rumor had it that Kelly and Milo had spent the previous night in a nearby motel and that Milo slept like a log while Kelly stayed awake all night, worrying.

Just before taxiing out to the runway, Burcham took some short, unhelpful orders from the boss.

"Just fly her, Milo," Kelly said. "Treat her nice, and find out if she's a lady or a witch." He then stepped back, joined a small group of top generals—including Hap Arnold—and marinated in the same thought he had every time one of his planes took off into the air for the first time.

What did we forget?

Spectators arrayed on the hills and along the runway stuck index fingers into their ears as Burcham increased power to the XP-80's jet engine. Sand and dust spun up and blasted the crowd as Burcham firewalled the throttle and raced down the runway, from west to east, and took off.

It was not a beautiful start. Almost immediately, the plane began to wobble like crazy, because Burcham wasn't used to having such sensitive aileron boost controls, and the pilot made just a single pass, turned around, and landed.

America's first-ever fighter jet flight lasted all of six minutes. "Great show," Robert Gross cracked to Kelly.

Typically, this would be the end of the day. The plane would be packed up and shipped back to the plant for a postmortem, followed by fixes. But Kelly had a good idea what the problem was here.

An engineer scrambled out to the plane, popped a panel off, and turned off the hydraulic system that controlled the plane's aileron.

Then Burcham hopped back into the cockpit, roared down the runway a second time, and took off toward the San Jacinto Mountains on a very low pass—maybe fifteen feet off the ground.

This time, the plane purred, and the crowd on the hill erupted with cheers as if watching their team's running back break into the clear. Burcham pushed the nose up, swung around, pointed the plane toward the Mojave Desert, and—in the colorful words of Charles Van Der Zee—"came barreling hell-for-election" over the airfield.

Maiden flights tend to be conservative, but this one was anything but. Burcham barrel-rolled the plane. He pointed the nose and flew for the heavens at full throttle. "I can simply say that it was the first time in my experience that I ever saw an airplane go straight up out of sight," one engineer recalled. "You just couldn't see it, it disappeared."

Then Burcham put the jet into a dive and roared toward the lake so fast that most spectators didn't even see the plane until it was past them.

"I have never seen a crowd so excited since my barnstorming days," said test pilot A. M. "Tex" Johnston, who went home and sent a wire to the chief test pilot at Lockheed's rival Bell Aircraft: "Back to the drawing board," it said.

According to the official log entry for the twenty-minute flight, Burcham "climbed, stalled, rolled, zoomed, and buzzed the field to everyone's delight. He reached 490 mph at one point, and everything felt good."

Test pilot Tony LeVier described a "tiny speck on the horizon" moving at a speed he couldn't process. "Before we had time for another thought that speck was a full-sized airplane that roared over us with a swish that became a green blur and disappeared. We were jumping up and down."

According to LeVier, Kelly Johnson cried at the sight. The jet fighter he built in a half year inside a circus tent had not just flown; it had become the first plane in history to approach 500 mph.

The miracle he'd promised the Air Corps, and his bosses, had occurred.

Kelly would later describe the jet's roar with near reverence—as "a blast of sound that surrounded us without seeming to originate anywhere. It was a new sensation."

And Burcham, who'd had the pleasure of flying her, was ecstatic. As soon as the plane came to a stop, he hopped out of the cockpit, threw his helmet on the ground, and screamed in glee: "Jesus Christ! What an airplane!"

Afterward, everyone piled back into those buses and headed to Mint Canyon, where the beer bash began in the shade of some oak trees. They'd barely settled in for an afternoon of revelry when someone noticed a sound, a faint drone that got louder until it was unmistakably the roar of a P-38 ripping through the canyon—the same P-38 that Tony LeVier had used to fly chase on the test flight.

LeVier soared and then dived from the heavens toward that oak tree.

"I thought he was going to prune the doggone oak tree with the props," said Van Der Zee, and customers inside the Mint Canyon Store, a nearby snack shop, came bolting out to take cover, assuming a crash was imminent.

LeVier pulled out of the dive, flew straight up into a hammerhead maneuver, and dived back at the tree, probably roaring with laughter inside the cockpit. His boss—Kelly Johnson—had done it again.

Lockheed basically took over the North Base at Muroc to get the XP-80 ready for delivery to the war's two fronts. For most of 1944, Kelly, the test pilots, and some flight-test engineers—no more than forty people total—squatted in a large hangar on the site and worked on improvements to and iterations of America's first jet fighter.

During that time, part of Kelly's team worked full-time in the hangar at Muroc, sleeping at the base from Monday to Friday and then going home Saturday morning—often with an FBI tail to be sure they weren't consorting with spies.

Kelly and his test pilots commuted in daily. They'd hop over the mountains on a company plane from Burbank every day at 4:00 a.m., then return home at 8:00 or 9:00 p.m. He used these trips to talk to the pilots about problems from previous flights, concerns about future tests, and ideas they might have for improvements on the craft.

Milo Burcham flew the first batch of flights, followed by some Army Air Corps pilots, and then, with flight 36, test pilot Jim White took over. "It was just work, work, work," White recalls. "Everybody did everything"—including, at the end of every day, cleaning the hangar floor.

Pilots flew all day. At night, Warren King would take the engine apart, check its various components, and put it back together for the next day.

White appreciated such intimate exposure to the boss and felt he could ask him anything. One morning, he asked Kelly if he thought humans could reach the moon, which at the time seemed like science fiction.

Kelly didn't laugh. His answer: It's not that hard. He'd just need the resources—"enough money to build a big rocket to get us up there."

White found Kelly's jet surprisingly easy to fly. He was the first pilot asked to attempt a spin in the plane, and he had no reservations doing it, because he'd seen film of a one-third-scale model of the XP-80 do some spins, and not crash, in Lockheed's wind tunnel.

White took off and ascended to 22,000 feet, so that he'd have plenty of cushion, because a plane in a spin loses altitude fast. He executed the maneuver, landed, and explained to the flight engineers in a postmortem data review what it was like to spin Lulu Belle. Those observations then went straight into the XP-80's manual.

Over the summer months, White did speed runs, and despite reports that the plane topped out at 502 mph in level flight, he later reported that he reached 515 mph on a speed run at 35,000 feet. With nearly every flight, a pilot pushed the boundaries further.

It was eighteen months before Lockheed unveiled its first jet to the public, and Lulu Belle—redesignated the P-80 for production—blew the public's mind. Warren King recalled being at Muroc on the day the press was first invited out to witness this exotic new plane. Jim White was the pilot, and he attempted to brief the photographers about his flight plan prior to taking off. *You should listen to me if you want a shot*, he said. The photographers, however, didn't want advice. They assured White they'd be ready.

Warren King warned one of the more prominent photographers that this wasn't hyperbole; he needed to be ready to act when told, and the advice was roundly ignored.

"I'll get the picture," the man replied.

When King saw the dot on the horizon, he knew White was eight miles out, and he told the photographer to have his camera ready. And when the man was slow to act, King warned him again, but by the time he finally did duck under his hood to take the shot, it was too late.

White screamed past, and the photographer tumbled over backward, breaking his camera and hitting his head.

White later asked if "the SOB" got his shot.

"No," King replied. "He got a knot on his head, though."

7

ITERATIONS

Once the P-80 was finished and flying, Kelly acquired bigger, better space for his fledgling operation, but it was—for now—still temporary, a rapid-fire engineering team that existed only to finish Lulu Belle's baby brother, the Gray Ghost, so named for its off-white lacquer paint.

The British engine was a stopgap. Kelly had planned from the start for his new fighter to have an American-built jet engine, and that's what the Air Corps wanted, too. So as soon as the first prototype, with the Goblin, was flying, Kelly and his engineering team went back to work on its successor.

On January 31, 1944, the project log was initiated with this note: The so-called L-141 (its internal Lockheed name) was "officially designated XP80A according to CLJ."

Shortly thereafter, two men were pulled off their jobs at Lockheed proper and asked by Dick Boehme to join this mysterious group that would work on the new variant. Both men—Ed Baldwin and Doug Wakefield—were under twenty-six, and likely to be drafted very soon, so Boehme—by putting them on Lockheed's single most important wartime project—was saving them from that fate.

The pace in the shop was frenetic again. Every entry in the log illustrates how iterative this fledgling shop was.

Kelly was not the only one keeping the log in those days because he was mostly up at Muroc, supervising the flight-test program with a crew of five engineers, but he appears in the notes regularly, lighting fires, making decisions, and solving problems. Some days, though, it seems to be him holding the pen.

Like on March 21, when the log notes that "Hall Hibbard assures me that there would be work for the project to do when XP80A is done. . . . He said project would be left intact."

This is an important note, and one worth pausing on. Because the project Kelly's talking about here wasn't just the jet and its spin-offs; it referred to the experimental shop that he'd built to develop that jet. The one with flat management, no walls, and minimal oversight—the project that would become the Skunk Works, but which was for the moment still just a concept and not yet an official branch of Lockheed.

There was no guarantee in 1944 that the not-yet-Skunk-Works would live beyond the XP-80—or at least beyond its iterations—but Kelly desperately wanted that. This experimental workshop, whatever it might become, was his dream. It was an operation built entirely on a framework of his ideas. An operation that had done exactly what he promised, on time and on budget, proving to him that it should continue as a permanent wing of the company.

And this is something Kelly was telling the troops. That the group would stay intact, and more prototypes would come. On March 31, day 68 of the program, Kelly told his team precisely this—that they would not be disbanded. "There may be a slack period and we will have to make the best of it," the log writer reported. "The group will be kept together. May have to subcontract some work on other projects."

Once again, Kelly set an ambitious target for a first flight and pushed his team to make that deadline, even if he himself was being pulled in more directions, often against his will. (On June 7, the log reports that Kelly, on show pony duty, "wasted entire afternoon in a radar meeting and inspection tour with HLH and 15 radar big shots.")

Finally, on June 10—log entry number 139—the YP-80A (as the XP-80A was rebranded after two early prototypes) was ready. This time, Tony LeVier had the honors, and after a "wobbly takeoff downwind" toward the dry lake, he flew a Gray Ghost with GE's new I-40 radial inlet jet engine around for thirty-five uneventful minutes—hitting 502 mph, 80 mph faster than Lulu Belle—and then landed with just one flap down in front of a large crowd that included both Gross brothers.

The size of the crowd, according to Kelly, "upset" the Air Corps contingent in attendance, but that didn't stop him from hosting another celebratory picnic that was "enjoyed by all."

Six days later, on June 16, the baby was born. It would go into production as the P-80A and then, straightaway, to the field—to Europe. "Latest dope is that the [Gray Ghost] is to go direct to combat zone," says the log's final entry.

Lulu Belle's brother had heavier weapons and wingtip fuel tanks, and was 80 mph faster. And from start to finish, the Gray Ghost prototype took 135 days.

Kelly named Harvey Christen his production superintendent and set him up, according to a company history, "in a room about the size of a large closet—a factory within a factory—to prepare for all-out production."

Robert Gross nicknamed the production jet the Shooting Star, and Hall Hibbard predicted that it would "open new horizons in aviation."

Kelly was an aerodynamicist at heart, and he obsessed over airflow. He wanted his planes to have as little drag as possible. Rivets were smoothed and painted over, and screw heads were filled with putty, which drove the maintenance guys nuts. They had to pick the putty out of the slots in the screw heads just to turn the screws and remove a panel.

Once production was underway, the Air Corps told Kelly to stop filling and painting the screws because it just created work in the field. The second prototype, nicknamed Silver Ghost, was the first plane to be outfitted with fuel tanks on the wing tips, because Kelly learned from wind tunnel tests that putting the tanks there, as opposed to under the plane's wings or belly, created less drag and improved overall performance.

Pilots loved the new jet, but it wasn't perfect. There were growing pains. Tony LeVier was flying a high-speed run between 15,000 and 20,000 feet when, according to Kelly, "there was a booming sound so loud that we could hear it on the ground." He looked up in time to see the plane disintegrate, and then "a parachute blossomed."

LeVier was not injured and explained in his debrief that the accident had been abrupt and catastrophic. He reported that he was just flying along—"sitting there all fat and happy"—when the plane "suddenly flipped over on its back, the wings broke off, and I was sitting out in space."

Men combed the desert to recover the many thousands of pieces of metal that had scattered in the crash, and, using that evidence, engineers identified the problem in the postmortem. The plane's jet turbine disc broke into three parts and sliced through the fuselage. This, Kelly realized, was a welding problem. America didn't have a single press large enough to forge the largest metal parts for planes, like these turbine discs, which meant that they had to be forged and welded. But even the best welds can fail.

Flying a jet was a completely different experience, and Kelly recognized that this would require some new skills for pilots. Even before the jet was finished, he and his test flight engineers created training protocols to show pilots how quickly and badly things could go wrong.

On the one hand, a jet was easy to fly. It had "radically simple controls," in Kelly's estimation. It also moved so fast that pilots could easily get into hairy spots and leave themselves no time to recover. Especially in dogfights. Jets flew so fast that a pilot flying head-on at another jet could easily run out of time to change course. So the jet pilot had to change his battle tactics entirely, attacking from the rear or side instead of from the front.

Not that it mattered much in the short term. Because these shiny new toys had limited use in combat; there weren't enough of them to make a difference in the waning days of World War II, and America's fighter force was doing just fine in planes like the P-38. But having these newfangled jets to fly in still helped American pilots.

The Air Corps used them in exercises at Muroc, to "develop tactics for flying against German jets"—defensive tactics, for bomber pilots, and offensive ones, for fighters who might have to face off against Me-262s in-theater.

A P-80 would make head-on passes at bomber formations, roll over, and pass underneath, inverted. Lateral and tail attacks were also practiced.

And Kelly was there, assisting this program, too. "I spent more than five hours each day, at 25,000 feet, wearing tennis shoes, shorts, and a parachute—riding 'piggyback' in a modified P-38 with Tony LeVier, watching them try to gun down the jet," he recalled later. "We go into some very fancy maneuvers; all of us spun in trying to turn into the P-80. I must confess I enjoyed it."

Useful data emerged from these exercises. For instance, while it was basically impossible to stop a frontal attack by a jet, "it didn't matter much," Kelly reported, because the pilot of that jet couldn't possibly be accurate with his cannons. "At a closing speed of more than 700 mph, the probability of an Me-262 pilot's scoring a hit in the few seconds' firing time he had while avoiding collision with the bombers and fighters was so remote that we decided not to worry about it."

The primary concern, analysts learned, was a rear attack. German jets could easily pursue and catch American planes, then slow down to match the speed, giving the pilots "considerably more time to aim and fire."

In response, the Air Corps taught fighter pilots to pay close attention to the rear view. "These exercises saved the 8th Air Force from having to discover in combat the characteristic of the German jet fighter," Kelly said.

Lockheed's test pilots generally loved the jet, which they found shockingly easy to fly. Herman "Fish" Salmon said, "I push my throttle ahead and I go. I pull it back and I don't. It couldn't be much simpler than that." Tony LeVier called the Shooting Star a "great airplane because there's practically nothing wrong with it."

But tragedy struck the program on October 20, 1944, when Milo Burcham took off from the east-west runway at Lockheed's Burbank field in the first-production Shooting Star and lost power at 200 feet when his fuel system failed. The plane crashed into a gravel pit at the end of the runway and exploded on impact, killing Lockheed's most experienced pilot.

Kelly was devastated and called immediately for the design and installation of an emergency fuel system. Every subsequent Lockheed plane—up to the present day—would have a standby fuel system, which was not an industry standard.

"Redundancy in systems since then became a mania with me," Kelly wrote. "With everything we build, we make sure that we can relight, restart, and keep flying if the main engine pump fails." Losing Burcham wrecked Kelly, but his tragic death definitely saved others.

That system, unfortunately, did not save Major Richard Bong—the legendary World War II ace who scored forty kills of Japanese planes over three combat tours in his P-38. Bong returned from combat and joined Lockheed as a test pilot, and was on his thirteenth P-80 test flight when his plane crashed into a Burbank intersection.

Bong had forgotten to activate his backup fuel system, even though it was on the preflight checklist, and died, like Burcham, shortly after takeoff, when his plane rolled over. Bong ejected but was too low for his chute to open.

In January 1945, Kelly elevated Tony LeVier to chief engineering test pilot. Kelly had offered LeVier this job twice before, and LeVier had turned him down because he didn't think that he'd earned it yet. Finally, on the third try, LeVier said yes, and a day later all the other engineering test pilots—who were furious at Kelly's choice—transferred to production.

"They didn't want to fly for me, and that's the truth," LeVier later said, and for two days, until Kelly recruited some flyers, he was the only active pilot in the department.

Regardless of these not insignificant growing pains, the Army was voraciously hungry for Kelly's new jet. "The Army has descended upon us and again completely disrupted the entire company with a mandatory demand for more P-80s quicker than we ourselves think is possible," Robert Gross wrote to his brother in February 1945. The Army was threatening to "eliminate any or all" of Lockheed's other production if it did not agree to meet this sudden thirst for jet fighters. "We are being asked to build up to a higher monthly rate than has ever before been attempted or achieved anywhere in the industry."

Specifically, the Army wanted triple the number of jets than it had originally requested, which is not the kind of change a manufacturer can easily accommodate, at least not immediately. When a top general came to Burbank to break the news, he said this to Gross: "I don't know whether you boys at Lockheed are the luckiest people in the world or the unluckiest."

A month later, the Army announced its new jet to the world. Hap Arnold did the honors.

Arnold introduced the "whispered-about, rumored-about, guessed-about, secret P-80" in a War Department press conference on March 1, 1945. "It is the fastest fighter in the skies," Arnold said.

That day, *The Lockheed Star* was filled with stories about America's first jet fighter, including a short interview with Tony LeVier, who said that what he liked best about a jet was that it was more comfortable and simpler to fly than anything else in the skies. "You relax when you realize there aren't as many things to watch as in a conventional plane," he said. The jet had very little vibration and a pressurized cabin. And he especially loved the sound of the jet engine—"like a big blowtorch."

Lockheed prepared for mass production and began tooling for a second plant in Kansas City, with a goal of producing thirty a day. Four demonstration jets were shipped to Europe, to show Allied pilots what this new plane was capable of. But the Shooting Star never did see combat. By the time it was ready, the Nazis were on their heels, and the Air Corps didn't want to risk having a jet shot down and its technology stolen.

When the Germans surrendered, U.S. commanders invited top engineers to Europe to pick apart German planes. Kelly was busy, so he sent his assistant, Ward Beman. The biggest surprise: The Germans had a more efficient power plant, with an axial flow jet engine instead of a centrifugal compressor.

Still, the P-80—and its successors—were America's future. The next war, whenever it happened, would be won and lost with jet fighters.

In January 1946, Colonel William H. Councill took off in a Shooting Star from Long Beach and headed east, carrying extra fuel in wing-tip tanks that were ejected and—according to rumor—snapped up by farmers, who cut them in half and used them as feed bins. Councill landed in New York City 4 hours, 13 minutes later, setting a new cross-country speed record. He'd flown 2,470 miles at an average of 584 mph.

One of the most popular and lucrative adaptations of the P-80 was the two-seat T-33 trainer that the Air Force rejected when Kelly first proposed it. (Because, they said, the P-80 was so easy to fly that there was no need for a trainer.) Kelly built a prototype anyway, and the Air Force bought a pile of them. So did the Navy.

When North Korea crossed the 38th parallel and invaded the South in 1950, sparking the Korean War, the Shooting Star was drafted as America's primary fighter in the first war that included air-to-air jet combat.

In early November 1950, a Shooting Star won history's first jet battle, when Lieutenant Russell Brown shot down a Soviet-built North Korean MiG-15 during a brief dogfight over Korea. Brown's P-80 was chasing the MiG in a maximum-speed dive when he fired with the only one of his four cannons that wasn't jammed and hit the enemy jet.

The Shooting Star was another cash cow for Lockheed; it would bring the company $213 million from the delivery of seventeen hundred jets, and subsequent versions "kept production lines humming for 13 years." Counting all the derivatives, including the T 33 trainer, Lockheed built more than six thousand planes in the Shooting Star line, and when the company passed that number, Kelly went to see Robert Gross, who had been skeptical of the program's potential.

"Bob, you said nothing much could come of this, but something has come," Kelly recalled telling his boss. "He was in no way convinced that the turbine engine was here to stay."

On October 6, 1948, Lockheed's publicity manager, John Tower, wrote to *Life* magazine to nominate thirty-eight-year-old Kelly Johnson for the magazine's "proposed story on men under 40 who have made important scientific contributions."

Kelly, Tower wrote, had led the development of "such forward-looking aircraft" as the P-38 Lightning, the Constellation, the Navy's 180-passenger Constitution, and especially the Air Force F-80 Shooting Star fighter jet. "Under his direction, Lockheed has established probably the largest independent research laboratory in the industry" and "the first privately owned wind tunnel in the industry" while building a stable of engineers who innovated daily. When test pilots of the P-38 "encountered the phenomenon of compressibility," Kelly solved the problem with his famous dive flaps, since "adapted by the Air Force for use on all high speed aircraft."

But his crowning achievement to date, in Tower's opinion, was the original Shooting Star jet fighter, designed and built under a circus tent in 143 days. "Although most Lockheed developments are the result of a conference system of developing ideas, the F-80 was actually a one-man job," he wrote.

The aviation world celebrated Kelly's achievement, and it was remarkable. But he knew even before the production model of his jet was flying that his plane was obsolete; all of them were by the time they were in production. Good engineers, Kelly thought, should be thinking years ahead—ideally 100 years ahead, but at the very least two or three years down the road.

"The P-80 is an interim airplane," he said at the time. "It is the best we can produce now with what is available."

Still, the plane's legacy was enormous. Wars change societies, in bad ways, obviously, but also for good. They force radical change. And one of World War II's greatest legacies, Hall Hibbard would argue, was the jet. "There will be new facets to our lives," he wrote in an essay titled "The Thrust Power of the Shooting Star" that argued jets would transform travel. In "10 or maybe 15 years . . . every capital on the globe will find itself within one day's flying time of every other capital."

The period following World War II, he said, would be one "at least as spectacular as the development period of the airplane in the 20 years after the first world war."

What began as a challenge to Kelly Johnson ended as a clear triumph. His experimental shop had been so effective that the P-80 prototype—America's first jet fighter—cost only 63 percent of what it cost to make the XP-38 prototype, which used conventional propulsion. It leapfrogged America's Air Force to the head of the global pack, without costing a fortune.

The other legacy of the program—arguably the greater one—was the unit that produced this jet, the hub for lean, nimble innovation that Kelly Johnson had been dreaming about almost since he joined Lockheed. It was essentially the legendary Skunk Works he'd run for the rest of his career, just without the name. That part came about by accident, and it took years for Kelly to accept that it was here to stay.

The origin of the Skunk Works name goes, roughly, like this. Secrecy during World War II was so strict that engineers in Kelly's group were forbidden from even identifying their department when the phone rang.

In fact, there was a single phone, and one day Irv Culver answered the phone and said, just off the cuff, "Skonk Works, inside man Culver."

A group of Pentagon officers were on the line, expecting to connect with a conference call. These stiff men, flustered, asked the guy who'd answered to please repeat himself.

"Skonk Works," Culver replied, and the rest is history.

It was, Culver later explained, not the first time he'd used the term. It's something he thought about, because he was often answering the phone and didn't know what to say. He needed a bit.

Culver's inspiration was *Li'l Abner*, a popular cartoon strip of that time that featured a drink called "kickapoo joy juice"—a non-alcoholic moonshine made in a shack out in the woods, from a secret recipe of ingredients, including old shoes and skunks. Inhabitants of the cartoon strip called it the "skonk works."

Culver probably chose those words because Kelly's shop smelled bad. Remember the plastics factory across the tracks?

Word of this "skunk works" got around the plant, and mystified employees of other departments. When it reached Kelly, he was not amused.

According to Culver, the boss was furious "and fired me twice that day. It didn't bother me because he always fired me at least once a day anyway. He fired everybody else too, so it didn't matter much—we got used to it. If you didn't get fired at least once a day you weren't working."

Over time, Kelly Johnson would accept the Skunk Works name. He even came to like it. But you know who didn't? Al Capp, creator of *Li'l Abner*. Capp notified Lockheed that Skonk Works was a copyrighted name and Kelly, rather than create a public fight, changed the *O* to a *U*.

Skonk Works became Skunk Works.

And thus was born one of the most famous names in engineering history.

8

YOU CAN'T WIN THEM ALL

Dick Heppe came to Lockheed in the summer of 1947 as a twenty-four-year-old kid who, like Kelly, had done both his bachelor's and master's in aeronautics—in Heppe's case, at Stanford. He was hired as a junior aerodynamicist, on a salary of eight dollars per week. Two years in, Heppe was assigned to a new jet fighter prototype project run out of a special division of the company that people were calling the Skunk Works.

By the time Kelly's Skunk Works handed off the P-80 Shooting Star to Lockheed proper for production as a fighter, there wasn't another big job to turn to immediately. Most of his band of nimble geniuses dispersed back to their regular departments to work on projects for the larger company, and Kelly resumed his role as Lockheed's chief engineer.

He did, however, keep the spirit of the Skunk Works alive and made it known to his superiors that his rapid innovation lab, which had built a jet from scratch in 180 days, could be quickly and easily reconstituted for work on any special projects that might arise.

Kelly's successor, Ben Rich, would later describe the Skunk Works between 1945 and 1954 as a "fast action shop" that could design or pitch in on exotic projects and prototypes that needed to happen fast, and quietly.

One such project was a twin-engine delta-wing plane that Kelly called the L-153. Papers liberated from Germany after the war showed that the Germans were bullish on swept-wing fighters—meaning, they had wings that angled back and didn't stick straight out the side of the fuselage like the P-80s did—and Kelly spent a lot of time considering this design.

In the summer of 1946, the Air Force gave him a contract to build two prototypes of a swept-wing fighter concept. This time, the job wasn't necessarily Lockheed's to lose. Both McDonnell, with the XF-88, and North American, with its XF-93, were also asked to make swept-wing prototypes. May the best contractor win.

Lockheed really needed the business. After the boom time of war, business had crashed, and these were grim days in Burbank. A new transport, the L-146 Saturn, failed, and one of the company's most promising planes, the Model 749 Constellation (the latest version of the passenger plane Howard Hughes had first commissioned), suffered a devastating blow when an electrical fire led to a crash in Reading, Pennsylvania, creating a PR disaster for a plane Lockheed hoped would have broad appeal with airlines.

That year, 1946, the company lost more than $21 million.

A new fighter contract could save the day, and Kelly settled on Willis Hawkins's delta-wing version of the XP-90. But this plane, too, was troubled. Hawkins's first design was too heavy and a subsequent version was unstable in wind tunnel tests.

Dick Heppe, the new kid, was asked to analyze the directional and dynamic stability. This wasn't something he'd learned in college, but Heppe picked up some books and taught himself the necessary tricks, then worked for several weeks and found that the vertical tail designed for the prototype was far too small—by as much as 25 percent.

Heppe presented the results to his boss, who did the math himself and agreed. Which meant that it was time to tell his boss's boss—Kelly Johnson—that a key component of his prototype, already under construction, needed a redesign.

Heppe was in awe of Kelly, who was already, by this point, "a legend" to young engineers like him. And suddenly he had to sit in front of the man himself and tell him that he needed to make his plane's tail much larger.

"I was, frankly, trembling in my boots," Heppe recalled.

Kelly did not dismiss the young engineer or get angry at him for bearing bad news. He listened, asked questions—prodding for mistakes—and then accepted Heppe's conclusion.

"It's a bitter pill," Kelly said, "but you're right."

He assigned an engineer, on the spot, to go redesign the tail.

"So that taught me the first lesson about Kelly," Heppe recalled, "which was that, if you knew what you were doing and you can satisfy his probing to a reasonable degree, he would trust you."

This trait—that a confident, intimidating, often intractable genius could accept bad news, even from underlings—instilled confidence in his workers. Which only made them want to work more, and harder.

Unfortunately, fixing the tail did not solve the L-153 program. Kelly's prototypes were plagued with issues. There was little optimism of winning the contract.

On May 19, 1947, Kelly admitted defeat. He wrote a memo to Hall Hibbard recommending that Lockheed abandon the delta-wing design and shift to a fixed-wing plane with Fowler flaps.

The decision did not go over well with Kelly's team. Hawkins and a few other key engineers fought to continue work on the plane and ran additional tests, but the data did not change Kelly's mind.

He dreaded going to Wright Field, he told Hibbard, "to change our mind for the third time on this configuration," but it was the only path forward.

The Air Force was not angry. The news that Kelly and Hibbard delivered—that they were changing course yet again—was "well received," Kelly noted in a log, "and in spite of its being an unpleasant job to admit our early mistakes, they took it in grand fashion."

Instead, he was given a green light to pursue a swept-wing version.

After a string of rapid and remarkable hits, Kelly's Skunk Works struggled with the XP-90 more than anything else he'd worked on. Numerous experiments failed and the U.S. government's needs required the team to pursue multiple variants—including a ground attack and reconnaissance version—at once. The strain was, for the first time, making Kelly doubt whether his Skunk Works system could handle this particular project.

Morale suffered, and Kelly was having to deal with key engineers who requested transfers to other departments at Lockheed. "Pressure has been so intense for so long," he noted in the project log, that some of the "old gang" wanted to make changes. "This has been a long tough job."

Through the end of 1949 and into 1950, Kelly reduced staff and stated openly that he doubted the F-90 would ever reach production. But work continued anyway.

In April 1950, Tony LeVier went supersonic for the first time, in an F-90. This was a single-seat plane, so Kelly couldn't ride shotgun. Instead, he was on the ground, watching LeVier go into a supersonic dive, and the sound of the plane breaking the sound barrier—causing sonic booms—frightened him.

He worried that LeVier had crashed and rushed inside to mission control, only to be told that LeVier was just fine.

Still, Lockheed didn't win that contract. In September, three months after war broke out on the Korean Peninsula, the Air Force picked McDonnell's XF-88 (which evolved into the F-101).

Increasingly, Kelly's reputation around Burbank and the industry was that he could do no wrong. But his shop was not infallible, and this was proof. It's hard to see the XF-90, for instance, as anything but a total failure. Heppe recalls the prototype as "a disaster"—grossly overweight, underpowered, and unstable, with a center of gravity that was too far aft.

Which was a valuable lesson for Kelly and for those who looked up to him. Proof that "he put his pants on one leg at a time," Heppe said. "He didn't jump into them." And that he could learn how to deal with failures, to step up and take the lead role to recommend cancellation once it was clear that a particular design just wasn't going to work.

Like Heppe, Kelly came to see that the project was doomed from the start. It was an inferior design, based on outdated specs, and incapable of performing with the underpowered J34 jet engines currently available to him. "Its military requirement concept was poor, and it was ten years before the engine for which it was designed was available," he later said.

He also called it "the only airplane to defeat the atom bomb," which refers to the bizarre afterlife of one of his two prototypes.

For some reason—perhaps because of its airframe's strength, or maybe just because it was a useless relic—that F-90 prototype was taken to Frenchman Flat, Nevada, and, in 1952, subjected to three separate atomic bomb tests.

It survived, the author Jay Miller notes, "with remarkably little damage."

Kelly would say at the end of his career that he never worked on a plane he didn't believe in. To do so was to invite failure, because there's no way to do the kind of work he was doing—ambitious and fast—without believing in it. There were, however, some projects that he regretted a little in retrospect.

One was the XFV-1, a concept cooked up at the Navy's behest that is sometimes remembered as the "Pogo Stick." The Navy wanted a Frankenstein of a machine that could take off from a stationary position, with no runway—like a helicopter—and yet fly like a plane. This machine should sit on its tail and take off and land "from a small area on any Navy ship," Kelly explained. It was also, he said, "the only airplane we ever built which we were afraid to fly ourselves in the final tests."

The concept wasn't bad; arguably, it was important. The Navy had been interested in vertical takeoff and landing—or VTOL, as it is commonly known—since 1947, when the first feasibility studies were conducted on a plane that could protect wartime convoys and provide close support. And the Korean War provided plenty of incentive for the Pentagon to invest in novel concepts that might increase U.S. air superiority.

So on April 19, 1951, Lockheed got its first-ever contract from the Navy, to build two prototype VTOL aircraft, which it called the XFV-1.

Kelly put a guy named Art Flock on the job as project engineer, and Flock began to design around an Allison XT40 turboprop. With Kelly working over his shoulder, Flock created a plane that sat on an X-shaped tail—each point of that X having shock absorbers and a load-bearing wheel.

Starting very early in a project's development, Kelly would invite his test pilots to come and look at the plans. The men who would ultimately fly a plane could, and often did, see things in a design that engineers wouldn't, or couldn't. Test pilots Fish Salmon and Tony LeVier kept a close watch on the project and were excited about it.

"It was such a radical departure from anything that had come before," Salmon later said. So radical, in fact, that whoever tested the plane would earn a bonus, for hazard pay, which was a bit of a red flag. Among the things Salmon helped Kelly figure out was how to position the pilot so that he wasn't uncomfortable in either vertical or horizontal flight.

What worried Salmon most was ejection. If the plane were to malfunction while it was in vertical mode, the pilot would be blown out sideways and not up. The solution, developed in cooperation with the Navy, was an extremely fast-acting parachute.

One day, Salmon was in a hangar at Edwards Air Force Base, watching mechanics tinker with a prototype in preparation for some testing, when some Navy pilots came through and were mesmerized by this bizarre aircraft. One of them cracked a joke: "I wonder where they keep the cage for the fella who's gonna fly that thing?"

By early 1953, the first prototype was ready, with a temporary engine and a temporary strut so that pilots could take off and land normally, for testing. Fish Salmon was at the stick on December 23 for the first high-speed taxi test and took off by accident but avoided catastrophe and set back down as quickly as he could.

Seven months later, Salmon flew the plane on purpose, and made twenty-one subsequent test flights, all in conventional mode, through March 1955. A major focus of those tests was the transition from horizontal flight to hovering, and vice versa. Salmon experienced buffeting during those transitions but worked through it.

The prototype experienced a number of complications, including regular and major breakdowns of the propeller and engine, which had heating problems, cooling problems, and lubrication problems. The temporary landing gear became a problem for high-performance testing, so Kelly had no choice but to get rid of it. That meant the pilots would have to get comfortable flying in vertical mode.

Getting into vertical mode while in flight, Salmon said, was easy, but once a pilot was in that state, things got confusing. It was difficult, he said, to know if you were hovering, rising, or falling.

Kelly's engineers found a solution for this in a headset that provided continuous voice transmission. It said either "up up up," "hold,

hold, hold," or "down, down, down," depending on the orientation of the plane. But this only solved half the problem. It was at least as important for the pilot to know how fast he was climbing or descending.

The latter was especially concerning. Pilots had to feather power as they descended. Go too slow and the plane would just "start to lean over and fall out of the sky," Salmon explained. To fix that, "you simply add power and slide sideways through the sky for a while." The closer a pilot got to the ground, though, the dicier things were. He had to look over his shoulder while adjusting an engine that wasn't very sensitive to slight adjustments. Mostly, the pilots tested "landing" on clouds, far from the ground, and no one ever did touch down vertically on land.

In the end, the concept of an interceptor that could take off and go straight up just didn't pan out. To take off and shoot straight up, fast, required way too much fuel. The top speed in vertical flight was also 60 mph. This meant that Salmon, who was also testing the F-104 at supersonic speeds, was flying the fastest and the slowest airplanes in the world at the same time.

Even in level flight, the plane's top speed was just 580 mph—too slow for battle, especially considering that a plane like this wouldn't be terribly nimble; it was likely to get crushed by Soviet fighters in combat.

The XFV-1 was simply a dog. And Kelly killed the program on June 16, 1955.

"Dear Navy," he wrote. "We're afraid to fly this thing." He later claimed it was "the only time I ever had to eat that much crow on an airplane."

Arguably, Kelly's most famous mistakes occurred around the same time. But the other one was a failure of his own imagination—a rare example of Kelly just being wrong—and not a project that went sideways. It happened when the Air Force came looking for a solution to a problem that arose during the Korean War. There just wasn't a cargo plane big enough and strong enough to move large volumes of personnel and war machinery.

In February 1951, an RFP (a request for proposal) went out to manufacturers calling for a medium-weight transport that could meet the branch's ambitious demands. It needed to carry ninety-two infantrymen up to 1,100 miles, or 30,000 pounds of cargo 960 miles, take off and land from short, crude airstrips (paved and unpaved), fly as slow as 125 knots for paradrops, have a rear ramp that could open in flight for dropping equipment by parachute, carry heavy equipment (like trucks and artillery), and be capable of flying on a single engine, should the other one fail or get shot out.

Willis Hawkins took the lead and drafted a team of engineers to work on a design for what Lockheed was calling the Model 82. Four months later, he brought his concept to Hall Hibbard, who asked if Kelly had seen it. Kelly had not, so Hibbard summoned him. Kelly took a look at what Hawkins was presenting and apparently scoffed. He was, the historian Walter Boyne writes, "perhaps not entirely immune to the 'not-invented-here' syndrome."

"If you send that in," Kelly told Hibbard, "you'll destroy the Lockheed Corporation."

Hibbard ignored his ambitious and bullheaded protégé, and history shows that he made a wise decision. The Air Force selected Lockheed's proposal and approved two prototypes, kicking off development of one of the most lucrative programs in the company's history.

This was a goliath of a plane—one that seemed to people who didn't understand aviation like a thing that couldn't possibly fly. The C-130—as it became known—could swallow Jeeps and tanks and bulldozers. Its floor was built to hold up to 300 pounds per square foot, and had tie-downs designed to keep a 25,000-pound vehicle in place even during a crash landing, to keep things from snapping free and crushing the cockpit.

It was not, to Kelly's chagrin, beautiful. Nor could it be. This was a utilitarian plane designed to be powerful and durable. It was "squat, square, and rugged," and when the company's top officers gathered in Hall Hibbard's office to review the first model, they were all speechless.

Finally, Hibbard cracked a joke. "Beautiful paint job, don't you think?"

The production model was renamed the YC-130—or Hercules, the constellation chosen as its nickname—and it was not sexy in any way. Its own designer, Willis Hawkins, later referred to the plane as a "terrible boxcar type machine" esthetically. But its utility, not beauty, was the point. "It wasn't a pretty airplane, but it did its job," he said.

For Hawkins, the ends justified the means. Sometimes, a great plane isn't beautiful. "But Kelly had a feeling that unless you were the fastest thing in the skies, you weren't really contributing. And so he had a little difficulty getting himself oriented to something that was functionally a truck, whether it looked like a truck or not."

The C-130 *was* kind of a flying truck, a flying truck that would become one of the most versatile military planes in history. The prototype flew in August 1954, on a short sixty-one-minute hop from Burbank to Edwards Air Force Base, with Kelly flying chase in a P2V Neptune.

The first production model flew in 1955, and in 1993 Lockheed delivered its two-thousandth production model, a C-130H, to the Air Force.

Since its first flight, in 1954, the Hercules has been everywhere and done everything—twenty-nine versions have been built and sixty-three countries operate the plane. "Aircrews have flown it to both poles, landed or airdropped military supplies to hot spots from

Vietnam to Afghanistan and performed countless relief operations around the globe," the company's internal history reports. "The Hercules has been used to drop bombs, retrieve satellites in midair, conduct reconnaissance and attack ground targets with cannons. Some models are flown as commercial transports. The C-130 has the longest continuous military aircraft production run in history and one of the top three longest continuous aircraft production lines of any type."

It's one of the most effective planes ever built. And the rare example of Kelly Johnson being extremely wrong.

9

THE MISSILE WITH A MAN IN IT

In the middle of 1952, the Air Force once again came to its close collaborator Kelly Johnson for help. Lieutenant General Benjamin Chidlaw, chief of Material Command, asked Kelly if he could talk to Korean War pilots in order to study the performance of U.S. fighters in combat. What, the Air Force wanted to know, do our guys *really* need?

There was reason to worry. North Korean pilots in Russian jets were consistently outperforming American fighters, including Lockheed's Shooting Star. And the enemy's strategy was easy to parse: Any pilot in trouble was instructed to point his plane straight up and ascend to 52,000 feet, where he was untouchable, because the Americans couldn't reliably get over 40,000 feet. The message, Kelly reported back, was universal: "For God's sake, when we start the next war, give us something that is faster and higher than the enemy!"

Kelly came back from Korea, assembled some of his favorite engineers, and announced that the Skunk Works was back in business with a new goal: We're going to Mach 2, and we're doing it in an operational fighter.

This was an audacious challenge. It had been only five years since Chuck Yeager broke the sound barrier, and that was in an experimental plane. Other manufacturers were chasing Mach 2, but they were doing it in odd, purpose-built designs that would never work as production fighters.

Lockheed had never built a supersonic plane (the F-80 could only go supersonic in a dive) or even conducted a supersonic wind tunnel test. "And we were to go out and get to Mach 2 with an operational fighter," Dick Heppe recalled. "Quite a challenge, but one which inspired all the people working for him."

The mood in Washington was right for this kind of leap. Inadequacies in U.S. firepower in the Korean War, especially in air-to-air combat, showed the Pentagon that the United States wasn't keeping up with the Soviet Union. Research budgets skyrocketed, from an average of $164 million per year during World War II to nearly $1 billion in 1951. "We are neither at war full-out nor at peace full-out," Robert Gross said, and there was a lucrative sweet spot there for his company.

In December 1952, Kelly, Dick Heppe, and Bill Ralston flew from Chicago to Dayton aboard a Lockheed Constellation operated by TWA on a dark, snowy night in order to visit Wright Field and present, Heppe recalled, "our concept of what was to become the XF-104."

The concept lacked specifics. It was, Heppe said, very much "Kelly's way of operating" to present basically jazz hands—a broad overview and general guidance of his concept. Kelly was too busy himself to get caught up in details and believed that his word—and the reputation of the Skunk Works, his impossible factory—was enough to sell programs, even unprecedented ones.

Heppe recalled Kelly's pitch as this: "We've never been supersonic before. And neither has anybody else, except in rocket-powered and research airplanes. And there are so many things we don't know. But nature has taught me over the years that symmetry is one thing you can always count on as being safe, conservative and right."

Kelly told the generals that Lockheed's XF-104 "Starfighter" would have a mid-wing. In other words, its wings would meet the fuselage right in the middle, which was an unusual and surprising choice. Putting the wings smack in the middle is not ideal on a structure that will be subjected to very high loads. But Kelly was confident that his structural engineers could figure it out.

The Air Force liked his pitch, at least well enough to say that if Kelly reached Mach 2 in his design, the program would go forward; if he didn't, it was dead.

So that was the bar that had to be reached, and Kelly would not let Heppe forget it.

The early tests were encouraging—enough for Heppe to assure his boss that they were going to make it. Heppe actually did believe this, but it was a bold position to take for a guy who was only five years out of college, and whose only primary Lockheed experience to date had come on the F-90, a failed program. Still, Heppe told Kelly he had "absolutely no doubt," and Kelly took the man at his word.

"Who am I? I'm nothing," Heppe later recalled. "I've done a little bit of work on one other airplane. Nobody has ever done a supersonic airplane like this in the industry before. And I'm standing there saying we're going to make it. And he trusted me."

That moment stood out to Heppe as a signature of Kelly's management style. He couldn't deliver miracles unless the men who worked for him were capable of making them, and the way to motivate these men was with responsibility and belief. Heppe called it "extreme motivation," which enabled people to "do things maybe they themselves didn't know they could do."

Kelly didn't just entrust Heppe with delivering that Mach 2 plane. He made Heppe Lockheed's principal contact with Wright Field. He, and only he, would fly to Dayton once a month to assure the Air Force that progress was being made.

"Routine as the clock," Heppe said, "the very first of the month, I was on the airplane."

The Starfighter—as it became known—wasn't technically a Skunk Works project, because there wasn't technically a Skunk Works at that point. Once the various XP-80 programs wound down, the Skunk Works ceased to exist as a physical space and moved instead to its temporary residence in Kelly Johnson's head.

He would, when presented with a problem, form a mini Skunk Works group to tackle experimental programs that required a rapid prototyping team. That's what he did for the XFV-1, C-130, and a corporate jet called the JetStar.

Kelly was also by this time critical to the larger company, overseeing numerous programs and thousands of engineers on big commercial jobs that earned the company enormous troughs of revenue. He had no time anymore to lean over shoulders and dig into specifics the way he had on the XP-80.

Heppe described Kelly's role in those days as "broad supervision." He didn't make the critical design decisions, but he did decide who would.

For instance, when it came time to choose the shape and thickness of the Starfighter's wing, or the size and location of the horizontal tail, Kelly delegated to engineers he knew and trusted to be the subject matter experts.

An issue that troubled the XF-104's aerodynamic design team was that the only place the horizontal tail could go was up high on the vertical fin, which was fine until the plane encountered a very high angle of attack, in which case it would often pitch up and lose control.

Heppe's team invented a mechanical system that could detect the problem before it began and take over to prevent a stall. Kelly heard the idea and immediately approved it.

Designers of the system considered it so critical—it would save pilots' lives—that they insisted the wiring be put in metal conduit. This was more expensive, more complex, and especially heavier. Engineers argued that regular wiring would be fine. But Heppe pushed back. It wasn't fine. The argument escalated until Kelly was dragged in.

The two sides made a case to the boss in Kelly's office. He listened, absorbed the arguments, and told the production guys to listen to Heppe.

"Put it in just like he wants," Kelly said.

That was that.

The year 1953 was an eventful one. On August 12, the Soviet Union raised the global stakes, and increased panic of an apocalypse, by detonating the world's first hydrogen bomb. It was also the most lucrative sales year in Lockheed's history to date, and the one in which Kelly was promoted to become Lockheed's chief engineer, succeeding his boss and mentor, Hall Hibbard.

Construction of the first F-104 began in 1953, too. Specifically, on February 27, Kelly's forty-third birthday. Six months later, Lockheed announced its contract to build prototypes of a new air superiority fighter known as the XF-104 for the Air Force, which would only allow Lockheed to describe this secret new plane as a "piloted jet airplane."

The F-104 was to be an evolutionary leap, a plane that, like the F-80, would eventually do all sorts of jobs. "We look to the F-104 to maintain our record for building the best and fastest fighter planes," Willis Hawkins said.

Kelly's new jet was unlike any of its peers. North American, McDonnell, Grumman—the other major U.S. military aircraft suppliers—were all at work on supersonic fighter concepts with swept-wing designs. But Kelly, never one to follow a trend, stuck with stubby, straight wings that helped give the plane its incredible (and mildly terrifying) nickname, "the missile with a man in it."

The F-104 would be capable of hitting 1,400 mph and reaching 103,000 feet.

When Tony LeVier saw the design for the first time, he was skeptical.

"Where are the wings?" he asked.

"You don't need much wing if you're flying at Mach 2," Kelly replied.

Ben Rich, then a young engineer, later described the wings as "thin and sharp as a Gillette razor blade" and said that they had to put red paint on their leading edges when the plane was on the ground so that people walking into them didn't cut themselves.

Whatever swagger Kelly may have lost after the XF-90 and XFV was back with the F-104. Upon returning from a three-week tour of Europe's aviation industry, he proclaimed confidence in America's—and especially Lockheed's—dominance. Kelly told reporters that he "saw nothing in Europe that approaches the concept of the F-104" and that his plane was "evidence that American designers are substantially ahead in supersonic aerodynamics." In fact, he said, Lockheed had the technical know-how from the F-104 program to build a Mach 2 jet transport—a passenger jet!—that could fly 1,250 mph and travel 1,860 miles in ninety-two minutes.

Two years after the F-104's first flight—with Tony LeVier at the helm—Lockheed and the Air Force finally showed this plane that was twice as fast as any previous operational fighter at what *The Lockheed Star* called "a Hollywood-type preview at the Palmdale jet center."

A giant curtain inside an enormous hangar opened to reveal a shockingly bizarre sight—a shiny silver plane with a pencil-thin fuselage and a "straight, thin wing, its leading edge almost razor sharp." They called it the Starfighter.

The F-80 had provided a huge leap in speeds, but after that, fighter speeds increased only modestly, until the F-104 Starfighter arrived. For years, its top speed was classified. One pilot called it "thermo-sonic," because the plane was so fast that it approached the thermal barrier, a zone where heat from friction could create structural issues. According to Kelly, "every time the F-104 flies it breaks the so-called world's speed mark."

The Starfighter program solidified Kelly Johnson's standing as the boldest man in aviation, but it does have a notable asterisk. Lockheed had tremendous success selling the F-104 to U.S. allies, but its performance in those foreign nations was mixed at best. West Germany's air force, in particular, had problems—118 pilots were killed flying Starfighters over a twenty-two-year period.

The problem? Primarily that the Germans nearly doubled the plane's weight, from 17,500 pounds to 31,000 pounds, by loading bombs, cameras, and ammunition onto a plane with tiny wings. They then tasked pilots with a very dangerous mission—extremely low-altitude flights, on the order of 50 to 100 feet, in bad weather and low visibility. German pilots just weren't trained for such harrowing flights, and also didn't fly the jet enough to be adept in those conditions.

This, Kelly explained, is why they kept flying into terrain.

10

LITTLE GREEN MEN

Kelly Johnson absolutely loved his work, but the stress of it also wore on him, and he escaped, when he could, to the ranch he'd bought with his wife, Althea—the one place he could actually relax. In the way that Kelly relaxed, anyway. Which mostly meant engineering, but for leisure—building roads, constructing barns, and fixing tractors.

Kelly never lost sight of the beauty of this land, on what is today a large plot of prime suburbia in the San Fernando Valley, and he often just admired the vistas at this place he loved.

On December 16, 1953, he drove north from the Encino house, where the couple stayed during the week, and arrived at the ranch at around 4:45 p.m. This is the golden hour in Southern California at that time of year, when the days are short and the low winter sun casts the hills and valleys in an almost Tuscan light.

Shortly after pulling in and almost certainly pouring himself a drink—"at approximately 5 o'clock," he later wrote, "within two minutes' accuracy"—Kelly was taking in—ho hum—yet another spectacular sunset over the hills through the plate glass window of his living room when he noticed, above a mountain, what looked to him like a black cloud.

"The whole western sky was gold and red," he recalled, and this dark object, whatever it was, didn't move.

Kelly assumed it was a lenticular cloud—a trippy, somewhat rare lens-shaped cloud that was not uncommon to spot over the hills along the Pacific Ocean, but something about this particular object wasn't . . . normal; in fact, it looked to him like "a so-called saucer" and not just a saucer-shaped cloud.

Kelly asked Althea to fetch his eight-power binoculars so that he could keep his eyes locked on this anomaly. Then he ran outside for a better look, but now the object was moving, and very fast—which Kelly recognized, he later wrote, "because of the rate of foreshortening of its major axis."

Within ninety seconds, the object had completely disappeared, in a long, shallow climb heading north and west.

The object, whatever it was, had been too far away for Kelly to accurately estimate its dimensions, but it definitely had an elliptical shape, "with a fitness ratio of the larger axis to the minor one of about 7 or 10 to 1."

He estimated the position to be roughly over Point Mugu, a cape near the city of Oxnard. And the whole encounter lasted five minutes, at most.

Kelly stayed on the ranch for a week, then returned to Lockheed. That first morning back, two members of his flight-test group stopped by his office to catch the boss up on the week he'd missed, including a test flight of a Navy Super Constellation WV-2 over the Pacific.

The flight had been uneventful for the most part. Rudy Thoren, Lockheed's chief flight-test engineer, ran Kelly through the technical details and then, at the end, added a surprising story. Thoren shared it hesitantly, fearing ridicule.

Around sunset, he said, he'd chased . . . a flying saucer.

Kelly snapped to attention, and he wasn't laughing.

I know exactly when and where you saw this, he told Thoren. *It was around 5:05, over the ocean near Point Mugu.*

Thoren couldn't believe what he was hearing. Because that's exactly where the object had been when he and his crew encountered it.

He proceeded to tell Kelly the story of their encounter. He'd been at the controls, ascending from 15,000 to 20,000 feet, when Roy Wimmer, the pilot, pointed out the plane's windscreen.

"Look out, there's a flying saucer," Wimmer said, making what he thought was a joke.

Thoren, Wimmer, and flight-test engineer Joe Ware all stared at what seemed to be a very strange cloud, as Thoren turned the plane right to get a better look.

The closer Thoren got, the more he saw that the edges were too defined to be a cloud; it looked, to him, like a flying wing. But he couldn't seem to get close enough for a good look.

This was around sunset, and the sky was a fiery red. Thoren was flying at about 225 mph but, while the object *seemed* to be stationary, he wasn't gaining on it. It started to move away, due west out to sea. In the matter of one minute, the object had traveled so far away that it was now barely a speck in the sky. And then it was gone entirely.

When Thoren looked back at the flight log, he saw that the plane had leveled off at 20,000 feet around 5:10, so the sighting—which occurred between 16,000 and 18,000 feet—would have been shortly after 5:00 p.m. The crew didn't speak any more about the encounter, or whatever it was. They just completed the test flight and landed at about 6:00.

But when Thoren got home, he told his wife about it, and even drew the thing he'd seen out there over the ocean.

Kelly took this all quite seriously. On January 20, he submitted four copies of a memo titled "Sighting of a Flying Saucer by Certain Lockheed Aircraft Corporation Personnel" to the commander of the Air Force's Air Technical Intelligence Center, at Wright-Patterson Air Force Base, in Ohio. The memo was co-authored by Thoren and his team—Wimmer, Phil Colman, and Ware—and it explained what Kelly and the crew had seen.

You might assume Kelly would be a skeptic. That this extremely logical brain would not have room for something so . . . fantastical seeming. Instead, Kelly wrote, "for at least five years I have definitely believed in the possibility that flying saucers exist—this in spite of a good deal of kidding from my technical associates."

He was even more curious now. "Having seen this particular object . . . I am now more firmly convinced than ever that such devices exist, and I have some highly technical converts in this belief as of that date."

The report was subsequently investigated—as part of the famous Project Blue Book—and ultimately dismissed as what Kelly had originally thought it to be: a "lenticular cloud."

Kelly never spoke of the incident again, and this is the only alleged alien sighting he ever reported.

But sightings of UFOs were about to spike, as Johnson and the Skunk Works moved into a new era cloaked in secrecy—building spy planes.

PART III

A VERY SPECIAL NEED

11

INTO THE STRATOSPHERE

5 November 1954

MEMORANDUM FOR: Director of Central Intelligence
SUBJECT: A Unique Opportunity for Comprehensive Intelligence.

For many years it has been clear that aerial photographs of Russia would provide direct knowledge of her growth, of new centers of activity in obscure regions, and of military targets that would be important if ever we were forced into war. During a period in which Russia has free access to the geography of all our bases and major nuclear facilities, as well as to our entire military and civilian economy, we have been blocked from the corresponding knowledge about Russia. We have been forced to imagine what her program is, and it could well be argued that peace is always in danger when one great power is essentially ignorant of the major economic, military, and political activities within the interior zone of another great power. This ignorance leads to somewhat frantic preparations for both offensive and defensive action, and may lead to a state of unbearable national tension. Unfortunately, it is the U.S., the more mature, more civilized, and more responsible country that must bear the burden of not knowing what is happening in Russia. We cannot fulfill our responsibility for maintaining the peace if we are left in ignorance of Russian activity.

Kelly Johnson understood that engineers, especially those in defense contracting, had a responsibility to understand and predict the market. No commercial enterprise that lives at the bleeding edge of technology can survive for long if it doesn't anticipate the needs of its buyers far in advance. So, Kelly was constantly talking to his contacts in the Defense Department and in the national intelligence establishment about the global chessboard and the challenges that lay ahead.

He caught wind of a "desperate need" for a new type of American aircraft before anyone asked him for it—one that "could safely fly over the USSR" and bring back critical information on Russia's missile capability and other details about its defenses and military infrastructure.

On May Day 1954, the Soviets unveiled their latest nuclear bomber—the Myasishchev M-4 "Hammer"—which soared low over Red Square, creating quite a stir in Washington, especially because it hadn't even been a year since the USSR had detonated the world's first hydrogen bomb.

Top officials—including, if not especially, President Dwight Eisenhower—were particularly worried about the Soviet Union's strategic bombers, like the Hammer, which could carry nuclear weapons and which U.S. military and intelligence leaders knew almost nothing about: a type of plane that would potentially allow for a Pearl Harbor–style sneak attack, but far worse.

In the spring of 1954, Eisenhower asked James Killian, president of MIT and a key scientific and military adviser to the president, to form a committee to make recommendations for how the United States could leverage its tremendous base of scientific and technological firepower to determine what the Soviet military was capable of—and as a result, how much danger America was truly in. It was thought of internally as the "Committee on Surprise Attack."

A subcommittee known as the Land Panel, chaired by Edwin "Din" Land—co-founder of Polaroid and another key civilian scientific adviser to the president—was told to "find ways to increase the number of hard facts upon which our intelligence estimates are based, to provide better strategic warning, to minimize surprise in the kind of attack, and to reduce the danger of gross overestimation or gross underestimation of the threat." No pressure there.

The United States and the Soviet Union were still in the early days of a nuclear arms race, and any edge in that race created leverage in the battle for global supremacy. But there was a fundamental imbalance when it came to intelligence gathering. As a free and open society, the United States was susceptible to on-the-ground spying. But the Soviet Union, being a closed, authoritarian state, was nearly impossible to infiltrate with spies, to gather what's known as HUMINT, or human intelligence.

To make up for that, the United States had to be creative. It would need to use science and technology to out-spy the Russians. And if American spies couldn't get into the Soviet Union to gather intel, they'd have to fly over it, which presented its own challenges. Like being shot down.

The design challenge facing Kelly Johnson, then, was daunting: To safely overfly the Soviet Union and take high-quality photos undetected required a plane that could fly more than four thousand miles without refueling, and reach at least 70,000 feet—beyond the reach of Soviet air defenses and so high that it wouldn't create vapor trails, thus revealing itself.

In short, this plane would need to be extremely light, while carrying an array of the most advanced cameras, sensors, and navigational gear available.

The need was also urgent. Existentially so.

Kelly asked some of his most trusted engineers to determine if it was possible to use any parts of the F-104 Starfighter as the backbone of a new, experimental concept plane he had designated the CL-281. In particular, Kelly wanted his engineers to study whether they could modify the F-104's wing area while also cutting weight in order to greatly expand its altitude ceiling.

"It soon became obvious," he later wrote, "that the only equipment we might retain from the F-104 might be the rudder pedals."

In other words, back to the drawing board. This would have to be an entirely new plane.

That March, Kelly submitted Lockheed Report #9732 for an ultralight, high-flying surveillance plane with an enormous wingspan to the Air Force and was told the branch was "extremely interested" in such a concept and would like a formal proposal.

A month later, he handed over a full proposal to build thirty planes, with Lockheed taking full ownership of the program, including the then-novel idea that it would also service the airplanes in the field.

The pitch was also radical in that the plane Kelly was proposing had no landing gear, to save weight. Being as light as possible is mission critical for flying high, so Kelly was looking for weight savings wherever he could find it and decided that his plane would drop its gear upon takeoff and land on a reinforced belly—an idea he'd actually cooked up way back in the 1930s for Wiley Post's modified Vega, *Winnie Mae*.

The pitch was not a hit. Kelly received a letter from the Air Force declining the proposal "on the basis that [the concept] was too unusual" and had just a single engine, which seemed overly risky for a plane that would overfly enemy territory and be so far from help.

But one Air Force official loved the idea: Trevor Gardner, the "technologically evangelical" assistant secretary of the Air Force for research and development. Gardner knew of a different buyer who might be interested and summoned Kelly to Washington in November for an urgent meeting.

On November 19, Kelly met with a group of officers, engineers, and scientists, including Din Land, and endured a grilling that reminded him of his college days. A select group of America's most brilliant and accomplished minds wanted to know if what Kelly Johnson was suggesting was truly possible and why he was so confident that he could pull it off.

Later that day, Kelly met with Allen Dulles—director of the still-nascent Central Intelligence Agency, which had been formed just seven years earlier—and was told that certain top officers of the defense and intelligence establishment were convinced that Kelly was the man for this job. They thought, in fact, that he was the only one who could do it quickly, and quietly. And the proof was in his record—in the P-38, P-80, P-80A, and F-104.

Shortly thereafter, Allen Dulles, his brother John Foster Dulles (who was secretary of state), and three other top officials took the proposal to Eisenhower in person, because the president feared leaks and the subject was considered too highly classified to be put in a written report.

Eisenhower listened, asked many hard questions, and approved the plan, with a stipulation. "It should be handled in an unconventional way so that it would not become entangled in the bureaucracy of the Defense Department," according to Killian. It was given instead to Allen Dulles and the CIA.

"Well, boys," Eisenhower said, "I believe the country needs this information, and I'm going to approve it. But I'll tell you one thing. Someday one of these machines is going to be caught, and we're going to have a storm."

In advance of his trip to Washington, Kelly had been warned by his bosses not to commit to anything and he worried that he might have to take a leave of absence from his job at Lockheed proper to take on this project.

Lockheed's capacity was straining, especially in engineering, and the company was carrying a commercial backlog of $202 million, the most in its history. Still, when Kelly met Robert Gross and Hall Hibbard—the only two men he was cleared to tell—on November 21 at Gross's home, he told them that this was a job Lockheed had to take. And that—despite the large number of airplanes wanted by the CIA—secrecy demanded that he run the entire program, from design to manufacture, in his Skunk Works.

The two bosses heard him out and agreed.

This was, arguably, the biggest single moment in the history of the Lockheed Skunk Works, in that Kelly now had approval for something more than an experimental design shop. He was given the green light to run his own production, too. Which meant that he wasn't just building prototypes. He would oversee full production of all planes built for the program.

What's more, the government was willing to hand him unprecedented control. Lockheed, the program's outline stated, was taking "full responsibility for the design, mock-up, building, secret testing, and field maintenance of this unorthodox vehicle."

Kelly spent two days redesigning the concept himself, to accommodate a new, detachable landing gear, a different engine, and a camera bay. Then, on November 29, he summoned five key Skunks to his office for a meeting.

Kelly looked at the assembled talent and spoke of a new program, one more secret than anything any of them had ever worked on, and then, without revealing any details about what they'd actually be doing, asked if they were willing to commit eighteen months to such a project.

All five said that, yes, they absolutely would. And then Kelly leveled with them. He'd sold the CIA a high-altitude reconnaissance airplane. They could have a few days to wrap up their current work but should be ready to go full bore on Monday, December 2. It was time to make history.

Kelly's reconstituted Skunk Works began with twenty-five engineers, with his trusted shop man, Art Viereck, in charge of production. Kelly assigned Ed Baldwin, better known as Baldy, to handle the traditional three-view drawing (the plane as seen from top, bottom, and side) and laid down some basic parameters.

The plane's front section, forward of the cockpit, should be the F-104, but with a rounded nose, and a 600-square-foot wing with an aspect ratio in the range of 10 to 11. It should have an off-the-shelf engine requiring minimal adjustments, a bay in the belly of the fuselage to handle a very specific package, and no landing gear. Instead, just as he'd pitched to the CIA, the plane would take off on a dolly, drop the wheels, and then land on a reinforced belly.

Four days later, Baldwin had the first drawings completed, and by December 10 the basic design was frozen. It was, more or less, the configuration that would go into production, with only minor changes to the horizontal and vertical stabilizers after wind tunnel tests. Which is fairly astounding to consider once you know what the plane Kelly laid out and Baldy sketched—all within a few short weeks—would become.

Shortly after beginning, Kelly prepared a twenty-three-page report for the CIA with his updated thoughts on the plane his Skunk Works would build. Among them, that the Angel, as he was calling it (because it was going to fly so high), would have a maximum speed of Mach 0.8 (460 knots) in level flight, with a ceiling of 73,100 feet—an absurd altitude that had only been reached at this point by research balloons and a few highly experimental one-off aircraft.

He promised to have the first plane flying by August 2, 1955, and all Angels finished and delivered to an as-yet-unchosen test site by December 1.

When Kelly selected his design team, he announced that they'd all work forty-five-hour weeks. That number quickly rose to sixty-five, and the actual schedule, once the project was fully running, required more like one hundred hours a week. It was the only way Kelly could hit his audacious eight-month target.

The work was fast and furious. "Working like mad on airplane," Kelly wrote in the project log. So mad that Kelly began work before he had a contract or any idea of how the money would flow from the government to Lockheed.

Government contracts were sometimes paid upon completion or on delayed schedules. Kelly insisted on splitting the tab up into smaller payments, made regularly, so that he didn't have to "go running to the bank to carry the government." When he submitted the first two invoices, for $1,256,000, Kelly had to have the checks sent to his house in Encino. After that, he set up a special account for the project at a nearby bank.

It is almost impossible to believe that a company as large as Lockheed could charge forward on an experimental program without a contract from the government, but this combination of mystery and subterfuge only assured Kelly that no bureaucracy would stand in his way. As for the government, this unconventional method of paying a contractor—in secret, out of oversight of even Congress—wasn't illegal.

So-called unvouchered funds were allowable for covert projects, according to a law passed by Congress in 1949, which stated that only the director of the CIA could access them. This was the only way a program could control secrecy, by avoiding things like competitive bidding and public procurement of parts.

On December 15, 1954, Kelly was back in Washington to meet with Trevor Gardner, as well as Colonel Osmond "Ozzie" Ritland and Richard "Dick" Bissell, the "special assistant" to Allen Dulles handpicked by the director to oversee this audacious program.

Kelly presented the group with an estimated cost of $22.5 million for twenty planes (plus a two-seat trainer)—far cheaper than the $35 million Eisenhower had already approved for a spy plane project—and agreed to comply with unprecedented security parameters.

"A large amount of time," he wrote in his project log, was "taken on the optimum cover story for the project."

Bissell would become Kelly's foremost champion on this project and on others that followed. He was an impeccably dressed MIT-trained economist who had helped implement the Marshall Plan after World War II before joining the CIA's predecessor agency, the Office of Strategic Services (OSS). Bissell was a walking prototype of the Blue Blood Spy. He grew up in Mark Twain's former mansion in Hartford, Connecticut, and had an ancestor, Samuel Bissell, who served in George Washington's spy corps. Bissell summered in Maine and taught economics at Yale before being recruited over to help his country solve novel problems during the Cold War.

Bissell was summoned to see Dulles "with absolutely no prior warning" and was handed the shocking new assignment: Develop a revolutionary method to spy on the Soviet Union! Fast!

Eisenhower had just approved the development of a high-flying spy plane to take photos of "denied areas," and Dulles wanted Bissell to lead it. He was to proceed immediately to the Pentagon to meet with Gardner and several other top generals to decide how to organize and run a covert program in conjunction with the Air Force. By meeting's end, Gardner had called Kelly Johnson to approve the project and had given him the green light to build twenty planes.

The project was code-named AQUATONE and would be funded by the CIA's secret Contingency Reserve Fund. Herb Miller, chief of the Office of Scientific Intelligence's Nuclear Energy Division, was named the program's executive officer.

At Dulles's direction, Bissell drafted a memo for the president and then set about forming a "Project Staff" that included a former Marine pilot named James Cunningham as administrative officer. Bissell's staff was compartmentalized away from the CIA's main branch and given special cover.

For instance, Bissell was outside the CIA's traditional chain of command. He reported directly to Director Dulles, and his personnel and operating costs were not drawn from regular Agency accounts.

His staff moved three times in the first two years, starting in an old building on E Street, near the Lincoln Memorial. The premises, he later wrote, "were so rickety that the installation of even basic office machinery threatened to bring down the second floor." Within a few months, he managed to find a better space and moved to H Street, where there was less risk of imminent collapse.

There Bissell set up his small, secret program office. It was tiny—fewer than one hundred people—but functioned as a mini version of the CIA, with its own finance, procurement, and security staffs. Very few people at CIA headquarters, including most of the top officers, even knew of its existence because the program's security team was so concerned about leaks.

Bissell's so-called Development Project Staff was the only CIA section with its own communications office and operational cable traffic that transmitted to and from Lockheed and several other key contractors on the program. Only Bissell, who read every cable, could disseminate them, and they were the only cables at the CIA that didn't automatically get copied and distributed to the director's office. Bissell just sent him the most important ones.

Bissell was a brilliant man and an autodidact. He set out to learn all he could about making and flying airplanes. Fortunately, he had a good teacher in Kelly Johnson, who was, in Bissell's opinion, "an artist in his field, a superb technician, and a brilliant engineer."

Kelly's loose, often garrulous nature wasn't an obvious fit with Bissell's stiff, effete stoicism, but the two got along well. They grew close and would become friends for life. Bissell understood that the program would only succeed if it stayed small and moved fast, and Kelly was almost uniquely suited among defense contractors to follow that model. His decisiveness in particular—"which allowed him to take shortcuts and render quick judgments without jeopardizing safety"—impressed Bissell.

Bissell didn't give Kelly a deadline, but he imposed one upon himself. This new plane, which would fly higher than any in history, would be in the air by August 1—nine months after the project commenced.

Bissell doubted this was possible, but he worked with Kelly to strip as much bureaucracy as they could from the program. Kelly had just one point of contact—Bissell—who could answer his questions in a single phone call, and their monthly progress reports would be ruthlessly short, about five pages. Bissell could also approve any costs up to $100,000. Over that amount, he'd need Director of Central Intelligence (DCI) Dulles to sign off.

If this had been the Air Force, Bissell noted, that same report would be an inch thick, and every design change would require approval by "Wright Field, a couple of different laboratories, the budget office, the regulations office, and so forth."

The CIA project was really version two of the Skunk Works, and its new home, Building 82, was an upgrade from the lean-to, but not a big one.

Ben Rich was told to report there in December 1954. Rich was a twenty-nine-year-old thermodynamics expert who specialized in solving heat problems on planes and inlets. His first patent had been for a special heater that helped solve a painful and embarrassing problem for naval aviators: At higher altitudes, their penises would sometimes freeze to the side of the tube used for peeing in flight.

Rich had no idea what was happening inside that enormous assembly building by the runway before Kelly called his supervisor and asked "to borrow a thermodynamicist, preferably a smart one." The timing was fortuitous. Rich, in his first year at Lockheed, felt "creatively frustrated" and was on the verge of leaving the company. His father-in-law had almost convinced him to take over the family's deli bakery.

This job was a dream. The surroundings, not so much. Rich was surprised to find the company's brilliant star engineer, the venerable Kelly Johnson, tucked away in what felt like a warren. Desks were crammed together. The purchasing department had been stuck in the only free space available, on a small balcony overlooking the shop floor. "The place was airless and gloomy and had the look of a temporary campaign headquarters where all the chairs and desks were rented and disappeared the day after the vote," Rich recalled.

"Adding to the eccentric flavor," Rich later wrote, when the hangar doors were opened to get some air flowing, birds would fly in "and swoop around drawing boards and dive-bomb our heads, after knocking themselves silly" against the windows that were painted black, at Kelly's direction, for secrecy.

Rich asked one of the top engineers, Dick Boehme, how long he'd be needed. Probably no more than six months, Boehme replied.

"He was slightly off," Rich later said. "I stayed for thirty-six years."

Boehme assigned Rich to a desk in an office with six other engineers and gave him a copy of Kelly's ten basic rules. "For as long as you work here," Rich recalls him saying, "this is your gospel." Then he told the young engineer what he'd be working on—a jet engine modified to fly 15,000 feet higher than any engine had flown before—and showed him a picture of the plane it was to go with.

Rich was stunned. He'd expected a fighter, not a glider.

"What is this?"

"The U-2," Boehme replied. "You've just had a look at the most secret project in the free world."

Rich knew Kelly a bit from his previous job, working on unclassified projects in the cavernous hangars where Lockheed's biggest, largest-volume planes were designed and built.

That's how Kelly spent the first half of each day—lording over the rows of men in white shirts diligently hunched over drafting desks, making the commercial planes of the future—and though Rich had only minimal interaction with him, the big man's presence was always felt.

By this point, Kelly was a towering figure at the company, a "living legend" held in awe by young engineers like Rich. "All of us had seen him rushing around in his untucked shirt, a paunchy, middle-aged guy with a comical duck's waddle, slicked-down white hair, and a belligerent jaw."

Rich would go on to have his own legendary career at the Skunk Works, as Kelly's right hand, and the boss's determination was one of the first lessons he absorbed: "Once that guy made up his mind to do something he was as relentless as a bowling ball heading toward a ten-pin strike," Rich would say. But for that to work, you have to be willing to back it up with results, and spine. "With his chili-pepper temperament, he was poison to any bureaucrat, a disaster to ass-coverers, excuse-makers, or fault-finders."

The sum total of Kelly's attributes, Rich thought, was that you just wanted to make him proud: "We peons viewed him with the knee-knocking dread and awe of the almighty best described in the Old Testament."

Rich was among the younger engineers in the group, but Kelly quickly recognized his talent and agility at tackling problems. Rich also had a rare asset that helped him greatly around the Skunk Works. He had a healthy respect for Kelly, sure, but he wasn't intimidated by the man. "For whatever reason," Rich later said, "I really was not afraid of him."

Which is different from being above his scorn. Rich made mistakes, and heard about it, but Kelly's temper tended to rise fast and pass quickly, like an afternoon storm.

Kelly was arrogant but had the right to be. Rich was struck, regularly, by how much this man knew or could just pick up by osmosis. "He was so sharp and instinctive that he took my breath away," Rich later wrote.

Rich might, say, predict a shock wave hitting the tail of the plane and Kelly would consider the factors, think a moment, and guess that the temperature at that point would be 600 degrees Fahrenheit. "I'd go back to my desk and spend two hours with a calculator and come back with a figure of 614 degrees."

It wasn't that the Skunk Works was established as a set of rules and governed exactly that way for decades. The basic principles were set on the XP-80 program, by necessity, but Kelly was constantly refining his methods along the way.

One of the key elements of making an experimental shop work was that the company had to allow the person running it to—as he once explained—"tear down long-established empires."

Departments become entrenched and defend their responsibilities. It's hard to take them away once they've been established. Purchasing, for instance, gave Kelly fits. The Skunk Works needed its own purchasing, with its own rules, one of which was that the guys who worked there shouldn't also do engineering. But the engineers also shouldn't do purchasing.

Periodically, Kelly would mail a letter to his vendors with a reminder: "Do not take what the engineers tell you might be a good suggestion and do it."

Engineers want the perfect part, even if they don't actually need it, and they don't always know the cost. Often, something slightly less perfect, or less expensive, works just as well. And vendors want to make the engineers happy. They also like sales. The end result is higher bills—"about four times normal," Kelly suggested. So he told the vendors to ignore his engineers.

Secrecy was of paramount importance. This was the most secret defense program since the Manhattan Project. And Kelly took that very seriously. But, on the team itself, this was mostly about trust and understanding.

Keep the group small, make the stakes clear, and don't bother with an elaborate security apparatus. Kelly's philosophy about secret documents was that they should not be labeled. If you stamp secret on something, you're just asking for someone to try to read it. A document is far safer if it looks like any other boring old report. He felt the same way about locked drawers, and even doors. Any additional attention invites risk.

When a program was finished, he mostly just destroyed the documents. Years later, when the Air Force came in to perform a security audit, they asked Kelly where the files were. *I destroyed them*, he said. *When did you do that?* Kelly couldn't recall. *And where's the record?* He didn't have that, either.

A given of joining a Skunk Works project was that every worker would take a vow of secrecy. No one in any other Lockheed department was aware of what went on inside the Skunk Works, and even wives couldn't be told what exactly they were building. But there was more to it than that. Security officers investigated personal lives. *Where do you go after work? What are your hobbies? Who do you socialize with? God forbid you have an uncle who's a Communist!*

The phrase "need-to-know" is today a cliché of covert projects, but the concept was born on these early CIA black programs. Many workers even *inside the Skunk Works* didn't have the entire picture of what was going on. They might know that they were building a wing for a high-altitude plane, but they didn't know what that plane was being designed to do. That simply wasn't information you needed to know to do your job.

The CIA's security chief put rules on Kelly, too. He was instructed to vary his schedule and his route to work. He was to avoid certain intersections and keep an eye out for large trucks that might block his path.

Secrecy complicated everything, including the official name of the plane. Inside the Skunk Works, people tended to use the nickname Kelly liked—"the Angel"—while the small group cleared into the program at the CIA, being bureaucrats, called it "the Article."

The project was so closely guarded that in early 1955 the Air Force put a call out to contractors for a plane it called the X-17, and when Johnson saw the proposal, he was irate. It was, he thought, "a dead-ringer for our original presentation." The Air Force department that issued it had, in his opinion, clearly used his original pitch and somehow didn't know about the CIA's secret project, which was a good sign for secrecy but infuriating nonetheless.

Kelly called Dick Bissell on a Sunday, then flew to Washington to share the proposal with Bissell and Gardner. Their reaction, Kelly wrote in the project log, was "stark horror." The proposal was swiftly killed.

It was, to Kelly, yet another sign of the Pentagon's broken contracting process.

Throughout Lockheed's development process, tension simmered

among the small number of people within the Air Force and CIA who knew about Kelly's project. Some of those in the USAF leadership resented that it was in the hands of the Agency and sought to take over operations as soon as possible. And it's hard to blame them—this was an airplane, after all.

In March 1955, the Air Force chief of staff told DCI Dulles that he hoped to take over the program once the plane was flying, and he met stiff resistance. That debate simmered until Eisenhower declared that the CIA would remain in control even once missions commenced.

"I want this whole thing to be a civilian operation," Eisenhower said. "If uniformed personnel of the armed services of the United States fly over Russia, it is an act of war—legally—and I don't want any part of it."

This was a month of maximum stress for Kelly, because the U-2 wasn't his only responsibility. His daily duties included four different airplane projects, and even though the U-2 was his primary worry, Lockheed's executives insisted that his other duties not lag.

He'd spend 6:00 to 10:00 a.m. at the Skunk Works, overseeing the U-2, then from 10:00 a.m. to noon on the F-104, still in production, and from 1:00 to 5:00 p.m. he'd put on his corporate chief engineer cap, working mostly on the C-130 and the R-1049.

Which probably explains why, that March, the stress came to a head. Having four projects on his plate at once "almost nailed me," Kelly later wrote on a copy of a letter his personal physician at the Beverly Hills Clinic sent to Robert Gross on March 18, ordering Kelly to take three weeks off to rest.

"From a thoroughly physical standpoint his condition is satisfactory except for moderate overweight," the doctor wrote. "However, from an emotional and fatigue standpoint he is in a very precarious condition at the moment." The schedule Kelly was attempting to keep, in this doctor's estimation, was "unreasonable" and that if he couldn't find a way to give up "a large part of his responsibility, he is likely to collapse completely."

This seems to have been an early warning of what we might today call a mental breakdown. But in the 1950s, a man was expected to just suck it up. Emotional exhaustion was not a thing people talked about. And a man like Kelly—who projected strength, and for whom work *was* his life—wasn't really capable of throttling back. Or delegating command. So mostly he just powered through.

The accelerated schedule meant that there was no time for a wholly new engine. The Angel would have to fly with an existing product, and Kelly selected Pratt & Whitney's J57 turbojet, which he knew and liked.

The mock-up arrived from Pratt & Whitney on February 1, 1955, in a damaged shipping crate. It had a hole in the side, Kelly noted in the log, big enough that "anyone who wanted to try could see the engine." Even worse, Pratt & Whitney had distributed twenty-one copies of the shipping instructions, labeled with a code name, and this weak grasp of operational security drove Kelly nuts. "They just don't seem to get the idea on this job," he wrote.

Engineers who worked under Kelly often talk about how practical his genius was. Rather than obsess over innovation that might be possible, he'd focus on what he knew could be done, based on existing technologies. The U-2 is a prime example. "It's sort of a nothing, technically," is how Dick Heppe later described it, as a preface to explaining how impressive Kelly's design mind was.

It was, essentially, the fuselage of his F-104 fighter with subsonic inlets for the engine, because it didn't necessarily need to be fast. Its key novel attribute was the enormous wing, and extremely light wing loading, paired with a powerful engine. The result, Heppe explained, was a capability "completely unknown and unavailable in any other machine," the ability to fly long range at 70,000 feet or higher.

And by late summer 1955, the prototype was ready.

The plane had arrived on time and under budget—a lot under budget. By the time Kelly had a prototype flying, there was $4 million to $5 million in leftover funds. He used that, plus spare parts, to deliver five extra planes to Uncle Sam for free. This special bonus price was, Air Force liaison officer Leo Geary later said, "probably the finest bargain the American taxpayer has ever had under any circumstances."

12

NO-MAN'S-LAND

In January 1955, with work on the Angel well underway, Kelly called Tony LeVier to his office at the new Skunk Works base, in Building 82. LeVier was one of Kelly's favorite pilots because he was brutally honest and had a tremendous memory. If something wasn't right with a new plane, even a very small something, LeVier would notice it and tell the boss what he needed to hear.

"Do you want to fly my new plane?" Kelly asked.

"What's it like?" LeVier replied.

"I can't tell you," Kelly said. "So if you say yes, okay. If not, get the hell out of my office."

"Sure," LeVier said. "I want to fly it."

Kelly sat LeVier down and explained that this plane was the result of a top secret program, more secret than anything Lockheed had ever done.

"You can't say a goddamn word about it to a living soul," Johnson barked. "Not your wife, your friends, your mother."

Only then did he share a drawing of what he'd been up to. LeVier couldn't believe what he was seeing. *This* was the secret? The drawing on the paper looked like a giant sail plane. Like some kind of glider! It had a skinny body and enormous wings—85 feet from end to end.

This aircraft, Kelly told him, was designed to fly higher than any plane in history because its mission was so dangerous. And it was so secret that all testing had to be done in a remote location, where no human would see it.

"Chrissake, Kelly," LeVier said. "First you had me flying the F-104, which had no wings. Now you've got me flying an airplane with the longest wings I ever saw."

"Here's what I want you to do," Kelly said, and laid out a plan: *Get Dorsey Kammerer and the company plane and erase all of the Lockheed insignias. Then put on some old clothes, so that you don't look even remotely official, and fly around the western United States until you find us a test site that's far off the beaten path, but not so far that it's impractical for people to actually get there.*

"So," LeVier later recalled, "we put on roughneck clothes, took food and a camera, and off we went."

LeVier and Kammerer told co-workers they were going to Baja California to look for whales, then went scouting in a twin-engine Beechcraft, using air maps of the western United States, for dry lakes. They flew around Death Valley, skirting bluffs and buzzing the scrub, then moved to the northern Mojave, where LeVier identified the perfect site in a remote piece of Nevada desert—a round lake about three and a half miles across.

Kelly had favored the land near Death Valley, but the CIA's security staff didn't like that the land was so close to towns and to public areas that were heavily traveled by tourists. But Air Force Colonel Ossie Ritland—who had previously commanded the 4925th Atomic Test Group—proposed a safer, more remote location, one where his group had previously dropped atomic bombs. "We were looking for the most isolated part of the USA where you could fly this airplane and not cause a lot of curiosity," Ritland later explained.

And no area was more secure than this patch of desert where the United States had been testing its atomic munitions—the one Tony LeVier was now flying over. He circled the lake bed and landed on faith, hoping the ground below was as firm as it looked from above. It was.

LeVier rated the site a ten-plus and, upon his return, raved about it to Kelly, who ordered him to prep the plane so that he could go see it himself, along with some very special guests: Dick Bissell, the CIA's fastidious program lead, as well as Ritland and Herb Miller of the Air Force.

Kelly sat up front, in the right seat, with geological survey maps on his lap, and LeVier carried a picnic lunch his wife had packed.

The area LeVier identified was in an area known as Frenchman Flat, near Groom Lake, Nevada, approximately four miles in diameter and—Bissell would report—"as smooth as a billiard table." It even had what appeared to be an old landing strip, probably built for emergency landings by pilots who were rapid training for World War II deployments.

Kelly asked for a closer look and LeVier opted, for some reason, to land on the lake bed and not the airstrip. Whether it was instinct, or just a last-second coin flip of a decision, this was wise. Because after landing, the group walked from the lake to the runway—and their feet sank into the soft soil. What had looked from the air like old pavement was in fact just formerly compacted earth that had, over the years since World War II, turned into layers of dust.

"Had we attempted to land on it," Bissell later wrote, "we most assuredly would have crashed"—very likely killing four of the most important national security principals in America and perhaps altering the course of the Cold War and world history in one fell swoop.

So Tony LeVier intuitively and unintentionally saved his boss's life, along with the lives of three government VIPs who backed Kelly Johnson and his ambitions. And, sure, the dusty airstrip needed a refresh, but the location was otherwise ideal: just a short flight from Las Vegas (eighty-four miles north) but a long drive from the nearest town and any random passersby. Most important, it was adjacent to the Department of Energy's atomic testing range.

That was what had appealed to Ritland in the first place, and it was a critical detail because—as the site of actual nuclear tests—the airspace above this land was blocked. Private planes could not overfly it.

Kelly loved everything about it—he called it "a dandy" in the project log—but he worried that the atomic bomb ties, while having obvious benefits, would also potentially create a pile of red tape to navigate.

When Bissell investigated further, he learned that the Groom Lake area wasn't part of the Atomic Energy Commission's Nevada Test Site, after all. It was just adjacent. So he asked the president to add "this strip of wasteland," as the CIA's official project history called it, to the site.

Adding this small rectangular strip to the test site was a clever move, because this was some of the most secure airspace in America, highly restricted because it was part of the country's nuclear weapons program. This would secure the airspace without arousing any suspicion, and the fact that the U.S. government was actively testing nuclear weapons nearby was only occasionally a problem.

By month's end, that red tape had been cut, and Lockheed was clear to use this so-called Site II, which Kelly had taken to calling Paradise Ranch, so that it would seem like a plum assignment and not the desert exile it was.

"It was kind of a dirty trick, since Paradise Ranch was a dry lake where quarter-inch rocks blew around every afternoon," he later said.

On official reports, the location was labeled as Watertown, which added to the list of nicknames for the site that would become interchangeable (and confusing) in subsequent years: Groom Lake, Watertown, Paradise Lake, and a bit later—the most infamous nickname of all—Area 51.

Picking a test site was, in theory, one less pain point for the program. Kelly now had a place to fly his radical new plane—or, rather, he had a place that *could* work for testing.

For the moment, though, it was just a barren patch of desert with no amenities. No buildings. No roads. Not even a functional runway.

The pressure of the U-2 program was constant, and crushing. Kelly projected competence and calm. But the process he himself asked for and even preferred—full bore, with virtually no room for error—created massive amounts of stress. And that stress, as doctors had been warning, was literally wrecking him.

On March 25, 1955, Lockheed issued a memo to "all supervision" on the status of Kelly Johnson's health. "On advice of his doctors, Chief Engineer Clarence L. Johnson will take a complete rest," VP and General Manager Burt Monesmith noted. Kelly's health "has suffered from the unusually heavy work load pressures his position has forced him to carry," and when he was cleared by doctors to return to Lockheed, his load would be limited "for an extended period of time to top level technical assignments."

But, despite his doctor's orders, Kelly did not give up work that spring. Not even close. His biggest concern was that the Angel would be ready to fly before he had a test site that could accommodate it. And just because he'd picked a location, and Bissell had secured it, didn't mean that the ranch would be usable anytime soon. "Have a growing feeling that our government friends are not getting organized for the project in the same manner in which we are building the airplane," he wrote in the log in late April.

On a subsequent trip, Tony LeVier, along with Dorsey Kammerer and Kelly, surveyed the site and selected a spot for a 5,000-foot-long north–south runway, as well as a rough plan for the buildings, and Herb Miller of the CIA Special Projects staff allocated $800,000 in funds for construction.

The Agency also picked a firm to do that construction and, using the Las Vegas Field Office of the Atomic Energy Commission as cover, announced this new base to eighteen local media outlets, saying that the AEC had "instructed the Reynolds Electrical and Engineering Co., Inc. to begin preliminary work on a small, satellite Nevada Test Site installation . . . essentially temporary . . . a few miles northeast of Yucca Flat and within the Las Vegas Bombing and Gunnery Range."

The ruse didn't exactly work as planned. Local reporters got curious, began asking questions, and learned enough to be skeptical. That's how the *Las Vegas Review-Journal* ended up running a front-page story on May 19, 1955, under the headline "Secrecy Cloaks Satellite Test Site Near Yucca," which was the first of many reports suspicious about what would become a legendary (and globally infamous) U.S. black ops site.

With no time to wait for grunt labor, LeVier and another Lockheed test pilot spent weeks combing the lake bed for debris left over from World War II gunnery practice at the site, in order to clear it for use in emergency landings.

On May 4, Kelly went to Vegas to meet with the construction company that would build his base, which was to include the runway, three hangars, a control tower, and housing for workers, as well as a movie theater, a volleyball court, and a well to supply fresh water.

Afterward, he flew out to the site to mark out the location of those buildings and to realign the runway "in keeping with our best wind information." On the way home, his pilot took a shortcut, passing over the Nevada Test Site, and the plane flew directly over an atomic bomb sitting on its tower, about nine hours before it was to be detonated.

For obvious reasons, the CIA could not directly commission, or pay for, any of the work the Groom Lake site needed. That responsibility fell to Lockheed. But Lockheed wasn't licensed to operate on this government property, and there was no time to cut all the red tape required to fix that. Instead, Kelly passed the drawings off to a contractor who did have the license, and that license put the job out for bid, using Kelly's initials, "CLJ," as the client.

The site was perfect, in that it was remote and protected by airspace that couldn't be overflown, but that also made construction expensive. The budget swelled from an initial estimate of $200,000, before there was a location, to $832,000 once the site had been selected. Reynolds Electric took on general support, which everyone called "roads and commodes," and consisted of turning a barren wasteland into, as pilot Frank Murray later said, "a place where a few hundred people could survive in relative comfort," *relative* being the key word: Workers would be housed in mobile homes and travel trailers.

On July 15, Kelly made a very important entry in the project log. "Airplane essentially completed," he wrote. "Terrifically long hours. Everybody almost dead."

A few days of testing on the ground in Burbank followed, then final inspection, and on July 24, the prototype—which the CIA was calling Article 341—was taken apart and loaded into a C-124 Globemaster over three rainy hours in the very early morning.

To this point, the airplane had no official name. It was Article 341 to the CIA, and the Angel to Kelly. The plane was still top secret and needed to be disguised even in official records, so it couldn't be labeled B, as a bomber, F as fighter, or R for reconnaissance plane. Instead, Kelly chose to designate the plane as a U for "utility" aircraft and the idea, he said, was that "anyone checking the lists of new planes would pass it over as unimportant." The Air Force had only two utility planes in the books at the time—the U-1 and U-3—and neither of them was of much interest to adversaries.

Thus, Kelly's Angel was officially designated the U-2.

With the Angel ready to fly, on schedule and an unbelievable eight months after work began, Kelly and a small team of engineers and program managers took off first in a Lockheed C-47 and flew to Nevada, only to find that the lake bed (where they'd planned for the C-124 carrying the important cargo to touch down) was wet and likely unsafe for landing the heavier plane.

Kelly asked the pilot of his plane to land on the runway, so that he could walk over and assess the lake bed himself. He did, and decided that it was fine,provided they let out most of the air of the C-124's tires.

But the newly installed base commander, Dick Newton, disagreed. He could not allow any risk of an accident here, and refused to let the big plane land. Kelly had yet to meet Newton before this, and was angry to have a military officer pushing back on his own carefully considered judgment. ("It was real gory for a first meeting," he later wrote.)

This was *his* project, and his plane, and he had declared the land safe. Kelly suggested they call Dick Bissell in Washington and let him break the tie.

Newton resisted, then relented.

Bissell heard them out and said that he would let Kelly—the man who knew the equipment best—decide if it was safe to land the C-124 or not.

"Two hours later, it landed in a cloud of dust, making a beautiful landing."

Another important decision was made that summer, one that's largely lost in all the U-2 lore. A graphic artist named Bob Charlton, who'd been working on the program, showed Kelly a sketch that had nothing to do with Aquatone or the Angel. It was a drawing of a cartoon skunk—the Skunk Works skunk, to be used as a logo.

A different artist finalized the sketches and the skunk logo appeared in the wild for the first time on customer reports for the U-2, and then in the plane's maintenance manual. A few years later, when Kelly caught wind of a rival trying to use the name for its advanced project group, he had the name he once hated trademarked. And a few years after that, around 1960, a taxidermied skunk in a plexiglass box was hung on the wall in Building 311, the engineering headquarters.

The Skunk Works wasn't just a nickname; it was now an institution.

13

SUCH GREAT HEIGHTS

In advance of the test flights, Kelly Johnson had sent Tony LeVier off on another mission, this time to Wright Field in Ohio, for work in the pressure chamber. LeVier had been up as high as 50,000 feet in the F-104, but the U-2 would go far higher, which required new equipment to keep him and other pilots alive, because at altitudes above 65,000 feet, fluids in the body will literally vaporize—boil off—if the pilot is not kept under pressure. And not just that: The reduced pressure at those altitudes means that a pilot's blood is not properly oxygenated. He could, in several terrible ways, fly himself to death.

Colonels Don Flickinger and W. Randolph Lovelace, both Air Force physicians, were assigned to work with the Skunk Works on the program's medical challenges. Both men were experts in the perils of high-altitude flying and had completed a series of parachute jumps from B-47 bombers in the early 1950s in order to test pilot survival gear. Flickinger co-invented the Air Force's standard oxygen mask system.

First and foremost, pilots would need pressure suits to fly the U-2, as well as training on how to prepare for and manage the effects of operating in such a low-oxygen environment.

The David Clark Company, of Massachusetts, was chosen to design a complex, first-of-its-kind life-support system, and the result was the world's first pressurized space suit, which delivered a friendly 16 pounds per square inch of pressure across the body in order to keep the blood from boiling off at altitude. Clark's main business was "ladies foundation garments" (girdles and bras), but a specialized group inside the company had been making pressure suits for NASA from the onset of the space program.

And yet, the suit was a fallback. The primary pilot safety system was a pressurized cockpit that simulated an environment of 28,000 feet, but should that pressure drop—due to a system failure or cockpit collapse—the pilot's suit would automatically inflate, and because the oxygen was to flow directly into his helmet, and not through ambient air, his oxygen supply would remain constant.

At Wright Field, Tony LeVier was fitted for one of these pressure suits and put through tests that simulated 75,000 feet. The systems worked. And then, in late 1955, he was ready to fly Kelly Johnson's Angel for the first time.

When LeVier arrived back at the dry lake bed he'd first identified as a test site not even a year earlier, he found a very different place. The formerly barren patch of desert now had a 6,000-foot runway, two large hangars, a control tower, a mess hall, and numerous mobile homes for housing.

At dawn on the morning of his first test, LeVier got into a pickup truck and did one last safety pass on the runway, cruising slowly and picking up any last stones and debris.

The U-2 was unlike anything LeVier had ever flown. It had a high aspect ratio and was very light and fragile; the plane looked and felt almost flimsy.

Kelly was worried about how LeVier would land it without traditional gear, so Tony had been surveying the other pilots for advice and decided that he would not land on the nose wheel because the plane might porpoise, which means bounce back up, causing big problems for the pilot.

Kelly disagreed. He was certain that LeVier should land on the nose. He worried that trying to land on the tail might cause a stall, which would almost certainly lead to a crash that would destroy the plane and kill its pilot.

But before he went airborne, LeVier's first order of business was a taxi test. Kelly had instructed him to taxi the plane up to 50 knots, then 60, then 70.

On the first test, with Kelly bouncing along behind the plane in a chase car, the plan was to max out at 50 knots, and LeVier noticed the wingtips begin to lift at around 45 mph, but the special temporary pogos at the end of each wing—which had been bolted into place for the test—never left the ground.

On the second test, LeVier pushed the throttle a bit more, and LeVier noticed his wings riding up again. This time, though, when he reached 70 knots, it wasn't just the wings. "It was at this point that I became aware of being airborne," LeVier reported in the debrief, "which left me in utter amazement, as I had no intention whatsoever of flying."

LeVier scrambled to lower the plane from a height of about 35 feet, but with only the vast lake bed around him for perspective, he had no idea how high he'd risen or where, exactly, the ground was. He slammed the throttle to prevent a stall, then held the plane straight and level and waited for the bird to hit ground, which it eventually did, with a crunching thud.

Both tires popped and the plane bounced back into the air two times before finally staying on the ground. The brakes didn't really help LeVier much, and the plane rolled for almost a mile before it finally stopped, with both wheels on fire.

Within minutes, Kelly arrived in the chase car and lit into his test pilot. "Goddamn it, Tony, what the hell happened?"

"Kelly," LeVier replied, "the SOB took off and I didn't even know it."

No one had anticipated this outcome. But the U-2 was so light, and had so much lift, that it could—apparently—ascend at just 70 knots.

As a message wired to CIA headquarters, likely composed by Kelly himself, put it: "No ill effects except to Tony's ego."

Tony LeVier's first intentional flight of the U-2 was on August 4, when he took off and flew to 8,000 feet in a rainstorm. LeVier worried that a storm could tear the plane's wings off. The plane "flew beautifully," Kelly later noted, but he decided to cut the test short, for safety.

He instructed LeVier to land on the nose wheel and he did, gently, but the second the wheel touched down the plane began to porpoise, so LeVier pulled back, climbed, and circled around.

"What's the matter?" Kelly barked.

"She was starting to porpoise," LeVier replied.

"Okay," Kelly said. "Come around again and approach at an even shallower angle."

LeVier did just that, and the plane again began to porpoise. So, again, he lifted off and circled around.

Now Kelly was worried. He could lose the plane.

"Bring it in on the belly!" he ordered.

"Kelly, I'm not gonna do that," LeVier replied.

LeVier made five subsequent attempts before Kelly talked him down from his seat in the C-47 chase plane. Finally, as LeVier later recalled, he "put it down slick as a cat's ass."

The key, it turned out, was to put the tail wheel down at the same time or slightly ahead of the main gear.

Ten minutes after LeVier landed, the skies opened and two inches of rain—nearly half the average annual total for that area—fell and flooded the lake.

In a postflight debrief, LeVier told Kelly that the plane's giant wingspan—it was essentially a powered glider—made it very difficult for him to keep the craft on the ground. This plane just wanted to fly, and kept trying to get back aloft, even when the pilot no longer wanted it to.

That night, everyone on the ranch celebrated. "Everybody got smashed," LeVier recalled. Kelly, still irked by LeVier for refusing his order to land on the plane's belly, challenged his pilot to an arm wrestling competition and, LeVier said, "banged my arm down so hard he almost busted my wrist."

Later, as a safety measure, LeVier laid out eight different three-mile runways for emergency landings. The additional construction made Groom Lake, overnight, the world's largest airport.

Select "customers," including Ozzie Ritland and Dick Bissell, flew into Groom Lake for the U-2's first official flight, on August 8. This time, LeVier had no issues at all. He took off, flew to 32,000 feet, and landed easily, as if he'd done it a million times. The U-2 was accepted for duty. And Kelly Johnson had actually pulled off what at first seemed impossible: He designed and built the world's highest-flying plane, from scratch, in eight months.

LeVier made the first twenty flights—all of the Phase 1 testing—in the Angel prototype, including the first-ever flight over 50,000 feet in a pressure suit, and then went back to Burbank to resume testing the F-104 as Lockheed's director of flying.

Kelly replaced LeVier with two test pilots from the Burbank mother ship, Bob Matye and Ray Goudey, and by October, three more planes had arrived in Nevada and the U-Bird had reached its design altitude of 70,000 feet. A fourth plane followed in early 1956 and the testing program continued with "altitudes considered incredible in 1955" being reached "on an almost routine basis," according to Lockheed historian Jay Miller. "On one occasion, three consecutive altitude flights made by Matye broke, by significant margins, the then-extant world altitude record . . . of 65,890 feet," set by a British test pilot.

Flying at these altitudes, in secret, presented challenges for the CIA security team. Kelly had asked Kollman Instruments, a subcontractor, to supply him with altimeters calibrated to 80,000 feet, 35,000 feet above what its off-the-shelf models could handle. Naturally, this raised eyebrows, and security officers working undercover swept in with a cover story to explain this peculiar request: The altimeters would be installed on experimental rocket planes.

Fuel was another problem. Regular jet fuel wouldn't work in a plane that flew so high; it would boil off and evaporate above 55,000 feet, just like bodily fluids. Kelly called on an old friend, the World War I flying ace turned (now retired) Air Force General James Doolittle, to solve the problem. Doolittle was now a VP at Shell Oil, and he arranged to supply a low-volatility, low-vapor-pressure kerosene known as LF-1A, which had a boiling point of 300 degrees at sea level.

To make enough LF-1A to supply the program, however, Shell had to divert supplies of petroleum by-products it typically used in the manufacture of a bug spray known as Flit, causing a nationwide shortage of this consumer product in the spring and summer of 1955.

"The biggest and most disturbing" of the development problems, according to CIA project boss Dick Bissell, was "the unreliable behavior" of the U-2's first engine above 60,000 feet. Up there in the thinner air, Pratt & Whitney's powerful engine—designed in the early 1950s for fighters like the Convair F-102—experienced so-called flameouts. Basically, the engine would just quit from oxygen starvation.

During development, this wasn't a huge concern for pilots, because the engine could be "windmilled" by the pilot by simply descending to a lower altitude with denser air and restarting the engine at 40,000 feet.

This tactic would be a significant problem on operational flights. The whole point of the U-2 was to fly at extreme heights to steer clear of antiaircraft weapons. But a pilot who has to descend from that safe aerie—from 60,000 to 40,000 feet—over enemy territory is going to get shot down.

The U-2's incredible gliding ability was also a safety feature. If a plane were to flame out at 70,000 feet, the pilot could glide it another three hundred miles without power. So, during testing, pilots were told that, in the event of a flameout, they were to glide to the nearest Strategic Air Command base. These bases were scattered around the West, and commanders were given sealed orders that were to be opened only "if a mysterious aircraft appeared," Bissell reported.

The U-2 test flights over the latter half of 1955 resulted in one surprising phenomenon: a spike in UFO sightings. Reports of unexplained shiny objects by commercial pilots poured in to air traffic controllers working the southwest sector of the United States, especially in the early evening. People on the ground spotted the silver planes soaring 70,000 feet up, too, and could only assume that the objects were extraterrestrial, because at that time there was no known craft capable of reaching such heights.

Such reports found their way to the Air Force's Project Blue Book, the UFO investigation group that Kelly himself had once filed a report with, and once certain investigators were cleared into the U-2's existence, they began to call Bissell's office to check reported UFO sightings against scheduled U-2 test flights. It's estimated that half of the UFOs reported during the late 1950s were in fact U-2 flights.

All this testing at Paradise Ranch required more support and services, which meant more people, and Kelly realized he needed a way to transport workers back and forth from Burbank to this remote desert outpost.

Basically, he set up the Paradise Ranch Express. Starting on October 3, 1955, daily Military Air Transport Service began flying the route in an Air Force C-54 that Jim Cunningham nicknamed "Bissell's Narrow-Gauge Airline." But seven weeks later, one of the planes crashed en route to Groom Lake, killing all fourteen passengers, including a senior CIA security officer. It was by far the single greatest loss of life in the history of the AQUATONE program.

From the U-2 program's outset, Eisenhower worried about the political fallout should one of these planes be shot down over Soviet territory. It's why he insisted that the program be a civilian operation.

The CIA tried first to recruit pilots from abroad and even brought seven Greeks and a Pole to America for training, but the effort was a bust. There was the language problem, plus the pilots had limited skill sets; in the end only two of the Greek pilots qualified to fly the U-2 and were sent to Groom Lake, but they "made only a few flights" in the plane and were out of the program by the fall of 1955.

Instead, the Agency turned back to the Air Force, and the deputy chief of staff for personnel approved the use of fighter pilots from the

Strategic Air Command—but only those flying in the reserves, because any active-duty pilot would have to retire from the Air Force first, and then be hired into the program. Reserve pilots had to resign, too, but it was far easier, and the process of reservists doing this and then reenlisting as civilians became known as "sheep dipping."

It was a lot to ask of a pilot, to give up his commission, so the CIA had to pay well and arrange for the Air Force to promise that any man who went through the process could return to his unit, with no loss of seniority, if and when he left the program.

The selection process was rigorous. This was a physically punishing job—flying a plane for hours on end at high altitude, over dangerous terrain, in a pressure suit. "Painstaking efforts were made to exclude all pilots who might be nervous or unstable in any way," the program's official CIA history reported.

Francis Gary Powers, the most famous U-2 pilot of all time (more on him soon enough), recalled being pulled aside and asked if he was interested in a civilian job. If so, he should meet with the prospective employer at a local motel that night. Powers was intrigued. He drove to the motel, knocked on a door, and met a man who called himself William Collins, which was almost certainly not his actual name.

Collins shared few details but said that Powers was one of a precious few pilots chosen to be part of a very special project.

"It will be risky but patriotic," the man told him.

Powers eagerly accepted and was given a fake name—Frank Palmer—to travel under. He and other pilots selected to apply were excited at the opportunity. This was rarefied work, for the best of the best. And they had no idea what to expect. It wasn't always love at first sight.

Marty Knutsen was the first pilot picked and sheep-dipped to become a "driver," as they were officially known in CIA communications, for security reasons. Knutsen was twenty-six and had a thousand hours of flight time under his belt when he arrived at

Groom Lake and "almost died of disappointment" when he saw what he was going to be flying. He was a fighter pilot, and this was a glorified glider, with a yoke and no stick. Then he flew the thing. "It was a bitch to land, and easy to stall out, but I fell in love," he later said.

Learning to fly the fickle machine was just part of the challenge. These men would be flying higher than any men ever had, and for hours at a time. And these accomplished Air Force pilots, who had hundreds if not thousands of hours of flight time in regular flight suits and cockpits, struggled to adjust to the U-2's life-support systems. They especially hated the special survival suits.

Early models were hot, tight, and bulky—just uncomfortable. The system included thick gloves and a heavy helmet that was prone to fogging. It was also a work in progress and led to several crashes.

It was claustrophobic feeling the helmet's twelve-inch-wide rubber neck seal pulled down over the face to be snapped into place around the pilot's neck. Driver James Cherbonneaux says: "[I] had to learn to ignore the feeling of panic and suffocation that occurred each time it was over my face, blanking out my vision and breathing."

To prevent the bends during ascent, pilots had to do the reverse of what scuba divers do after surfacing from extreme depths and pressures—U-2 drivers would "rebreathe" on the ground as part of the preflight preparation, which meant putting the suits on and breathing pure oxygen for ninety minutes ahead of a flight so that their bodies could dissipate nitrogen in the blood. This made eating, drinking, and peeing problematic.

The first version of the pressure suit forgot to account for urination. That was fixed by adding a catheter to the next model, and then replaced by a bladder, which was better, but still not easy.

A pilot had to navigate three layers of zippers to free his penis and pee into a bottle. Defecation was not an option, which is why pilots ate only a minimum of high-protein food for a day before flying. "You got used to evacuating your bowels before takeoff," James Cherbonneaux said, "because a lapse in that area could turn into a painful experience indeed."

Drinking was by straw, through a tiny self-sealing hole in the visor, and scientists working to help feed U-2 pilots also pioneered many of the first ready-to-eat foods in squeezable containers. These grown-up squeezies packed 150 to 300 calories per tube and were primarily bacon- or cheese-flavored mixtures.

And yet pilots still lost an average of 3 to 6 pounds during a single eight-hour mission.

Lockheed test pilots trained the first six Strategic Air Command drivers, and these drivers in turn trained all the inductees who came after them. But even this training was unusually hard, because there was no two-seat version of the U-2, which would allow an experienced pilot to fly along with, and train, a newbie. Instead, trainees would get exhaustive instruction on the ground, and then, when they were ready to attempt a flight, the only option was to go with God while being coached via radio by another pilot flying in a chase plane.

All drivers struggled at first because the U-Bird was essentially a glider, which made it a challenge for fighter pilots used to operating planes built for speed, power, and maneuverability. The U-2 had tremendous lift, and took off easily, but it was also very fragile and could only fly within a narrow speed window. When a U-2 was climbing, a pilot had to watch his speed like a hawk. "The slowest it could go safely," Kelly's protégé Ben Rich later wrote, "was right next to the fastest it could go as it climbed steeply to above 65,000 feet."

At cruise altitude, that window was even narrower. Up in the stratosphere, only six knots separated the speed at which the plane would stall and lose lift from the speed at which buffeting could occur from going too fast. Pilots referred to this as the coffin corner. And because the plane was not designed to withstand even mildly strong G-forces, the pilots had to be vigilant about keeping the nose slightly up, lest it quickly gain speed and begin to break apart.

And then there was navigation. There was no GPS in those days. The primary navigational tool for a U-2 pilot was a downward-facing periscope, which the pilots used to follow the scrolling strip maps that emerged like ticker tape and plotted the plane's course. Once read, the pilot would place that strip into a small tank of water, where it would dissolve, so that if the plane were to crash, no one could tell where it had flown. (That soluble paper, by the way, was its own development project. "I guess we spent a year or more on this thing," Jim Cunningham said.)

Pilots could deviate no more than one mile off the preplanned course and a downward-facing tracker camera known as the "lie detector" took photos from start to finish so that the pilot's course could be studied for accuracy later. On cloudy days, however, that method was useless, so pilots carried a small sextant and used stars to pinpoint their location.

14

GO TIME

By the spring of 1956, the U-2 was ready to go operational.

Much consideration went into a plausible cover story to explain why this experimental American plane would suddenly be soaring extremely high in the skies over Europe. Dick Bissell and the CIA ultimately settled on the only story that made sense. This experimental high-altitude plane was flying research missions for the National Advisory Committee for Aeronautics.

On Monday, May 7, 1956, NACA announced the "start of a new research program" and a "new airplane, the Lockheed U-2 . . . expected to reach 10-mile-high altitudes as a matter of routine."

The U-2—according to the cover story—would study clear air turbulence, convective clouds, wind shear, and jet streams, as well as cosmic rays, ozone, and water vapor concentrations in the atmosphere, NACA Director Hugh Dryden said. "The first data, covering conditions in the Rocky Mountain area, are being obtained from flights from Watertown Strip, Nevada"—in other words, the final test runs from Groom Lake.

As Kelly Johnson later noted, this was "all true—in time."

What everyone, including Kelly, feared most was having a U-2 pilot shot down. Rumors persisted that the Soviet secret police tortured captured spies, and this worried Dick Bissell in particular. Bissell asked Dr. Alex Batlin of the CIA's Technical Services Division for a way to "help 'captured' U-2 pilots avoid such suffering." Batlin's solution, borrowed from the execution-eve suicide of notorious Hitler capo Herman Göring: a small glass capsule filled with liquid potassium cyanide.

If a situation was truly dire—if capture was inevitable—the pilot could just bite the capsule, and he would be dead within fifteen seconds. Six of these so-called L-pills were ordered and offered to pilots in advance of early flights, but carrying one was completely optional. Pilots could opt out, and many did.

A last-minute test of the plane's capabilities, even when damaged, occurred on April 14, when a U-2 flown by driver Jacob Kratt had an engine flameout over far western Tennessee. Kratt followed procedure and dropped to lower altitudes, but he was unable to restart his engine and told the project office he would need to make an emergency landing.

The CIA was prepared for just this possibility, which is why Bissell had asked that sealed orders be delivered by the Air Force to every base in the country. Those orders would remain sealed until the base commander was instructed to open them, in the event of an emergency they couldn't possibly predict.

Jim Cunningham took the call about Kratt's engine failure and asked for details on which base he'd need to call. Kratt was by this time gliding over Arkansas and did the math.

Considering the winds, and his plane's impressive glide ability, he estimated that he could make it as far as Albuquerque. Cunningham called Leo Geary, who called the Air Force operations office, which called the commander of Kirtland Air Force Base and explained that an unusual aircraft was incoming and would be making a dead-stick landing.

Kirtland's commander was to have his security staff isolate the plane and move it into a hangar until a rapid response team could get there and repair it.

Kirtland's commander was shocked to see what looked like a giant glider touch down on his runway, and even more shocked when he saw that the pilot who climbed out of it was wearing what appeared to be a space suit.

President Eisenhower gave his approval for overflights of the Soviet Union in late June 1956. Bissell would have ten days, starting immediately, to commence high-altitude surveillance flights over the country's most formidable enemy. At the end of those ten days, the president wanted detailed updates and a review of operational security. Basically, he wanted to pause the program and ask two questions: What did we get, and how much did the Russians know about it?

Bissell asked Eisenhower's staff secretary, General Andrew Goodpaster, if that meant ten days total, or ten days of good weather. "It's ten calendar days, period!" Goodpaster replied. "You'll have to take your chances with the weather."

Weather mattered a great deal. If a target area was overcast, the plane's trip would be useless. So at noon the day before a scheduled flight, Bissell would receive a weather report stating whether or not conditions looked favorable for flight the following morning. That same evening, at around 6:00 or 7:00 p.m., he would get a second report, and if the weather still looked good, word would be sent to the detachment that there was a "high likelihood of a mission."

By midnight in Washington, a final decision had to be made and transmitted in order for the first deployed U-2 detachment, six hours ahead—at Lindsey Air Station in Wiesbaden, Germany—to have a U-2 ready to go at first light.

Bissell would make this call personally, from his office, after one final review of weather and flight plans. Then, because he knew he wouldn't have any new information until the morning, he would go home and, as he later put it, "sleep (perhaps unwisely) the sleep of the just."

The first-ever operational overflight of a U-2 over the Soviet Union took off on Independence Day, July 4, 1956, from Germany, with the agreed-upon cover story should its mission be detected: This was a research plane, operated by NACA to study the weather and cosmic rays in the upper atmosphere.

Article 347, flown by driver Carl Vito, passed over Poznan, Poland, Belorussia, and Moscow, before turning north toward Leningrad. The primary targets were the naval shipyard where Russia built submarines, and a nearby airbase where the country was thought to keep many of its Bison bombers.

No one was 100 percent sure that the Russians' primary surface-to-air missile, the SA-2, couldn't reach the U-2 above 70,000 feet, but it was decided that the missile's fins were too small to accurately guide it in the thin air at those altitudes. In all likelihood, the drivers were safe. It was, Bissell later said, a "calculated risk."

Especially considering that, despite CIA assurances to the contrary, the Soviets tracked the plane all the way, from the moment it crossed into East German airspace. Premier Nikita Khrushchev was notified immediately, and within hours, a protest had been relayed by the Russian embassy in Washington.

But the mission otherwise went off without a hitch and came back with photos that Bissell described as "remarkable." The U-2, he marveled, had captured "perfectly beautiful" photos of Russia's two largest cities "in which one could literally count the number of automobiles on the street."

An assessment of the flight written by the CIA's Herb Miller two weeks later was effusive. "There can be no doubt of the value in terms of our national security of the photographic coverage obtained . . .

of five of the seven highest priority targets specified by the USAF," Miller wrote, clearly with the president—and his worry about whether or not to continue these controversial overflights—in mind.

Later photos, from subsequent iterations of the U-2's camera system, would be so clear that analysts could identify golf balls on the ground. Kelly hung one photo in his office, taken by a U-2 flying at 70,000 feet over Burbank. The image was so clear that you could read some of the street signs.

A full review of the photos from that first flight revealed that the mission had "much greater significance" beyond just its specific targets, Miller reported. "For the first time we are really able to say that we have an understanding of much that was going on in the Soviet Union. . . . We are no longer depending on an 'estimate' or a 'judgment' or an 'assessment' of what the situation is."

This single flight, conducted on Independence Day, provided American analysts with a "cross-section of a part of the entire Soviet way of life" that had previously been hidden from view—military bases, sure, but also farms, irrigation systems, factories, power systems, housing, recreation, railroads, the amount of traffic on roads and rails. The photos revealed previously unknown airfields, training bases, and even urban centers. The haul was just astounding.

A day later, on July 5, Bissell approved two nearly simultaneous flights. One went over central Russia; the other targeted Ukraine. In total, there were six U-2 flights in that first ten-day period, and every one of them, as suspected, was tracked by Soviet radars. But there was nothing the Russians could do about it.

Except complain, over diplomatic channels. Which they did. The Soviets said nothing publicly, which Bissell interpreted as a sign that they were embarrassed; they didn't want their citizens, and the armed forces, to know that American planes were overflying Soviet territory with impunity, and there was literally nothing they could do to stop them. You can imagine how certain paranoid people in the government could extrapolate from there. *What if these new, secret, high-flying planes that we can't reach were to show up carrying bombs?*

Instead, the Soviets worked back channels, and the angry complaints passed along via officials in Washington had an effect on Eisenhower. The president worried that diplomatic tumult over the flights could affect peace talks that were already scheduled. But as the former Supreme Commander of the Allied Forces, he also understood the importance of accurate intelligence.

The head of the CIA's Photographic Interpretation Division, Art Lundahl, played a big role in convincing Din Land's panel that a spy plane was the way to solve America's intelligence problems. So he was a natural choice to lead the new department that would handle the program's "take"—the spectacular photos that the U-2 drivers were suddenly bringing home.

Before the flights commenced, Lundahl was summoned by DCI Allen Dulles and briefed by Bissell, who told him to go create a "storefront" for photo analysis. It would have to be big, Bissell said, almost industrial in scale—to accommodate the thousands of gallons of processing material and chemicals required to develop the images. "This was not something that could be accomplished out of a locker in a bus depot."

Lundahl told Bissell that he needed 50,000 feet of space that no one would wonder about. So the CIA found him a spot above a Ford repair shop on Fifth and K Streets in downtown Washington. "The dilapidated exterior—and grim neighborhood—effectively masked the most high-tech photo-intelligence center in the world," Bissell later wrote.

The new analysis hub needed a code name, and Lundahl suggested Automat, because that was the model he had in mind—a highly automated twenty-four-hour processing factory that ingested film and put out some of the most valuable intelligence of the Cold War.

Unfortunately, Lundahl was told, Automat wouldn't work. According to CIA security rules, a code name couldn't start with a vowel. But Lundahl dug in. He wanted it anyway, and told his security officer, Henry Thomas, to figure it out. Thomas put his initials in the front, and Lundahl's secret photo analysis program became HTAUTOMAT, with a silent HT.

As soon as the U-2 was flying, the Soviets mobilized to unlock the secrets of this plane and its program. They knew where the detachments were based—at Lindsey Air Force Base in Germany and at Incirlik in Turkey—and could easily keep eyes on those airfields to know when a plane was leaving.

But intelligence is a chess match, and Stanley Beerli, commander of the Incirlik detachment, had an idea for disguising the flights. The Agency would create a mobile unit that could deploy rapidly from a more forward base—to allow the U-2 to take off in greater secrecy, since there wouldn't be observers posted nearby looking for it. Often, this meant Pakistan. He code-named the program QUICKMOVE.

It worked like this: Two C-130s would fly to whatever remote base had been chosen to rendezvous with a U-2 and a driver. One C-130 carried a twenty-man ground crew; the other was loaded with 55-gallon drums of fuel for the plane. This mobile support team would fuel the plane, then fly back home to their actual base, leaving just the U-2, and no trace of the operation, behind.

The C-130 was a perfect cover. It was among the most widely used military planes on the planet and there was nothing unusual about two of them flying in and out of any base, in any country. "The whole operation was compact and beautifully run," Beerli said.

A QUICKMOVE deployment from Peshawar, Pakistan, resulted in photos of new nuclear installations in Siberia and a new radar test site in Kazakhstan. Pilots weren't supposed to divert from their predetermined flight path unless there was an emergency, but it was an unauthorized diversion that resulted "among the richest intelligence finds of the entire program," Bissell wrote.

In June 1957, three months before the spacecraft *Sputnik* launched, a U-2 driver noticed something in the distance that looked curious. It turned out to be Tyuratam, Russia's first test site for ICBMs. "Departing from his course, he stumbled on the crown jewel of Soviet space technology, whose existence had not even been suspected," Bissell wrote. Within a week, U.S. analysts and photointerpreters had built a scale model of the test site.

The CIA's new toy would come in handy elsewhere, too. When war broke out between Egypt and a tri-national force from Israel, France, and the UK over the Suez Canal, Kelly Johnson's U-2 delivered again. This time, the mission had no geopolitical risk; the drivers could fly over the disputed territory without fear of being shot down.

One day, a U-2 flew over an airport and saw a whole line of Egyptian air force planes lined up. It made another pass ten minutes later, and the airport had been destroyed. British and French forces had attacked it—and the United States had an almost immediate damage assessment. Eisenhower couldn't believe it.

"Twenty-minute reconnaissance," the president marveled. "Now that's something to shoot for."

15

THE END IS NIGH

The fact that the Soviets were tracking the U-2 was a concern, but not a huge one, at least in the early stages, because there was nothing the Russians could do about Kelly Johnson's elusive "angel." Still, that knowledge bothered CIA leadership, the Pentagon, and Kelly, who all understood that the Soviets would be throwing every possible resource at finding a way to target the high-flying planes.

Within months of the plane's first flight, the Skunk Works settled on one primary strategy: They wouldn't push the plane to fly higher or improve its defenses, both of which would require major changes. They wanted to try to make the plane harder to spot by enemy installations, and so began to study ways to lower the U-2's radar cross section, or RCS.

RCS is sometimes called radar signature, and it's the measure of how detectable a particular plane is on radar. The higher the RCS, the easier the plane is to see and shoot down.

In October, Bissell noted in a memo that the "number 1 aircraft"—presumably Article 341—would be pulled from service for modifications.

Kelly contracted with the Lincoln Lab at the Massachusetts Institute of Technology to do this specialized research—code-named Project Rainbow—and a number of strategies were pursued to hide the bird from Russia's C- and S-band radars. Lincoln Lab engineers tried radar-absorbent paints; so-called Salisbury screens glued to the fuselage; and wires loaded with ferrite beads, strung around the entire periphery of the plane's wing and tail.

Each of these strategies did the job of reducing the RCS, but only at some specific frequency, or angle of attack, and every one of them also lessened the plane's performance, cutting its maximum altitude by as much as 4,000 feet. The radar-absorbent paint, for instance, added 100 pounds to the U-2's weight. So in every case, the negatives far outweighed the gains.

Test pilots nicknamed these experimental planes the "dirty birds" and were not excited at the prospect of flying them. One pilot, James Cherbonneaux, drew the unfortunate lot of flying a dirty bird—specifically, one of those with the ferrite beads and wires—out of Pakistan on an operational mission. "Part of my paycheck was compensation for high risk missions in a semi-experimental airplane," he told the co-author of Ben Rich's memoir, "but I had never before risked flying an airplane wired like a guitar."

On July 7, Cherbonneaux flew that dirty bird over the Black Sea, within twelve miles of the Soviet coast, in order to test the Russian defenses. The plane seemed to mostly work—it was pinged much less often than an unadorned U-2 would be—but the flight revealed two blind spots. Radar operators were "homing in on" the plane's cockpit and tailpipe, neither of which had been treated.

Two weeks later, Cherbonneaux was sent on another dangerous test mission. This one had been specially cleared by President Eisenhower. Cherbonneaux took off from a black site in Pakistan on a QUICKMOVE flight and headed straight into Russia to locate and photograph a missile site where the Soviets were thought to be prepping for an ICBM test.

The additional weight of his dirty bird's wires and beads meant that Cherbonneaux's maximum altitude was 58,000 feet, so he was more vulnerable than he would be in a regular U-Bird, but he also had a better-than-normal view of the land below when—just seventy-five minutes into his three-hour flight toward the target—he spotted something startling.

It was a nuclear test site previously unknown to the United States. "I brought my drift sight up to its maximum four-power magnification and focused on a large tower," Cherbonneaux later said. His pulse raced, and he felt "a chilling terror." At the top of that tower was a "large object," and he also spotted signs of activity around a huge blockhouse near the site. "And then a paralyzing thought slammed me: What if those bastards were getting set to let that nuclear weapon blow just as I was directly overhead?"

Cherbonneaux fumbled nervously to line up his cameras and then fired away.

"My heart was pounding in my throat," he said. "I just knew I was going to be evaporated in the next seconds. Five minutes later I was clear of the nuclear test site, laughing to myself for being so chicken."

Cherbonneaux continued his flight, photographed the primary target, and then returned home, nervous the entire way that he was vulnerable to antiaircraft defenses at this altitude. But no Soviet planes took off to chase him, and no radars activated.

He expected a hero's welcome for having uncovered a Soviet nuclear test site with a bomb in the tower. "Instead," he said, the team that debriefed him "scoffed incredulously," one debriefer having the gall to say, "There is no atomic test facility in that part of central Russia."

Still, the debriefing team relayed Cherbonneaux's news back to Washington by a special wire and rushed the film to processing. The CIA's immediate response, sent by coded cable before any results were back, was displeasure—annoyed that this driver had essentially cried wolf. "Their communication was stern, halfway between a personal rebuke and an official reprimand," Cherbonneaux recalled.

A day later, at lunch, the senior CIA officer on the U-2 program in Turkey pulled the driver aside. "Apologies, Jim," he told Cherbonneaux. "Collateral intelligence sources just reported that a nuclear bomb was detonated from that tower less than two hours after you flew over it."

Cherbonneaux's success in the dirty bird notwithstanding, no amount of stringing piano wire or hanging metal beads on the U-2 was going to reduce its RCS to a point where Dick Bissell could confidently assure Allen Dulles, or the president, that these spy planes could overfly Russia without risk. Given enough time, with no quantum leap in RCS reduction at the Skunk Works, the Soviets were absolutely going to shoot a U-2 down.

So, about a year into Project Rainbow, engineer Frank Rodgers broke the bad news to Bissell, who absorbed it and replied, "You're telling me the ship is still sinking, but maybe more slowly?"

What about starting over? Did it make more sense to design a plane from scratch, to avoid radar, as opposed to tinkering with the existing design? And could he do it?

Rodgers thought a moment and answered that it was certainly possible, but he wasn't confident enough to promise anything on the spot.

That was good enough for Bissell. He asked Rodgers to try.

"For the next month, I spent most of my time alone in the shack on the roof of the Lincoln Laboratory building, where our experimental program had begun," Rodgers wrote in an unpublished memoir. "There, I returned to basic research."

For starters, Rodgers wanted to understand "the nature of the radar return from simple geometrical shapes without regard to the aerodynamic practicality of such shapes."

He used a circular aluminum disk as his test vehicle and began to experiment with various treatments, and then compared his treated disks to the naked one. "After almost a month of frustrating failure," he hit on something—an idea that went back to World War II and that, to his great surprise, basically erased the disc's signature across a broad band of frequencies.

Unfortunately, Rodgers wrote, "a flat disk would be about as aerodynamically stable as a circular Persian rug." You could improve it slightly, by making it fat in the middle, like a UFO, and while these weren't viable plane designs, the UFO-type design did cause Rodgers to wonder: Maybe aliens had a similar radar-reflecting treatment, and this explained why their crafts were so rarely seen.

Frank Rodgers wasn't the only engineer to be asked to think outside the box. Or even the only Frank.

A thirty-seven-year-old control systems engineer named Frank Bullock arrived at the Skunk Works in 1956, when a guy he'd worked with on the Constellation commercial airliner invited Bullock to join him in a special group that operated in great secrecy. This friend, Victor Sorensen, had been called to join that same group back when Kelly's organization was developing the prototype for what became the C-130 Hercules.

This project was far more experimental. Kelly was in search of something radical to replace the U-2—something that flew much faster and much higher. He needed to establish new limits, and he had an idea.

Hydrogen.

Air-to-air fueling did not yet exist in 1956, so the only viable way to power such a plane at high speeds, to heights above 80,000 feet, and give it any useful range was to use liquid hydrogen as a fuel. This plan, Ben Rich later wrote, was "certainly ambitious," but "in those days Kelly seemed entirely capable of moving the world."

The idea wasn't new to Lockheed. In 1954 and 1955, the company had conducted a preliminary design study on the feasibility of using liquid hydrogen, which was harder to produce but safer to handle than gaseous hydrogen, as a fuel source for a supersonic, high-altitude airplane.

Two years later, in 1956, the Skunk Works got a contract from the U.S. Air Force to take things a step further: to design and build a Mach 2.5 reconnaissance plane that could fly at 100,000 feet—far beyond the capability of any Soviet antiaircraft weaponry.

Kelly promised a prototype in eighteen months and called the concept the CL-400.

The Air Force wanted the Skunk Works to do what it did best—work at lightning speed—so it cleared only twenty-five people and moved the program office for this compartmentalized, ultra-secret program—code-named SUNTAN—from Wright Field in Ohio to the offices of the Air Research and Development Command in Baltimore.

Kelly insisted that the Air Force grant him the same freedom he'd gotten from the CIA, including a waiver from typical procurement processes, a minimum of paperwork, and the freedom to contract his own suppliers. Inside the ARDC, the program was closely guarded. Project numbers were changed regularly and certain contracts were issued from different offices to disguise their true origins.

This was the first time a Skunk Works project would require development of both an airframe and an engine, and the latter was assigned to Pratt & Whitney, which built a prototype 304 "hydrogen expander" engine at a Connecticut factory and then moved the project to West Palm Beach, Florida. Among its innovations: This was one of the first engines to use fully electronic fuel controls.

Kelly summoned one of his favorite young engineers, Ben Rich, and asked him what he knew about cryogenics. This was still pre-*Sputnik*; very little knowledge about the handling of hydrogen (or LH) existed outside the extremely classified hydrogen bomb program. But Kelly understood that liquid hydrogen was only viable if he could find a way to make it as easy to use as traditional fuel, like gasoline.

Rich told his boss that he knew very little about cryogenics, aside from what he'd learned in college chemical engineering courses, so Kelly ordered him to go and become an expert, as quickly as possible.

As Rich later recalled in his memoir, Kelly dispatched him to travel the country in secret, learning everything he could about the manufacture and handling of LH without letting on what he was up to.

Rich traveled under his favorite Skunk Works pseudonym, Ben Dover, and pretended to be a self-employed thermodynamicist. He went to Boulder, Colorado, where the National Bureau of Standards had a cryogenics lab run by the world's foremost expert on handling liquid hydrogen, and to UC Berkeley, where a Nobel laureate worked out of a basement "dungeon" because of explosion risks.

Every expert Rich encountered warned him that manufacturing large quantities of liquid hydrogen was dangerous, and a bad idea, but Rich had his orders, and when he returned to Burbank, he told Kelly what he'd learned: that, in the opinion of America's top experts, hydrogen was too dangerous to store or handle.

"Goddamn it," Kelly said. "Liquid hydrogen is the same as steam. What is steam? Condensed water. Hydrogen plus oxygen produces water. That's all that liquid hydrogen really is. Now, get out there and do the job for me!"

Kelly arranged for a supply of hydrogen from the National Bureau of Standards lab to be trucked to Burbank in a 2,200-liter refrigerated dewar and set Rich up in a small, extremely secret test shop installed in the safest place on the Lockheed property—in some of the old concrete revetments built during World War II to shelter people and planes during air raids.

Rich's secret lab had explosion-proof electrical fixtures, spark-proof tools, a combustible-gas alarm system, and eight-foot-thick walls. And there, in his bunker, he set out to design an engine and storage/handling system for a hydrogen-powered plane, years before the idea would come to fruition for the space program.

Rich picked Dave Robertson, "one of the shrewdest, most instinctive engineers" at the Skunk Works, to be his deputy. Robertson, Rich thought, had "the right flair for these wild experiments," and the two men began to carry out a series of dangerous tests inside what a colleague nicknamed Fort Robertson.

The two men, and a small group of handpicked Skunks, created intentional explosions of various sizes to study and carefully analyze how to prevent and manage them.

To do this, they built a number of scaled-down models of the tanks and first punctured them, then blew them up, on purpose, to assess the explosion risk. They conducted sixty-one controlled explosions using a scale model of the test facility and learned that the potential damage wasn't as great as anyone feared.

Hydrogen turned out to have a rapid flame speed, which meant that its fireballs dissipated quickly. As a result, the explosions, according to a Lockheed report, "were generally mild and hydrogen fireball radiation was much less than that from a comparable kerosene fire." Only two of the sixty-one explosions, in fact, produced sizable booms, and in both cases oxygen had been deliberately mixed with the liquid hydrogen. And—by comparison—gasoline fires were "an order of magnitude more severe" than hydrogen fires. This, at least, was encouraging.

Within three months, Rich's prototype plant was manufacturing 200 gallons of liquid hydrogen a day, more than any other place in America, and virtually no one knew about it. To prevent sparks, workers wore grounded shoes and were forbidden from carrying metal objects.

There *was* one near disaster, when a wood-framed oven that Rich installed to superheat a prototype wing caught fire in the wee hours. Lockheed security panicked because the classified nature of this facility made it impossible to call the fire department. So they called Rich, who raced to work and, thinking quickly, ordered workers to bleed out the liquid hydrogen. The extremely cold liquid hitting the floor of that extremely hot room produced a thick fog.

When Burbank's fire department finally did arrive, Lockheed security still wouldn't let them into the secret facility, until Rich himself signed off, having decided that they wouldn't see anything through the fog anyway.

While Rich was attempting to pioneer a propulsion system without blowing up Burbank, Kelly worked on the airframe, which looked a lot like the F-104, by necessity. The accelerated schedule basically mandated that he and his designers build on existing work, using "known aerodynamic configurations."

But the design had problems.

The low density of hydrogen—a 50-gallon drum full of it weighed just 50 pounds—meant that any viable LH-powered plane would need an enormous volume of it. So much that this supersonic jet would have to be wide-body, with a fuselage diameter of ten feet.

The CL-400 design Kelly cooked up was massive all around: at

160 feet long and 48,515 pounds before fueling, it was twice the size of a B-52 bomber. With a full belly, the CL-400 would weigh 69,955 pounds, and the prototype alone used nearly fifteen miles of aluminum extrusion.

Fuel handling was another enormous challenge. Filling the plane meant delivering liquid hydrogen cooled to −425 degrees through a hot wing with an average surface temperature of 325 degrees. It was, Rich once said, "the world's largest thermos bottle."

Kelly was determined to either solve the problem or rule hydrogen out as a potential alternative for building extremely fast and high-flying aircraft. He modeled different variants, up to the gigantic CL-400-13, which looked like a manned missile and could fly at Mach 4 but at great cost—the plane, which had rear wings and canards (small wings at the front of the fuselage), was 290 feet long and weighed 358,000 pounds fully loaded, 50 percent of which would be fuel. Not exactly practical.

Ultimately, none of the CL variants Kelly drew up had enough range to match the U-2. Pilots flying from allied bases would have to stop for refueling en route to the Soviet Union, which meant the United States would need to build and maintain hydrogen tank facilities in Europe and Asia. What's more, the sheer volume of LH needed was almost impossible to supply. If Lockheed had built just sixteen of the planes, flying them would require 25 percent of the natural gas supply in Los Angeles.

In March 1957, Kelly delivered his verdict to Air Force Secretary James Douglas Jr. during a program review. He told Douglas that he had crammed as much hydrogen into the fuselage as he could and that he couldn't put more into the plane without storing it in the wing. At best, he said, the plane's range could be extended by 3 percent by making airframe changes. And Pratt & Whitney had no answer, either; their best possible estimate was a 5 or 6 percent improvement through engine modifications.

The Air Force pushed back. They insisted that 2,800 miles of range was possible and asked Kelly to continue. But by late 1958, the writing was on the wall. And Kelly, the guy who sold the idea to the Air Force in the first place, had seen it coming. Looking back, he'd known within six months of the program's onset that a hydrogen plane would probably never exceed 2,500 miles of range.

The CL program was not a complete failure, however. Vital data on the handling of liquid hydrogen, as well as tank construction and materials, were turned over to Convair, which had just won a contract to make rockets for the Centaur program.

"We showed that hydrogen as an aircraft fuel was feasible," Ben Rich later said in a speech.

The U.S. space program undeniably benefited; the CL-400 was, basically, the first liquid hydrogen space vehicle. And there was some other "excellent fallout," too, according to Kelly. "It showed that a large supersonic airplane and engine could be developed on a Skunk Works program basis." And it was, he pointed out, "the first time in which Skunk Work procedures were used to build a completely new engine." It also showed that a spy plane's mission range could be "almost doubled at some altitude loss" by using a hydrocarbon fuel that didn't have the "logistics and handling problems of liquid hydrogen."

This failure, in fact, would lead directly to Kelly Johnson's next great achievement—a spectacular new plane to replace the U-2.

16

MAYDAY, MAY DAY

Throughout 1958 and into 1959, as the U-2 was penetrating Soviet airspace with some consistency, President Eisenhower worried a lot about the overflights. Soviet antiaircraft defenses were improving rapidly, and the president had diplomacy on his mind, too. He was anticipating an important summit with his volatile counterpart, Soviet premier Nikita Khrushchev.

Eisenhower wasn't alone in worrying. Kelly, Dick Bissell, really everyone who mattered around the program talked openly about when, not if, the Soviets would finally shoot a U-Bird down.

But Eisenhower was also still very worried about ICBMs, especially about the possibility that the Soviets could move and launch the missiles using flatbed trucks. So, despite all the concern over advances in Soviet radar and air defenses, the president approved one final flight of the U-Bird—a QUICKMOVE operation—to be carried out no later than May 1.

Weather foiled numerous attempts in late April, but the skies finally cleared on May 1. And late the night before, Bissell approved "the most daring" U-Bird mission yet—the first ever to traverse the entire Soviet Union. The mission was code-named Operation Grand Slam.

A few days ahead of the flight, driver Francis Gary Powers flew into Peshawar with a U-2 and awaited orders.

The mission plan called for Powers to take off from Pakistan in a U-2C that flew 3,000 to 5,000 feet higher than the original U-2A, cover a huge swath of the Soviet Union, and then land in Norway.

Powers took off and began his mission without incident. He was cruising comfortably at 68,000 feet about 75 miles southwest of Sverdlovsk when he heard a "thump," followed by an orange flash, causing his plane to jerk and throwing him back in his seat. Powers tried to regain control of the plane, but nothing worked.

The fragile aircraft convulsed again, and both wings broke away, causing the U-2 to plunge and spin toward rural Russia, 50,000 feet below.

The plane spun so violently that Powers could barely tell up from down. It was impossible for him to reach the self-destruct switches, so Powers could not do what he was supposed to do in the event before bailing out—destroy the U-2 to prevent its secrets from being pillaged.

Instead, Powers popped the canopy loose, and, as the frigid air caused his visor to ice over, threw himself away from the plane. He saw hills, a lake, some buildings. He thought, as he floated slowly down toward the hostile territory below, that this part of Russia looked a little like Virginia.

Stan Beerli, the Detachment B commander who created QUICK-MOVE, had flown into the mission's chosen landing site at Bodo, Norway, with a team a few days before, and awaited Powers's arrival.

U-2 pilots flew silently. There was no radio communication until they were safely out of harm's way. So the recovery team waited for Powers at his estimated arrival time, which came and went. Once a full hour after that time had passed, it was clear that Powers wasn't coming back. He'd either been shot down or was out of fuel.

Bissell was out of town when word reached him that the program "lost a bird." He flew straight home to Washington and strolled calmly into a chaotic program office "as if he were about to assemble a Monday staff meeting," a staffer later recalled.

Kelly heard by phone, from Bissell. "Three-sixty is missing," he said.

There was a cover story for just this scenario: An experimental plane conducting research on clear air turbulence on a joint NASA-USAF Air Weather Service mission had strayed accidentally into Soviet airspace and crashed. In fact, a draft of the release explaining this was prepared two years before Powers's flight, and that release was quickly dusted off and distributed.

A State Department spokesman addressed reporters and said that the civilian pilot of this research plane signaled problems with his oxygen system somewhere over Turkey, near the Soviet border.

"It is entirely possible that, having a failure in the oxygen equipment which could result in the pilot losing consciousness, the plane continued on automatic pilot for a considerable distance and accidentally violated Soviet airspace," he said. "The United States is taking this matter up with the Soviet government, with particular reference to the fate of the pilot."

Eisenhower's worst nightmare, the thing he feared from the day he first heard about the program, was that a pilot could be shot down and captured. But most of those cleared into the program assumed that if a U-2 were shot down, it would almost certainly kill the pilot. And the fragile plane, without question, would disintegrate.

U-2 drivers were drilled, in the event of a disaster over hostile terrain, to flip two switches before bailing out. These switches would trigger a process that waited seventy seconds—to give the pilot time to get clear of the plane—and then detonated explosives to destroy the cameras, electronics, and other proof of the plane's true mission.

No one in Washington was yet aware that Powers hadn't been able to do that. And Nikita Khrushchev seized the moment. This was a publicity coup, a chance to boost Russian morale while publicly shaming the evil American empire. Russian media published pictures of some wreckage, with children swarming over the remnants, and stated that these were pieces of an American "pirate plane."

The Soviet premier was playing a game, trying to get the United States to admit what it had done. But the photo only heightened the confusion in Washington. Maybe there was no wreck? Could Powers have defected? Or crashed into the ocean?

Back in California, Kelly watched the footage and knew immediately that this wasn't his plane. There was just no way the Soviets would let kids climb all over the wreckage of the most important enemy plane shot down over Mother Russia in the Cold War. "That's no damned U-2," he said. More likely, he thought, these were pieces of Soviet fighters that had been accidentally shot down *by other Soviet fighters* while pursuing Powers in his U-2.

Hearing this, Bissell asked Kelly to perform a new patriotic duty, as a pawn in the intelligence game; he was to thumb his nose at the Soviets in order to poke Khrushchev. "I was given the job of insulting them to the point where they would show us what they had, because we did not know whether Powers had just defected," Kelly explained decades later.

So Kelly declared publicly that the photo was a fake. He said that he knew his plane, even in pieces, and that pile of twisted metal wasn't it.

"The minute I saw the photography, I knew the whole thing was phony," he told a reporter. Kelly expressed skepticism that the Russians could have shot down a U-2, and said that it was more likely that the plane—if the Soviets really did have it—had experienced mechanical problems.

This was by far the most public event of Kelly's life. It was, arguably, the thing he would be best known for, at least until much later, when some of his secret designs were made public. And speaking publicly during this tense time raised Kelly's profile to an uncomfortable degree. He was, by this point, a critical asset to the national security of the United States, which also made him a significant target for America's adversaries, and CIA security advised him to take extra precautions. He should start varying his arrival and departure times from Lockheed, as well as the routes he took. Maintain extra vigilance, and watch for unusual activity. At home, Kelly began to keep a pistol near his bed.

But the ploy worked; Khrushchev took the bait. The Soviet premier produced real photos, and then the wreck itself, even inviting foreign journalists to come see the great American failure in person, in Moscow. And then, Kelly said, "they put on such a fine show with all the airplane pieces and all the part numbers, cameras, and they did it in Moscow and LIFE took some very good pictures and we could derive from that exactly what happened."

The U-2 had been driving the Russians nuts for four years. And it wasn't until they saw the wreckage of Powers's plane, with its lightweight frame and wings filled with fuel, that they understood how it flew so high.

"Simple but very clever," a general from the Soviet air force, who saw the wreckage in situ, told a reporter two decades later.

Powers was lucky to have survived. Had the Soviet missile hit the cockpit, or even the fuselage, he would have been killed. Some in the public, and within the government, would later question whether the pilot had done the right thing by bailing out and allowing himself to be captured. A real patriot, they said, would have swallowed his cyanide pill. Kelly disagreed. "He did exactly what he should have done," he later said.

A twenty-three-year-old Russian farmer named Mikhail Vasilyev saw a man in a green flight suit—dangling from a red-and-yellow parachute—float down to Earth in a field outside a small village. The farmer assumed, at first, that this was a Russian pilot, until he tried to communicate with the muscular man in the green flight suit. He and a friend helped detach the chute, then got Powers up and out of his sweat-drenched pressure suit, which wasn't easy, considering all the laces and wires.

Vasilyev gave this strange American a cigarette and some water, and later recalled him being quiet but friendly. His plan was to deliver Powers to the local airport when two KGB officers rode by on bikes and took this surprise American prisoner into custody.

Powers was seen by a doctor, then kept under guard until a KGB general arrived to question him. He said nothing, answering only one question, according to that general's wife. It was after he found the cyanide pill.

"Why didn't you kill yourself?"

"I wanted to live," the pilot replied.

Pilots who did not choose the pill were ordered to tell the truth, at least about certain things. They should lie about the plane's ceiling but inform their captors that they worked for the CIA. It was of utmost importance to the president that these flights were not seen as military operations. The CIA, in turn, would say nothing, and the U.S. government would make no public comment.

And when Powers went down, that was Bissell's plan. But the State Department panicked. They wanted to make a statement, that Powers had been on a weather mission and had veered off course. The

CIA reminded State that the plan was to stay quiet and communicate only via back channels. Also, they knew that Powers's flight had been tracked from its origin. He clearly wasn't off course.

The State Department prevailed. At State's urging, NASA released that statement that one of its "research" planes had flown off course and crashed in Turkey after the pilot—an employee of Lockheed flying for NASA—reported problems with his oxygen.

When the Russians announced that they had the pilot, these oxygen problems provided an explanation. Probably the guy blacked out and the plane just kept flying on cruise until it ran out of fuel.

Everyone had a theory. And even at Lockheed, there was debate. Robert Gross wrote that "we do not believe that the plane was shot down by a rocket while flying on its mission." The prevailing opinion at Lockheed, Gross said, was that the engine failed while Powers was at "extremely high altitude where the air is rare and the operation of the engine at best is sensitive." With no other choice, Powers then glided down and eventually was within range of Russian air defenses, where he was easily shot down.

Kelly wanted answers. He ordered a study into possible causes. He asked engineers to simulate Powers's mission and study "what aircraft components could fail at altitude to cause him to lose cruising altitude."

One theory was that a hydraulic system failure could have led to a loss of boost pump pressure and caused an engine flameout that would have forced Powers to "descend to a vulnerable altitude," where the Soviets just shot him down. But tests proved this theory wrong.

"We found nothing basically wrong in the aircraft or its systems which would have been likely to cause the 1 May incident," Kelly wrote. "After these studies, my conclusion was that we had to assume that the aircraft had been hit at high altitude by a missile, as stated by the Russians at the time."

The actual crash photos, ultimately, provided his answer, which allowed Kelly to tell Powers what happened when he was finally returned: "They knocked off the right-hand side of the stabilizer, the airplane cartwheeled over, and broke the wings off."

Allen Dulles offered to fall on his sword. The CIA director told the president that he'd take the blame and resign. But Eisenhower insisted that *he* should assume responsibility. On May 11, the president of the United States admitted to authorizing the U-2 flights, calling these missions "distasteful but a vital necessity" and forced by the Soviet Union's "fetish of secrecy," which kept the United States in the dark on advancements in weaponry.

This wasn't just rhetoric. At a summit in Geneva the previous year, Eisenhower had proposed an "open skies" agreement that would allow both nations to safely overfly each other's territory, to observe and photograph nuclear facilities and launchpads. This would provide mutual assurance, objectively, that neither side was ramping up for war.

Khrushchev rejected the proposal. He would not open his skies. He also boasted that the Soviets had built even more advanced ICBMs, which in turn worried the U.S. leadership that it could be falling behind.

"No one wants another Pearl Harbor," Eisenhower said when he claimed responsibility for the downed spy plane.

Khrushchev had been willing to let the incident pass quietly, without a crisis, as long as the U.S. government stayed silent on it, allowing the Soviets to save face. He announced that the president was lying.

But when Eisenhower owned up to it, Khrushchev was livid. Francis Powers, he said, was from the CIA.

This sudden flare-up in tension put in jeopardy a long-planned summit between the two leaders in Paris. Khrushchev would go forward, if his counterpart apologized and promised to discontinue the flights.

Eisenhower refused. And the Soviet delegation bolted from Paris before the talks could even begin.

"The political implications of the flight were extremely major and President Eisenhower was really beaten down by Mr. K at the summit conference," Kelly, saddened by the events, wrote in the U-2 log.

Based on the crash intel he was given, he and his team agreed with Bob Gross's assessment: that Powers had been shot down. "The airplane was hit at less than normal altitude by a rocket," Kelly wrote in the U-2 log. He also speculated that the plane "disintegrated completely enough" that Powers didn't have time to use the self-destruct button, or even his ejection seat.

The image of Francis Gary Powers among the U.S. public was that he was a traitor. It turned out that he wasn't. He'd followed his orders exactly. He admitted to working for the CIA and lied about the plane's capabilities. Certain top officials, though, never forgave him.

Finally, in February 1962—eighteen months after the U-2 went down—the U.S. government worked out a trade and Powers was returned. When he got back to America, Powers was moved to a safe house outside Gettysburg, Pennsylvania, and Kelly flew immediately to see him.

"Before we start, I want to tell Mr. Powers something," Kelly said. "No matter what happens as a result of this investigation, I want you to know that if you ever need a job, you have one at Lockheed." Kelly paused to let that set in, then asked the question that had been bothering him for nearly two years: "What happened to my plane?"

After speaking to Powers, and analyzing all available reports,

Kelly concluded that he had been correct all along. He presented his conclusion to the CIA on February 21: The plane was shot down by an SA-2. The missile never hit the plane, but it exploded close enough that the resulting shock wave blew off the plane's right-hand stabilizer.

"I was so impressed by the very clear description of the incident by Frank, and having direct knowledge of what he was ordered to do in case of capture, that I will gladly contribute to a fund for decorating this officer for the fine job he did under the most difficult circumstances," Kelly wrote in his report. "He satisfied me, by detailed questions, that the Russians could not have brainwashed him on detail matters of his escape from the aircraft."

Still, America wasn't quite sure how to receive this infamous pilot. Some portion of the population just couldn't shake the idea that he'd possibly turned traitor and had told the Soviets all kinds of secrets. Even the government doubted him. But Kelly protected his pilots, and he believed Powers. If no one else would stand up for him, the Skunk Works would.

The CIA offered Powers a job but that didn't last. So he quit and headed west, to Lockheed, where Kelly honored his promise and gave Powers a job as a company test pilot. Powers flew U-2s that had been repaired or modified, and he declined the many interview requests that poured into Lockheed's PR office.

In 1969, Powers hired a ghostwriter and began work on a book called *Operation Overflight*, being careful to omit any classified information on the U-2 or the overflight program. He did, however, lash out at the public for tarring him as a traitor, and the CIA for doing nothing to defuse that rumor. "A scapegoat, by dictionary definition, is one made to bear the blame for others or to suffer in their place," Powers wrote. "It would be tragic if, in the process of trying to protect our government, we forgot that it was founded on the concept of the worth of the individual."

As required by law, Powers submitted the book to the CIA for review. Not long after, Kelly told Powers that Lockheed was running out of work for him. When would his job be eliminated? Powers asked.

"Yesterday," Kelly replied. And it wasn't his call. Turns out, the CIA had been paying Powers's salary all along.

Operation Overflight was published on May 1, 1970, exactly ten years after Powers was shot down. He made the publicity rounds and gave Kelly a signed copy, with this inscription: "It was a pleasure working for you and flying your plane. The U-2 will always be my favorite. Best wishes and continued success. The U.S. needs your planes."

Powers did ultimately find a new pilot gig—as a flying traffic reporter for KGIL radio in Los Angeles. He loved the job, soaring over the L.A. sprawl, and bought a new house with his wife in Sherman Oaks.

But there's a sad, somewhat ironic, coda to the Francis Gary Powers story. On August 1, 1977, Powers covered a fire near Santa Barbara, then flew on to grab some footage of a softball game, just three miles from the Skunk Works, in Burbank. Powers was hovering over the field when something went wrong. One boy reported hearing a pop, and the chopper just fell out of the sky, breaking into pieces on impact. Meaning that this man who survived being shot down while flying one of history's greatest planes on an extremely dangerous mission was killed while flying a news helicopter on a sunny day.

An investigation revealed no sign of mechanical error. And wind or weather wasn't a factor. "We found no evidence of fuel on board and the craft didn't burn," a police spokesperson said. "So from all appearances, it would seem he ran out of gas. That's the unforgivable pilot error."

Powers certainly understood the range of an aircraft, but he had apparently tried to stretch that range and didn't make it.

17

THE U-BIRD SOARS ON

The U-2 wasn't grounded after Powers went down in Russia. Its utility as an intelligence platform continued to be an asset for the CIA and the U.S. military, adding to its legacy for decades more. Each of these high-flying planes cost less than $1 million and the total government personnel on one of America's great Cold War intelligence programs, at its peak, was only five hundred to six hundred. Kelly Johnson estimated that "95% of our hard information on the missiles in Russia came from the take of that plane."

Some of America's most influential Cold War leaders, including Dick Bissell, credit the U-2 with helping to lower global tensions by proving that the "threat from the Soviet long-range Bison bombers was considerably less than had been estimated." That lessened anxiety in the White House and enabled Eisenhower to reject an Air Force request to increase production of B-52s, "secure in the knowledge that national security was not being compromised."

And the Powers crisis only ended U-2 overflights of Russia. It didn't diminish at all the importance of this plane in overhead reconnaissance. Starting in 1957, the Air Force also began flying U-2s as part of the Strategic Air Command's 4080th Strategic Reconnaissance Wing, based at Laughlin Air Force Base, in Texas.

Air Force pilots selected for the 4080th ran into the same problems their sheep-dipped predecessors had dealt with in Nevada. This was a hard plane to fly and the first year was plagued with accidents. In the most famous of those, squadron commander Colonel Jack Nole ran into trouble at 53,000 feet, when his U-2 pitched down and went into a dive.

Nole tried everything he could think of to recover control of his plane, but its engine was on fire and the U-2 rolled over onto its back.

Nole had to abort, which was not easy even in stable flight, since the U-2 had no ejection seat. He popped off the canopy and bailed out from ten miles up, an action that no pilot was trained for or had even tried.

Nole struggled to activate his emergency oxygen and, because he feared he was losing consciousness, he opened his chute far above its recommended altitude of 14,000 feet—which meant that his troubled descent from there would be painfully slow, and possibly fatal.

"His emergency oxygen supply was exhausted before reaching 20,000 feet, requiring him to remove his mask and suck in deep gasps of air to remain conscious as he descended into more oxygen-rich altitudes," according to a report on Laughlin's website.

Somehow, Nole didn't lose consciousness, and he approached the ground in some arid wilderness outside Del Rio, Texas. Part of his rig caught on a flat-topped rock on the side of a gentle hill and he landed far more softly than he'd been bracing for.

Nole struggled to his feet, unbuckled his harness, and clambered to the top of a nearby hill just as a rescue helicopter touched down nearby.

No one needed to tell Nole that he was lucky to be alive, but he didn't know how lucky until the postcrash debrief. In particular, he would probably have suffocated if he hadn't been swinging around, like a pendulum, on his descent. This caused his parachute to bleed air, speeding him up. If not for that, the ride from ten miles up could have taken more than thirty minutes. In the official report, doctors stated that "Colonel Nole survived through an act of God."

Eventually, these Air Force pilots got the hang of flying the so-called Dragon Lady (as the bird would become known) and became the primary unit for U-2 surveillance flights. They were put on alert.

And when intelligence reports out of Cuba began to suggest that the Soviets had brazenly moved intermediate-range ballistic missiles with nuclear warheads onto the island, just ninety miles from the mainland United States, the White House needed proof—proof that the U-2 could provide if the job was safe. President John F. Kennedy did not want to see a military pilot shot down over hostile territory at a time when tensions were so high.

So in early October 1962, the CIA asked Kelly Johnson for an important favor: to get "a bird out of the North Base" at Edwards for a critical mission on October 16.

This mission became a top priority in Burbank, as Kelly's workers installed new cameras and electronic countermeasures on U-2s for delivery to the 4080th. His crew "stole a tail gear off" one of the Air Force models, changed out the engine, and delivered the plane to help the president determine next steps.

Planners named the program Operation Brass Knob and, for a week, this special U-Bird maintained an almost constant patrol above Cuba, capturing photos that were rushed to analysts as soon as the plane was on the ground.

It was an Air Force U-2 flying over Cuba that, on October 14, provided some of the most important intelligence of the Cold War: proof of Soviet intermediate-range ballistic nuclear missiles just a long swim from the U.S. mainland. "This aircraft took the pictures that made the basis for our move on Cuba," Kelly wrote in the U-2's project log.

A day later, the photos were on Kennedy's desk, and discussions began over what to do. Some of the president's advisers argued for airstrikes to take out the bases, and even for a large-scale invasion of Cuba.

But reason prevailed, and Kennedy merely threatened offensive action while instigating what he called a naval "quarantine" of the island. Semantics are important in diplomacy. Calling the action a quarantine and not a blockade differentiated the move from a military action.

This incredible provocation—nukes on America's doorstep—had been made at the invitation of Fidel Castro, who was worried about the stability of his government and realized that the best way to protect himself was to pull the Soviets closer. He invited Khrushchev to build up forces and put a missile site at San Cristóbal.

These Soviet actions were a serious escalation of the Cold War, and tensions between the world's two nuclear superpowers rose overnight. Apocalyptic nuclear war, for the first time, felt like a genuine possibility.

Meanwhile, U-2 overflights continued. And the Cubans were tracking them. Their Soviet-trained radar operators locked onto the flights from the outset and worked with SA-2 ground-to-air missile batteries to target the birds.

Finally, on October 29, the Cubans struck back, as eleven different radar sites tracked U-2 343, flown by Major Rudolf Anderson Jr. on his sixth mission of the operation.

"Our guest has been up there for over an hour," Lieutenant General Stepan Grechko is said to have told a deputy. Grechko's boss, the commanding general of the Soviet detachment, was the only man authorized to order a strike on the plane, but no one could find him, so Grechko made the call himself. "Destroy Target 33," he said.

A pair of SA-2 antiaircraft missiles were launched toward Anderson's plane and one exploded alongside it, sending shrapnel through the plane's fuselage and into the cockpit, where it likely killed the pilot instantly.

Anderson, a thirty-five-year-old father of two boys, was the only casualty of one of the tensest moments of the Cold War. His wife was newly pregnant with their first daughter. "Your husband's mission was of the greatest importance, but I know how deeply you must feel his loss," President Kennedy wrote to her in a letter. Anderson became the first-ever recipient of the Air Force Cross, the branch's highest award.

This was the third U-2 to be shot out of the sky, and when news reached Kelly, he knew that this was probably the end of this special plane's life overflying Soviet territory. "It is apparent that the U-bird has just about reached the end of its reconnaissance capability," he wrote in the program log. Crews continued to test modified versions with reduced RCS at Groom Lake, but "the U-2 program at this time is in quite a shambles."

And yet, the U-Bird kept flying. Until something better was in the air, the plane was just too important to mothball, especially with tensions rising across the Asian continent.

PART IV

BLACKBIRD

18

THE IMPOSSIBLE IS POSSIBLE

Ken Collins was a star pilot with 113 Korean War combat missions under his belt when he was asked, in the spring of 1961, to volunteer for a classified space program. He reported to the Lovelace Clinic in Albuquerque for what he was told was an "astronaut physical" at the same facility where the first Apollo astronauts were cleared. Over five intense days, a series of doctors and lab technicians X-rayed him from head to toe, probed his orifices, flushed his system, took numerous tissue and fluid samples, and measured his body and its capabilities in every conceivable way.

Next, Collins was sent to the nearby Los Alamos Lab, where more scientists analyzed his body's fat versus muscle composition, then put him in a soundproof black box for twelve hours in complete darkness and subjected him to a polygraph.

At this point, Collins still believed that he was being auditioned to serve as an astronaut, even more so when he learned that he'd survived this gauntlet and was one of five pilots selected to officially join the program.

It wasn't until a team of government movers arrived at his house to pack him and his family up that Collins got orders to attend a briefing in Washington at which he was told, finally, that he wasn't in fact going to space.

At least not in a spaceship.

Instead, Collins had been selected to "test an exotic new airplane for the CIA." There were no pictures or other details.

Collins didn't see that exotic plane—the Lockheed A-12—in person until December 1962, when he flew to a remote desert outpost known as Groom Lake and was led to a hangar that was completely dark inside.

The only light shone through some upper windows, and Collins could see where this mysterious plane sat, but he could only make out a few details—mainly, the plane's enormous nose and spikes.

But as his eyes adjusted to the low light, Collins began to take in the remarkable craft parked on the glossy floor of the hangar. The A-12 was huge—nearly twice the length of any fighter he'd seen or flown—and jet black. Everything about this plane looked novel, futuristic, almost alien: "Its sleek length, the massive twin rudders, and its total blackness," Collins later recalled. "It was a vision I'll never forget."

One of the many famous sayings attributed to Kelly Johnson is: "One miracle per program." But the project that would arguably most define him—the one that would become known by its clunky code name, OXCART—is where that golden rule didn't apply.

As soon as the CIA had signed off on Kelly's plan for the U-2 successor—to design and build a Mach 3 plane that could fly at 100,000 feet, with minimal RCS—reality set in at the Skunk Works that now they had to do it. As Ben Rich later put it, with the deadpan of an engineer: "All the fundamentals of building a conventional airplane were suddenly obsolete."

Years later, when one of Kelly's favorite and longest-serving engineers, Dick Heppe, was asked to reflect on the miracle of the A-12 spy plane and its successors—colloquially known as the Blackbirds—he noted that "everything was at or beyond the edge of the state of the art."

But what stood out as much to him was that Kelly Johnson wasn't at all daunted. The big boss reveled in the challenge and trusted his people to just "do a job where nobody had ever been. And as history books show, that trust, in fact, paid off."

The shape Kelly's team landed on for the A-12 just sort of happened. It was a gradual evolution, informed almost entirely by wind tunnel testing. "The result, head on, looks like a snake swallowing three mice," Kelly later wrote, attempting to compare the plane's unconventional shape to something an imagination can grasp.

Even in 2025, more than half a century after it first flew, the Blackbird looks futuristic. It has been shocking since the day it was built, and the notion of how that must have looked in its era boggles the mind. To calibrate its sheer originality, one of the program's young engineers, Steve Justice, recommends looking at other innovations of the time.

Check out a 1958 car or even a jet. Calculators were giant and mechanical, with gears. If you asked the enormous Friden desktop calculator—then state of the art—to divide 1 by 0, it crashed, because it tried in vain to calculate infinity. "The disconnect between the level of technology expected to create the Blackbird and the tools we had at the time was absolutely immense," Justice says. "And Kelly Johnson challenged his team to have it flying in two years. . . . In half the length of a normal college education, they were going to have to invent everything."

The second half of that quote may not be literally true, but it sure felt that way to Kelly and his engineers.

Chuck Yeager had broken the sound barrier for the first time in 1947, in his Bell-X1 rocket plane. But nearly a decade later, when the OXCART project began, the only planes that could get close to Mach 3—or three times the speed of sound—were, like Yeager's craft, experimental planes that used rockets, not jet engines. Even the highest-performance jets on the planet could do only Mach 2 or 2.5 for short bursts, and here Kelly Johnson was asking for an *operational* production plane that cruised at Mach 3-plus.

The list of complications in achieving such a feat was immense, beginning with the heat such a plane would generate. The friction on the body of a craft flying that fast heats the surface to over 500 degrees, which, for comparison's sake, is the temperature of an oven broiler. So the maximum temperature of an oven, which scorches meat within minutes, is what the A-12's outer skin would reach at cruise.

One problem with this: Metal expands quite a bit at that temperature. The A-12 would get longer—by inches—in flight because of thermal expansion, which meant that everything inside the plane, including the wires and cables, had to account for that.

The engines would be absolutely cooking, too; at full power, the temperature around the A-12's two J-75 engines would be 800 degrees, which is about the temperature of the lava flowing out of a Hawaiian volcano.

Arguably the greatest challenge of Kelly's plane, then, was managing the thermodynamics. Engineers needed to build a machine that could handle 800 degrees, and also 0, which would be the outside air temperature (and temperature of the metal) when the A-12 was flying slowly at lower altitudes to refuel.

This process bedeviled the Skunk Works. Even parts that were allegedly designed to withstand extreme heat failed when put to the test on OXCART.

For instance, numerous vendors claimed to have transducers that worked at up to 1,000 degrees, but tests inside Building 82 found them to have extremely short lifespans; they were built for rockets, which burn very hot, but not for long.

How about fluids? Oil turns to tar under extreme heat. Lubricants? There simply weren't any in existence that could withstand heat like this. In search of a hydraulic fluid for the plane's actuators, Kelly put a call out to the industry and to research institutions, and one of the first samples to arrive in Burbank came in an envelope, as a powder.

It was an excellent fluid at high temperature, but you had to heat it up with a blowtorch to get it there. "Not too useful on an airplane," Kelly cracked. So he sent out a second request, this time noting specifically that the fluid has to *be a fluid*, all the time.

No control cable on the planet could stand up to the sheer number of cycles required on this plane. So Lockheed commissioned its own, using Elgiloy, a rare metal typically used for watch springs, not a character from *The Lord of the Rings*.

Finally, there was the matter of the metal that would make up the plane's skin. Aluminum, the standard for planes, wasn't going to work. It just wouldn't stand up to the extreme temperatures of Mach 3 flight.

Stainless steel seemed like a good option. Until Kelly paid a visit to Lockheed's Georgia plant, where work on the supersonic B-70 bomber was underway. There Kelly saw that it took a clean-room environment—essentially a big pressurized airbag, with pressure locks at the entrance, and everyone inside wearing white clean-room suits—just to make the basic honeycomb panels for that plane's fuselage.

He thought back to his original Skunk Works axiom: Keep it

simple, stupid. The more complexity you introduce, the more potential for problems you create. Working with stainless steel was just too sophisticated for the Skunk Works.

That left only one good alternative: titanium. Titanium was on par with stainless steel for strength but weighed half as much and could handle extreme temperatures. Problem was, no one had ever built a plane with titanium, which meant that expertise in handling this exotic metal was scant, as was the supply.

One of the first major tasks the OXCART crew undertook was building a mock-up of the fuselage and cockpit that could be placed into a giant oven for heat testing the more than six thousand parts that would comprise this section.

Unfortunately, the titanium purchased for this job was basically useless; it was so brittle that pieces would often shatter when dropped, and during those early months, 95 percent of the B-120 titanium that arrived in Burbank was rejected. The reason? Poor quality control at the supplier.

The solution for this problem nearly drove Kelly nuts. The only way to tackle it was with process, and paper. "For an outfit that detested red tape," Ben Rich wrote, "we found ourselves wallowing in bureaucratic procedures." Skunk Works employees tracked every titanium part—millions, Rich says—so that a single bad piece could be traced back to its original pour. That way, any other parts from that same pour could be removed before they, too, broke.

"We put into effect a quality control program that I believe was and is unequaled anywhere," Kelly wrote. For every ten parts made, the Skunk Works manufactured three samples that would be heat-treated and run through a battery of stress tests before any of the other pieces were considered good enough for production and placed into storage. "There were times when I thought we were doing nothing but making test samples. But the effort was worth it."

Working with a new metal revealed numerous quirks. "Everything wants to poison it," Kelly said. Titanium did not work well with fluorine, cadmium, and chlorine—the last of those incompatibilities was found when, during the summer months, spot welds on the prototype's fuselage began to fail, and the process of elimination determined that the culprit had to be Burbank city water, which was more heavily chlorinated in the summer to retard algae growth. From that point on, the A-12 prototype got expensive baths, using only distilled water.

And then there was cadmium. Shopworkers used cadmium-plated wrenches to tighten bolts, and some bolts began to pop off under extreme heat. It took a while for engineers to sort out the cause, but they finally homed in on the wrenches, and that was the end of cadmium-plated tools in the shop.

Tooling, in general, was a headache. Drill bits that had been used for years at Lockheed on aluminum designs did not like titanium. Previously, a worker could drill one hundred holes before his bits needed to be sharpened; with titanium, it was every few minutes, and often the bits just snapped. So the bits, and the drills, were reengineered.

It was frustrating, one step forward, two steps back, and it weighed on everyone. Especially on Kelly. In one particularly exasperated moment, Kelly barked at Ben Rich that "this goddamn titanium is causing premature aging. I'm not talking about on parts. I'm talking about on me."

The United States didn't have any forging presses strong enough to form the titanium parts, and Kelly couldn't convince the government to pay for one, so most of the parts had to be fixed in machining, which resulted in 90 percent of the titanium going to waste.

But titanium—being expensive and difficult to work with—was also in relatively short supply because the United States didn't produce much of the ore required to make it. So where did Uncle Sam get enough ore to make titanium for the A-12?

From the world's largest supplier: the Soviet Union.

The CIA used a network of shell companies, working through Third World countries, to purchase the ore. According to one former pilot, the Soviets thought they were selling these large quantities for the manufacture of pizza ovens. "The Russians never had an inkling of how they were actually contributing to the creation of the airplane being rushed into construction to spy on their homeland," said Ben Rich.

To handle all this finicky material, Kelly's crew had to invent new drills, cutting machinery, power heads for profilers, and cutting lubricants to increase the rate of metal removal. It set up training classes for machinists, established a complete research facility for developing tools and procedures, and issued research contracts to the most trusted outside vendors to develop improved equipment.

But this was an incredible group, and they iterated fast. By the time the program concluded, Skunk Works manufacturing had improved the rate of metal removal from three times the industry average rate to ten times, and had increased the life of a drill bit from an average of 10 holes to over 119.

The A-12 "practically spawned its own industrial base," Dick Bissell later wrote. The 2,400 machinists, mechanics, and fabricators working on the project eventually even did their own milling and forging.

The A-12 prototype was mostly drawn and built by hand, but it was also the first Lockheed design to utilize newfangled contraptions known as computers. The few engineers in the shop who knew how to work these high-tech machines—which at that time were massive and filled entire rooms—used them for stress analysis and radar work, plus antenna and landing gear design. And during the busiest design period, 57 percent of the total computer capacity at Lockheed's headquarters was used for OXCART.

The pace of things required relentless improvisation. Working fast means cutting steps, which results in unexpected mistakes. When engineers were assembling the fuselage of the prototype, they found an inexplicable gap. Ben Rich figured out where the problem was: They'd forgotten to do drawings for this part of the fuselage. So one of the structural engineers grabbed a clipboard and some graph paper, put a drawing number on top of the sheet, and drew this portion of airplane skin, along with instructions for the stringers who would attach it. This was not unusual. In many cases, "official" drawings were done after the fact.

Rich, arguably Kelly's most valuable engineer, was also put in charge of designing the plane's inlet—the opening where air was sucked in to feed the engine. In the case of the Blackbird, this was a very complicated feature. Those giant spikes that stick out from each nacelle move in and out to control the flow of air into the engines. And what drove Rich most crazy were the electronic controls for these inlets.

Very late into the design, the controls were unreliable. They often didn't respond. "We tested it from hell to breakfast," Rich said, and no one could figure out the problem. Then, one day, someone was testing the high-frequency radio, and the inlets popped out. "We checked it for all sorts of things, but we just overlooked one circuit," he explained. And it took some random person, talking on a radio, to solve it.

At the peak of basic design, Kelly had 135 engineers on the A-12, including himself. They were crammed into every corner, and especially early on, the wind tunnel was in constant use. Tunnel testing on arguably the plane's most complicated feature—the engine inlet—resulted in 2 million data points alone. By the time the final design was completed, engineers had run more than 250,000 tests of OXCART models in the wind tunnel.

OXCART would trust test the capabilities of the Skunk Works. From the outset, there was doubt from some leaders at the CIA that Kelly could pull off what he was proposing—this incredibly complicated machine—while sticking to his unusual methods.

By the end of the first year, the program's costs were running wild, at least by Kelly's standards. He was running 24 percent over his projections, mostly because of unexpectedly high material costs. He also got bad news from Langley. CIA analysts had discovered a new generation of Soviet radar, the Tall King, which was capable of spotting an incoming plane much farther out than the radars it replaced. This meant that Lockheed would have go back to the drawing board on the A-12's RCS.

19

TOO MANY MIRACLES

The road to success is paved in failure, especially on a project as complicated as OXCART. Pretty much every day brought a new frustration, many of which were out of Kelly's control. And operations were no less fraught in Florida, where a Skunk Works–style group within Pratt & Whitney was facing enormous complexities of its own in the development of the powerful engine that would propel the A-12.

The idea was to use a prototype engine built for the Navy to power a variety of experimental concepts. It could generate up to 26,000 pounds of maximum takeoff thrust and provide dash capability of Mach 3, but only for a very short time. The OXCART version needed to *cruise* at Mach 3.

This engine had to be technologically advanced, and also a tank. In the end, not one part was made from materials used on previous engines. Pratt & Whitney, like the Skunk Works, had to reinvent fabrication. Its engineers learned to make sheet metal from materials that had previously been used only for forging turbine blades.

Each company embedded personnel with the other for the duration of the program and William Brown, Pratt & Whitney's integration engineer, later said that he had "difficulty differentiating between 'we' Pratt & Whitney and 'we' Lockheed." That's how closely the teams worked.

Brown, like Kelly, credited the customer. This entire operation only worked because the right people at the CIA believed in him and trusted the engineers. The lack of friction allowed smart people to find solutions quickly and with no oversight, and the process would have been impossible, Brown thinks, "under a more cumbersome management system," which would have demanded more reviews and compromises. "The result was an operating system incorporating a quantum leap in state of the art at an earlier time and at less cost to the government than would otherwise have been possible."

Still, development of the J58 lagged behind the airframe. It was clear that the new engine would not be ready in time for the A-12's first flight. Kelly didn't want to wait. He could learn plenty about his plane with a temporary engine, so he asked to use Pratt & Whitney's less powerful J75—the U-2's engine—as a stopgap until the J58 was tested and ready to fly.

Building the A-12 wasn't just hard for Lockheed and Pratt & Whitney. Every contractor on the OXCART program had enormous challenges to work around. Contributions often required great leaps in technology but simultaneously had to work within the very specific and difficult confines of this wildly complicated airplane.

Rus Daniell was Kelly's point man for these so-called interface agreements, and the job was not a fun one.

Kelly put enormous trust in Daniell, a quiet, unflappable man who'd been with him for many years. Daniell was given great responsibility, as the single point for engineers for all the subcontractors who needed to know, for example, how much space they'd have, what kinds of wires they needed to use, and what power would be allocated to them for their component.

Daniell had a team of engineers, and his superpower, according to Air Force Colonel Donn Byrnes, was an ability to "translate this seemingly impossible and endless series of hardware and electronic questions and definitions into written documentation."

The paperwork in this job was overwhelming, and Daniell managed the waterfall of incoming memos with a very specific strategy. He kept two inboxes. Any new memo or request went into box one, and if the person who sent it didn't follow up in four or five days, he'd throw the memo away. If the person did follow up, Daniell moved the memo to box two. "This was the one he worked from," Byrne said.

Daniell stood at arguably the worst intersection in the entire organization. He had to answer to Kelly and to the subcontractors. His clients—who supplied things like navigation, cameras, and cooling systems—needed power, which comes from the engine. But Kelly also needed that power, for thrust, in order to fly faster. Less thrust translates into less range, which means the plane has to carry more fuel, which increases the weight as well as the drag and changes the center of gravity. Performance suffers.

So Daniell was constantly upsetting either Kelly or his contractors. It was, Byrne said, "a serious lose-lose situation" and yet, somehow, "when you visited [Daniell's] office, he always appeared to be in complete control."

Which is probably why Kelly liked him so much.

The project's complexity surprised even Kelly, who had so far built a career on promising miracles that he actually did deliver. On OXCART, his Skunk Works was innovating at an unprecedented rate, but still, the delays piled up.

Of course, additional work, stress, and complexity weren't helping the ongoing drama of producing the world's most advanced plane. The engine, as always, was holding things up. Development issues on the J58 kept mounting, which meant that costs continued to climb. Development costs, Kelly noted, were headed for a 100 percent increase, while the cost of every engine was likely to climb 50 percent.

When his original delivery date came and went, Kelly wrote to Dick Bissell to tell his CIA benefactor that he was at least four months behind, sparking Bissell to fire off a response that was "needlessly rude," he later admitted. "I have learned of your expected additional delay in first flight from 30 August to 1 December 1961," Bissell wrote to Kelly. "This news is extremely shocking on top of our previous slippage from May to August. . . . I trust this is the last of such disappointments short of a severe earthquake in Burbank."

This was somewhat out of character for Bissell, who could be prickly but was famously stoic and one of Kelly Johnson's biggest fans.

But Kelly surely understood the pressure his friend was under. He had other issues on his mind. As the deputy director for plans—in charge of all CIA covert ops—Bissell was the architect of the infamous Bay of Pigs debacle, in which a CIA-trained paramilitary army of Cuban rebels invaded Cuba in April 1961 with the goal of overthrowing Fidel Castro. The rebel force was routed, and the Bay of Pigs became one of the most embarrassing failures of the Cold War for the United States.

Bissell had taken most of the blame for the disastrous invasion, both inside the Agency and in the media, and Kelly felt terrible for him.

At 7:00 a.m. on Friday, April 21—a day after the failed invasion ended—he sat down to handwrite a letter thanking Bissell for standing up for America, and for the Skunk Works. "If there is anything I can ever do to help carry part of your load, don't hesitate for an instant to call me," Kelly wrote. "I've thought continually of your problems of the last week to see what we can do here to assist. . . . [Please] accept a grateful US citizen's thanks for your splendid efforts in our behalf."

Four days later, Bissell responded: "I doubt if you realize how helpful it is to have your very nice note just at this moment in time," he wrote. "Last week was without any question the most unpleasant period of seven days I have ever experienced. It was even worse than the first week of May 1960"—when Francis Gary Powers was shot down over Russia in the U-2—"because this time there are hundreds of people's lives involved."

This letter was candid and confessional, the kind of thing a powerful man can only write to a friend he truly trusts, and feels unjudged by. Bissell wrote that he, more than anyone, deserved blame. "I certainly feel I have served my country ill, as it has turned out, in this major effort."

Bissell was under pressure, and this time he was less willing to just accept the delays. He needed to make some changes. He told Kelly that the CIA was cutting its order from twelve spy planes to ten, to save $161 million. Also, he was assigning a babysitter to Burbank. There was just too much at stake to trust even the inimitable Kelly Johnson's word that he'd make the next deadline and stay on the revised budget.

Kelly, as you might expect, refused, but he had no real leverage over Bissell. There wasn't a plane to build without the CIA's money. So he grudgingly agreed to let the Agency send an engineer—with his approval—and the two men settled on Norm Nelson, a CIA engineer Kelly had known for years. Even then, he insisted that Nelson know his place. He couldn't just show up and spy for Bissell; he had to work.

Norm Nelson was, according to Ben Rich, "the first outsider ever allowed a place inside Kelly's realm." And he was treated as such. The top engineers knew which meetings they should forget to mention to him, and across the board, Nelson was treated coolly at first. But he did find his place.

In fact, Nelson drank the Kool-Aid. He would become as much of a Skunk Works engineer as he was an Agency employee, and proved his value to Kelly within months of arrival, when the CIA allocated $20 million for the design of wing tanks to extend the A-12's range.

Nelson asked to take a look at the problem, ran some math, and came back to Kelly with the results. The tanks would add eighty miles of range, at best. It wasn't even slightly worth the effort, let alone Uncle Sam's money.

Kelly agreed and returned the $20 million to Langley.

The A-12 hadn't even flown yet when Kelly Johnson got some news that made him reconsider whether he was building the right plane. CIA analysts had received alarming new telemetry data indicating the test of a new kind of Russian missile, and sent it immediately to Burbank. Kelly's telemetry experts analyzed the data and spotted something even more worrisome: clear signs that the Soviets were flying a prototype supersonic bomber.

Rumors of this so-called Backfire bomber had been floating around for years—it was widely assumed that the Russians were also hard at work on a plane that could fly very far and very high, at Mach 2 or more. But this data confirmed that the Soviets were further along than anyone thought. If they could in fact build such a plane, this would tip the balance of power in the arms race decidedly toward the USSR because the United States had no plane, or missile, that could stop such an aircraft from reaching the continental United States with an array of bombs and missiles onboard.

Except, as Kelly quickly pointed out, the United States *did* have a plane that could intercept such a craft. Or at least it was working on one. Right in his shop. The A-12, with some tweaks, could easily be adapted into an interceptor capable of racing out from U.S. bases and shooting down a supersonic bomber well before it could harm the mainland. Such a plane could reach incoming targets much earlier than any other aircraft, and do that over an enormous range of altitudes.

An A-12 flown by Mele Vojvodich was scrambled from Groom Lake and vectored to the mysterious target by an officer who had no idea what kind of plane he was guiding; he could just tell by the image on his radar screens that it was fast.

Soon enough, Beale had its answer from Vojvodich: The unknown vehicle wasn't a vehicle at all; it was an escaped weather balloon caught in the jet stream. This bizarre test proved the value of a supersonic craft that could rapidly interdict objects inbound for the United States.

Thing is, Kelly couldn't just pitch the Defense Department's decision-makers on a major new fighter program based around the A-12, because only a tiny group of people at the Pentagon even knew about it. This was a CIA black op, cloaked in so much secrecy that very few Air Force generals had been cleared into the program.

Also, there was a new president, John F. Kennedy, who had a new defense secretary, Robert McNamara, who wasn't exactly a fan of the defense-industrial complex he'd inherited.

McNamara was a wonk, having just come from the Ford Motor Company, where, as president, he was lauded for implementing modern organization and management practices on an old and bloated business. McNamara distrusted contractors and, according to Kelly, disliked any program started under Eisenhower. And apparently the Skunk Works, which operated with almost complete independence, really bothered him.

Kelly flew to Washington to meet McNamara as soon as he was welcome, and returned from that first meeting unimpressed and worried. McNamara seemed "petty," he told Ben Rich, and distrustful of outsiders. "He'd love to stick it to an old-timer like me just to show the entire aerospace industry who's boss."

Kelly's pitch for an A-12–based fighter made sense; it was, arguably, a no-brainer if a major concern of U.S. homeland defense was stopping a Soviet supersonic bomber. But this was going to be an uphill fight.

He pulled Ben Rich aside and asked his protégé to help think of ways to sweeten the deal for the "blue suiters," as he called the Pentagon's generals. He asked Rich to study a few adaptations to the A-12, including whether or not the plane could carry an ICBM, which would greatly extend the range of such a missile (because so much fuel is used by a rocket in the initial seconds of lifting off). They also considered a so-called energy bomb—a 2,000-pound bomb made of high-penetrating steel that carried no explosive. A bomb that dense and heavy dropped from 80,000 feet at Mach 3 would, Rich said, "hit the ground with the force of a meteor," with 1 million foot-pounds of energy, creating a hole 130 feet deep.

The Air Force liked that idea but worried about the bomb's lack of guidance. At that speed, a bad miss seemed . . . very possible. And an unguidable meteor sounded like a disaster in the making. Most important, McNamara dismissed the energy bomb as "futuristic drivel."

Over time, a clearer divide developed. Air Force personnel—the military men—understood the necessity of a supersonic interceptor and badly wanted one. But civilians at the Department of Defense (DOD), Kelly vented in his log, "claim there is no threat," and he needed to find a way "to help break the logjam." The only hope, Kelly decided, was to find a way to make a universal plane that could be adapted into whatever the Pentagon needed it to be.

The A-12 design had shaped up to a point where Kelly could modify it easily. "By now we have a set of building blocks with which we can make just about anything," he noted in December 1962. "The airplane is truly universal!" Using the same basic airframe, Kelly wrote, he could make "a reconnaissance airplane, a reconnaissance/strike version, a bomber, a fighter, a drone launcher, or a satellite interceptor launcher."

Above left: Kelly Johnson with his mom and brother in Michigan, 1924

Above right: With a classmate at the University of Michigan

Left: A young Kelly during his early days at Lockheed

Below: Lockheed's entire staff, outside the Burbank headquarters, 1934

COURTESY OF JOHN HORRIGAN

The original home of the Lockheed Aircraft Company, Burbank, CA
COURTESY OF LOCKHEED MARTIN

Hall Hibbard (far left), Walter Jones (middle), and Kelly Johnson (far right) examine F-80 model, circa 1948 COURTESY OF LOCKHEED MARTIN

Going over test data with Amelia Earhart, 1936 or 1937
COURTESY OF JOHN HORRIGAN

Pilot Laura Ingalls with Lockheed Orion, 1935 COURTESY OF JOHN HORRIGAN

Kelly Johnson studio portrait, 1948 COURTESY OF LOCKHEED MARTIN

Right: Kelly's first wife, Althea Louise Johnson

Below: Althea, with dog and horse, 1930s

COURTESY OF JOHN HORRIGAN

Lockheed president Robert Gross, year unknown COURTESY OF LOCKHEED MARTIN

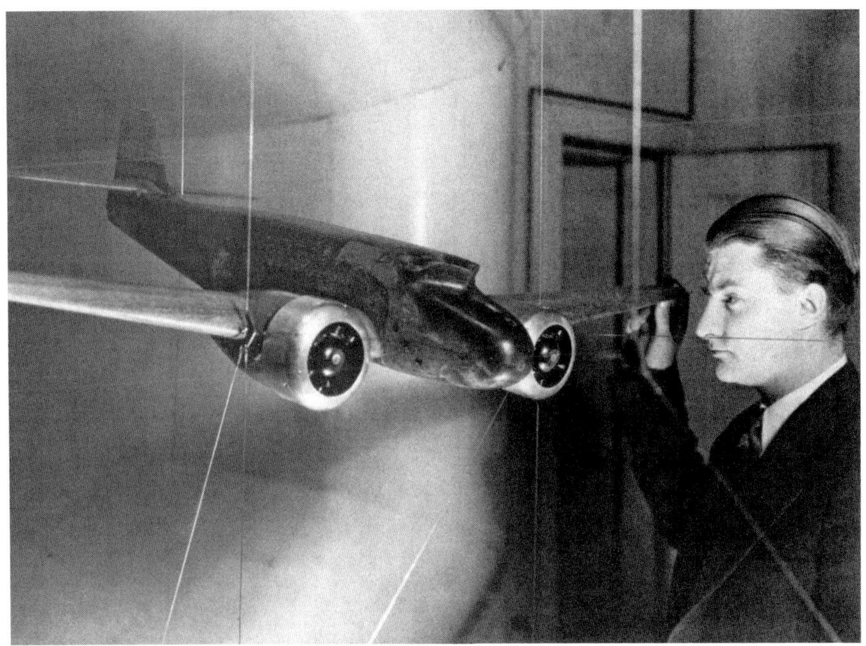

Kelly in the University of Michigan wind tunnel with early single-tail Model 10 Electra COURTESY OF LOCKHEED MARTIN

The P-38 Lightning, aka the fork-tailed devil, helped Allied pilots control the skies over Europe and the Pacific. COURTESY OF LOCKHEED MARTIN

Right: America's first jet fighter, the F-80 Shooting Star

Below: The F-104, nicknamed "the missile with the man in it"

COURTESY OF LOCKHEED MARTIN

Receiving the 1959 Collier Trophy from Air Force Chief of Staff General Thomas White and Vice President Richard Nixon COURTESY OF JOHN HORRIGAN

Left: Kelly with an F-104
COURTESY OF LOCKHEED MARTIN

Below: In his happy place, at Star Lane
COURTESY OF JOHN HORRIGAN

Looking northeast at Watertown, aka Area 51, during early U-2 operations
COURTESY OF JOHN HORRIGAN

The dining room at Area 51, U-2 testing years COURTESY OF JOHN HORRIGAN

U-2 “driver” reviewing weather, pre–test flight COURTESY OF JOHN HORRIGAN

Weather briefing at Area 51, prior to U-2 test flight, year unknown
COURTESY OF JOHN HORRIGAN

Left: An early U-2, circa 1955
COURTESY OF LOCKHEED MARTIN

Below: Lt. General Pat Carter giving Francis Gary Powers a CIA medal, April 1965 COURTESY OF JOHN HORRIGAN

Photo of U-2 wreckage, later proven to be retouched and fake, sent to *Life* magazine by the Soviet Union following the shoot down of Francis Gary Powers in May 1960 COURTESY OF JOHN HORRIGAN

Vol. LXXV No. 20 TIME May 16, 1960

THE WEEKLY NEWSMAGAZINE

THE U-2 IN FLIGHT

NATIONAL AFFAIRS

THE NATION

Cold-War Candor

"It is certainly no secret," said the State Department last week, "that, given the state of the world today, intelligence collection activities are practiced by all countries . . . The necessity for such activities as measures for legitimate national defense is enhanced by the excessive secrecy practiced by the Soviet Union in contrast to the free world."

With historic frankness, the statement went on to admit that "endeavoring to obtain information now concealed behind the Iron Curtain," an unarmed U.S. plane had flown over Soviet territory. Thus the U.S. told the world that a Lockheed U-2 brought down over Russia on May 1 was flying an intelligence mission, just as Premier Nikita Khrushchev said.

That admission stirred up a flurry of concern at home and abroad over the U.S.'s "embarrassment." The admission was embarrassing to the U.S. for one reason: it reversed the Administration's earlier claim that the U.S. was engaged in high-altitude meteorological research over Turkey and the plane drifted into Russia by mistake.

Open Skies. All the bored calm with which the world awaited an unproductive summit vanished in a new preoccupation: Would Khrushchev make use of his capture of the U.S. high-flying plane either to scuttle the summit or make unreasonable demands? Would allies be dismayed and neutrals angered?

The apprehensions, as they so often are, were exaggerated. The incident, coupled with Khrushchev's recent intransigence, has certainly heated up the cold war. But people everywhere have accepted the reality of the cold war, which has its own kinds of maneuvers, battles, tactics and weapons.

Faced with the unexpected, the State Department, after its manly candor, set out to make its own points about the U-2.

"One of the things creating tension in the world today," it said, "is apprehension over surprise attack with weapons of mass destruction. To reduce mutual suspicion and to get a measure of protection against surprise attack, the U.S. in 1955 offered its 'open skies' proposal—a proposal which was rejected out of hand by the Soviet Union. It is in relation to the danger of surprise attack that planes of the type of the unarmed civilian U-2 aircraft have made flights along the frontiers of the free world for the past four years."

Cleared Air. If the U.S. felt embarrassed, perhaps rocket-rattling Nikita ("We will bury you") Khrushchev must have found it embarrassing, too, to have the world learn that unarmed, big-target U.S. planes had been flying missions over Soviet territory for four years before his armed forces finally managed to bring one down.

For reasons of his own, Nikita Khrushchev chose to make a spectacular out of the U-2 incident (*see* FOREIGN NEWS). In Washington, there were some calls for a congressional investigation, and in both the U.S. and Britain some fears were expressed that the U.S., by risking the U-2 flight "at this time," had risked prospects for "agreements" at the summit. But if the shooting down of the U-2 dimmed summit prospects, they could not have been very bright beforehand.

Perhaps they were never very bright. President Eisenhower, Secretary of State Herter and Under Secretary of State Dillon have all made it clear in recent weeks that the U.S. will go to the summit determined to hold fast to its rights in Berlin, and Nikita Khrushchev has shown in tough-toned speeches that the U.S. firmness has undercut his hopes of making any headway at the summit.

The talk of endangered agreements at the summit showed a short memory of what the cold war was all about and how it got that way. Under standard Communist terms no agreements of any substance or durability were likely to be possible at the summit, before or after the U-2 incident, unless the U.S. and its allies would accede to Russian demands. By candidly admitting that the U.S. is flying intelligence missions over Russia, by vividly reminding the world that a cold war is going on, and by demonstrating that it reserves the right to defend itself in every way it can, the U.S. might have cleared the summit air for some hard talk on hard issues that could be a lot more worthwhile than vague, generalized agreements.

DEFENSE

Flight to Sverdlovsk

(See Cover)

The low black plane with the high tail looked out of place among the shiny military jets crowding the U.S. Air Force base at Incirlik, near Adana, Turkey. Its wide wings drooped with delicate languor—like a squatting seagull, too spent to fly. Its pilot seemed equally odd: a dark, aloof young man who wore a regulation flying suit and helmet but no markings, and had a revolver on his hip. Pilot Francis Gary

Time magazine article about the previously secret U-2, following the downing of Francis Gary Powers's plane, in 1960 COURTESY OF JOHN HORRIGAN

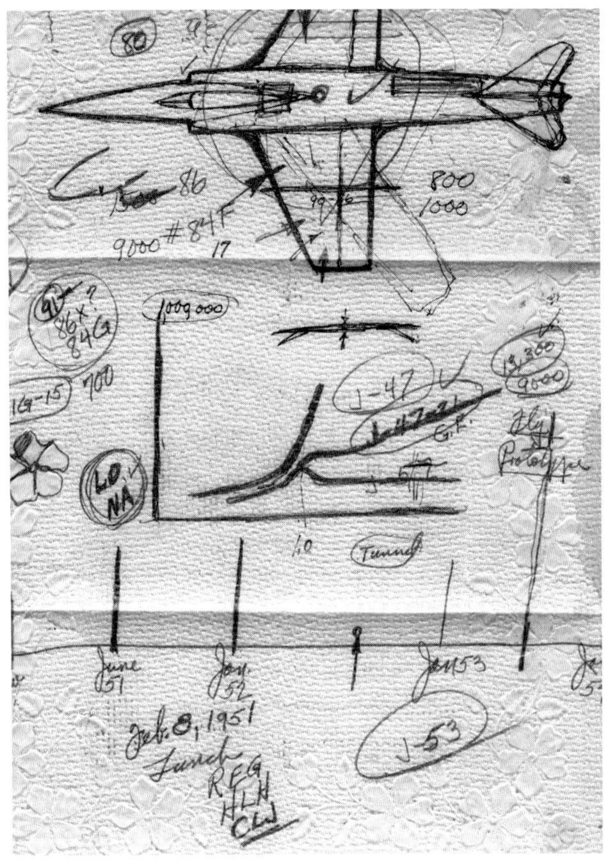

Sketch of a Soviet MiG fighter, drawn at lunch by Kelly in 1951 COURTESY OF JOHN HORRIGAN

U-2 in disguise as an experimental NACA weather research plane COURTESY OF LOCKHEED MARTIN

Above: SR-71 Blackbird in flight

Right: Receiving the Collier Trophy, for the second time, from President Lyndon B. Johnson

COURTESY OF LOCKHEED MARTIN

A-12 refueling in midair, 1962 COURTESY OF JOHN HORRIGAN

A-12 Blackbirds lined up at Area 51, 1964 COURTESY OF LOCKHEED MARTIN

A modified A-12 known as the M-21 carrying a D-21 drone during testing
COURTESY OF LOCKHEED MARTIN

Kelly Johnson in front of the legendary Skunk Works logo
COURTESY OF LOCKHEED MARTIN

Kelly riding shotgun in an SR-71 trainer COURTESY OF JOHN HORRIGAN

SR-71 at Beale Air Force Base in the late 1980s COURTESY OF LOCKHEED MARTIN

The F-117 Nighthawk in flight COURTESY OF LOCKHEED MARTIN

Above: Ben Rich, the second man to lead the Skunk Works, and his mentor in 1983 at Kelly's retirement ceremony COURTESY OF LOCKHEED MARTIN

Left: Sign of street named for Kelly at the U.S. Air Force Academy COURTESY OF JOHN HORRIGAN

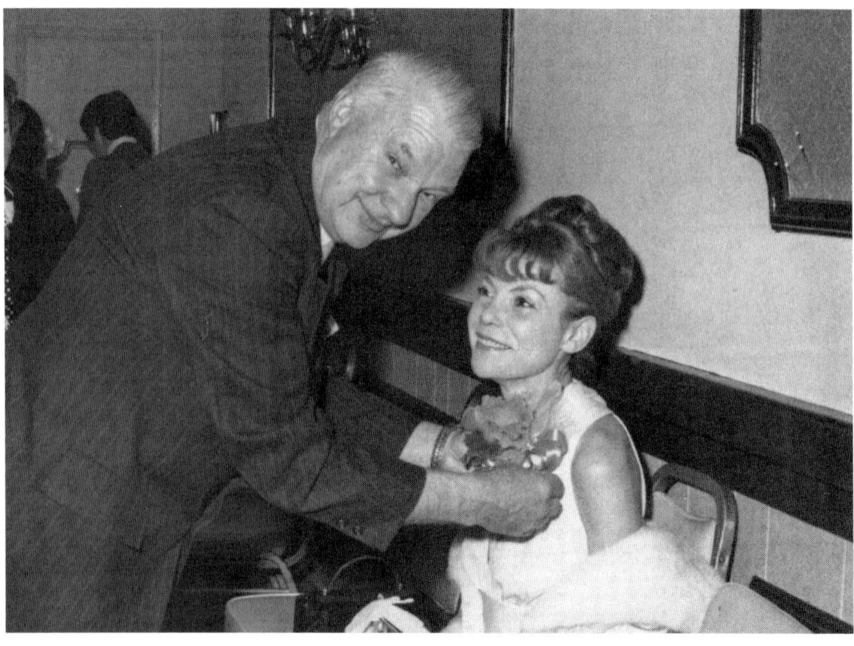

With Maryellen, his second wife COURTESY OF JOHN HORRIGAN

And in the end, it worked. Kelly didn't get his bomber, but he did get the Skunk Works a contract to build three prototypes of an A-12 interceptor.

Kelly put one of his best men, Rus Daniell, in charge of the project and slotted the interceptor—which he named the YF-12—right into his A-12 production line. The three prototypes weren't hugely different from the spy plane version and would be the seventh, eighth, and ninth planes built on the line. The YF-12 (known as AF-12 in the design phase) looked nearly identical to the A-12, with just a few additions—a modified nose to house the Doppler radar system, three air-to-air missiles, a second seat behind the pilot for a fire control officer, a ventral fin on the aft fuselage (that folded up for takeoff), and titanium in place of the radar-resistant materials on the fuselage, because this version of the plane wouldn't have to infiltrate enemy airspace.

Building an attack version of his supersonic plane was exciting to Kelly, but it also meant that he was inviting a new master into the Skunk Works during its most important and frantic period. While nine of the twelve planes were going to the CIA, under the deal to replace the U-2, these three fighter prototypes belonged to the Air Force—in other words, the Department of Defense, land of bloated bureaucracy and glacial timelines.

Kelly insisted that the Defense Department would have to accept and adapt to his methods, and the DOD agreed, at least in practice. But signs that working with the Air Force would be different from working with the CIA appeared almost immediately.

When a small group of overseers from the Air Force paid their first visit to Burbank, Kelly spent the first hour lecturing the generals—or, as he put it, "giving the new members of the group an outline of how we operate at ADP (Advanced Development Projects)." That whole day, Monday, went well, he reported.

The honeymoon lasted twenty-four hours.

"Tuesday morning the dam broke: Then they started throwing specifications at us," he wrote in the YF-12 log.

Kelly recoiled. In his own words, he "let fly" and made it clear to these blue suiters that the Skunk Works worked only one way—his way.

"If we don't stop the nitpickers now, at the beginning, we never will," he wrote in a log entry that oozes with snark: "Col. Nye was so shaken he asked four intelligent questions in a row, which were promptly answered. We went over our work statement and deleted such items as, making a specification for a plan on how we were going to do something. I refused to do either and put things in terms of our guaranty to make the hardware work." In other words, the only thing that mattered.

It wasn't just the Air Force that worried Kelly. The proposed armament for this interceptor—an air-to-air missile known as the GAR-9, made by Hughes Aerospace—wasn't going to fly, literally. At least not when fired from a Mach 3.2 airplane.

Kelly was "a little bit aghast" at what he saw in the briefing from Hughes (as in Howard), expressed his concerns to the Air Force in private, and heard that they, too, had concerns with Hughes's ability to make an effective guided missile to arm the YF-12. "I don't blame them a bit for being disgusted," Kelly said. "On the other hand, I know that a considerable amount of the failure on their previous fire control systems has been due to AF management." Subtext: If you back off, we might be able to fix it.

Kelly shared his concerns with the engineers at Hughes, ever so tactfully. "I told them this . . . and, after thanking them for our fine relations up to this point, told them that they had better lead a new life, or the whole project would go down the drain because of their failure to perform in the past."

Privately, Kelly doubted that Hughes could fix it. He assumed that he'd soon be overseeing a missile program, too.

One of the more annoying differences between the CIA and the Air Force was the reporting. Even under these special arrangements, which allowed for only minimal oversight, the Air Force wanted to keep close tabs on progress and liked to pay visits. And another entourage—fifteen blue suiters in total—came back to Burbank to review the interceptor mock-up.

The group's size worried Kelly, but the officers behaved themselves—which, to him, meant that they didn't nitpick or raise any objections that would cause him to go back to make any major revisions.

They did suggest a list of changes, but Kelly headed off any paperwork delays by incorporating eight of those changes overnight between the first and second days of their visit. "This seemed to shock them to such a degree that they decided to hold off writing comments," he wrote in the log, "because it was possible to modify the mock-up almost completely prior to their leaving."

This seemed to help Kelly's relationship with the Air Force. In fact, the political situation was beginning to thaw.

In July, a group of generals, led by Curtis LeMay, came for a visit, and LeMay "shook everyone" by requesting a proposal that would allow for the production of up to ten fighters a month.

Kelly was hoping, even banking, on the interceptor version going into production. It would likely result in an enormous order, and he was already thinking about how Lockheed could possibly handle it.

For starters, the Skunk Works could expand into Buildings 309 and 310, and would then comprise more than half of Lockheed California. But that was a Band-Aid at best. The logistics of delivering ten planes a month, Kelly wrote in the log, would require "such a staggering operation." It would mean, basically, producing one plane every other working day, and likely wasn't even possible given the facilities available in Burbank. "Our whole concept of operation would have to change. We are, nevertheless, proceeding to make a study, to see what it would take."

This year, 1961, was a whirlwind. Kelly had two versions of his supersonic plane in development for two exacting masters, and still had to prepare for what production of those planes would mean—a complete overhaul of production at Lockheed in order to accommodate a Skunk Works that would triple in size.

And he was still making regular flights to Washington, D.C., to meet with and placate his CIA and Pentagon contacts. On one of these trips, Leo Geary asked Kelly if he would consider taking "complete responsibility" for running OXCART, meaning that he would oversee Pratt & Whitney and all the program's contractors.

Geary was prepared to support this move and take it to his superiors, but Kelly wasn't enthusiastic. His workload was already "a platter," he said, and the travel such a job would require wasn't good for his health.

Geary told him that he'd made an impression on McNamara and Air Force undersecretary Joe Charyk; as a result, the two were considering him as a Pentagon consultant on "special jobs."

Something like that could work, Kelly said, "to a moderate degree." But he was currently stretched to the limit; it definitely wasn't the best time to talk about additional responsibilities, no matter how politically advantageous they may be. "Right now I don't feel like doing anything more than getting one of these beasts off the ground," he said.

Despite the hint from LeMay, there was still little clarity from

Washington about which variant of the A-12—the bomber or the fighter—was most likely to get funded. The Pentagon, Kelly noted, was "considerably interested" in the bomber version—an offensive weapon—while the Air Force, "from General LeMay down," seemed to favor national defense and wanted the fighter.

Things were unsettled in Burbank, too. Lockheed's founding leader, Robert Gross, resigned in September to fight pancreatic cancer, and his brother Courtlandt took over as chairman. Less than three months later, on December 3, 1961, Robert Gross died.

Gross had been one of Kelly's greatest champions. He showed trust in him straight out of college, and his loss hit Kelly hard.

Four days before Christmas, Kelly ended one of the most exciting and tumultuous years of his life to date with the annual Christmas speech. This speech was a tradition for Kelly, and he took great pride in writing it. Every year, he delivered it numerous times, to various departments, with subtle tweaks, and this time he began with the mechanics.

Kelly spoke from the production floor, in the shadow of the A-12 prototype. "As we look back over the year, we can see 1961 as a difficult, challenging, frustrating year, but, overall, one of great achievement," he said. "Standing here around this big bird, we can all be proud of our efforts to bring it into being."

Efforts were not always fun, even for a man who loved to be told that something was impossible. "We've had a terrible time doing it," he went on.

Kelly ran through some of the challenges that had been overcome, and then pointed to the remarkable work that had been done here, in this cavernous building, on the outrageous machine that lay before them. "We are being closely watched to see if we can do the job we promised in the time we said we could. As we stand here today, we are late. There is desperate pressure to get this one and those to follow it in the air. We have been plagued with material and part shortages; slow, cumbersome techniques called for by our high degree of quality control; and the need for invention at every turn.

"I can assure you that the potential of this aircraft is very great and it is worth your very best efforts to make it work. Our nation needs it for many things—as you do for yourselves and your children. In its defensive role, it can save our lives and our country."

Kelly rarely showed vulnerability, but you could hear it in his voice. Pride, too. "When things get tough and you are tired—when the holes don't match, the plumbing leaks, you can't get the parts out of the hot blocks, or when you wish you could stuff the engineers who drew it up underneath the tail chute box and seal 'em in—just think this—'I'm not building this for Lockheed or Art Viereck or Kelly Johnson—This is for me!'"

20

BACK TO THE RANCH

More dispiriting news arrived in Burbank on February 28, 1962. Dick Bissell, Kelly's dapper CIA advocate and one of his favorite humans, resigned from the OXCART project and was replaced by Herbert "Pete" Scoville, a PhD chemist who'd helped lead the U.S. nuclear weapons program before joining the CIA to work on its science team.

Bissell had done as much as anyone to create the Agency's overhead surveillance programs, and it was his faith in Kelly that allowed him to build the Skunk Works into the impossible factory it had become, but his benefactor was just under too much fire in the aftermath of the Bay of Pigs fiasco.

Bissell's departure was a blow for Kelly, personally and professionally, but Kelly had bigger worries in the early months of 1962. Foremost among them was how to get the A-12 out of development and into the air.

The A-12 was, without question, the most secret plane ever built by the United States—and the most secret thing, period, since the atomic bomb. But this highly secret plane was also not subtle in any way. It was wildly distinctive-looking and would be loud. Anyone who saw it flying was going to gape, which created a very big problem: What testing ground could possibly be used for its test flights?

There was really only one good option: Paradise Ranch.

This was a different Paradise Ranch from the one thrown together for the U-2. The CIA hadn't intended to use Groom Lake on OXCART; senior staff on the project knew how much work it would take to accommodate the program, including construction of new, far longer runways, and they also knew how much friction there had been between Lockheed and the CIA over who was in charge of the place and its operations.

The Agency had hoped to use one of ten Air Force bases scheduled for closure, but none of them worked, considering the secrecy and security required. In most cases, they were just too close to population centers. Someone was going to spot this radical-looking plane the minute it was airborne.

So, construction began on a newer, bigger, better-provisioned, top secret air base on the desert flats of Groom Lake. But before Bissell left the Agency, he did make plans to preemptively avert any repeat of tensions with Kelly. This time, the base would be set up by the Agency, and Lockheed would be a tenant for the purposes of testing the new plane.

Labor issues with a local sheet metal workers union caused construction at the site to briefly appear in the press, and security officers scrambled to cover for it, spreading word that the construction was part of an unnamed Air Force program. An article in the *Las Vegas Review-Journal* noticed, and didn't miss the mark by much, stating that Groom Lake "is ideally suited to secret projects because experimental aircraft can take off and land without detection from any outside point."

The location's previous connection to the CIA and U-2—a plane that was under extreme scrutiny following the shooting down of Francis Gary Powers—caused Agency security to worry that too much new activity on OXCART would draw attention to the site. To try to head off any speculation from contractors working on the program, previous names of the base—including Groom Lake, Watertown, Paradise Ranch, and the Ranch—were ordered to be "eliminated from the vocabulary of OXCART-cleared personnel." They were "henceforth"—per official memo—to call the base Area 51.

And where did that now infamous name come from? Some of the initial work done at the location for OXCART was on the radar cross section testing, which required construction of certain buildings and equipment on what was then Department of Energy Nevada Test Site land. That work needed to be assigned to a specific project for invoicing and a program accountant (for reasons unclear) chose Project 51. And the site became Area 51.

On the U-2 program, the site formerly known as Groom Lake was a ghost town. This time around, it would be a permanent installation, with facilities that were on par with the ones you'd find at Air Force bases. No more sleeping in trailers. There were houses for officers, storage buildings, a commissary, a control tower, and fire stations.

The project's fuel needs would be immense, but the existing facility didn't have fuel storage or an existing means of transporting the 500,000 gallons of PF-1 fuel (later renamed JP-7) required every month during testing. The project's logistics team studied airlift, trucking, and even a pipeline for delivery of fuel and decided, in the end, that trucking was the only feasible solution, but even that required the resurfacing of eighteen miles of roads to accommodate the heavy trucks. They also constructed a fuel tank farm capable of storing 1.3 million gallons on-site.

During U-2 testing there was almost nothing to do when the staff wasn't working. But the new and improved base, upgraded for the A-12, was a paradise comparatively. Its large gym had a basketball court, squash courts, an indoor swimming pool, and a six-lane bowling alley with automated pins and ball return. There was also a movie theater that played current films six nights a week, a softball field—where Groom Lake's highly competitive 8-Ballers team practiced and played—and a nine-hole golf course. The food was excellent.

Crews detailed to the site worked in shifts, five or six days at a time. The Skunk Works would fly its staff down every Monday, in three Lockheed Constellations, and flights also arrived daily from Las Vegas, carrying contractors doing short-term work.

Kelly assigned Dorsey Kammerer, his head of flight testing and a founding member of the Skunk Works, the task of figuring out how to move the Article from Burbank to Groom Lake. This wasn't a job that could be arranged at the last minute. And, unlike the U-2, the A-12 could not be broken down into pieces so that it could be transported in the belly of a cargo plane. Instead, they'd have to move this enormous plane more than three hundred miles, over public roads—in utter secrecy.

Even the basics were complicated. For example, the plane couldn't just sit on a flatbed, even under a tarp, because security staff feared that the shape itself might reveal its secrets.

Kammerer asked an engineer named Leon Gavette to design a transport system. Gavette's solution: two aluminum-framed, cloth-covered transport boxes that would be built alongside the prototype on the production line.

The larger box—105 feet long and 35 feet wide—would carry most of the fuselage and inner wings, while a second, smaller box carried the removable outer wings, nacelles, rudders, the plane's nose, and assorted bits and pieces.

A CIA security team scouted the transport route in advance, using a truck affixed with red-and-white-striped survey poles. One pole extended up into the air, the other off both sides, to approximate the height and width of the carriage trailer. The team drove the intended route and, when the poles struck something—signs, power lines, berms—these obstacles were noted and marked on a map.

Later, a second team, working with California and Nevada highway patrol escorts, came through and eliminated obstacles where possible. Trees were taken down, berms were shaved away, and, in some cases, signposts were cut and hinged so they could be temporarily lowered while the carrier passed, then raised back to normal height and reattached.

The larger of the two silver-sided boxes was as wide as a two-lane road; it was not unlike moving a house, and in some spots, the carrier cleared fences and signs by just a few inches. At one junction in Mojave, the advance team suggested a slight detour to avoid removing a cluster of obstacles. Instead, the carrier was instructed to go around behind a Mobil service station, which worked fine until two trailer wheels slipped off the pavement and sank into the sandy soil, causing a long and unexpected delay while a tow truck was improvised to yank the trailer out of the sand.

For the purposes of moving the Article, CIA security created a fake trucking company, Roadrunners Internationale—inspired by the nickname for members of the program's Special Activities Squadron, the Roadrunners. And the company's first convoy—led by a small fleet of rental cars carrying personnel—left Building B6 on February 26, 1962, at around 4:00 a.m. and arrived at Area 51 three days later, after overnights in Barstow and Shoshone. Winter weather, including snow, slowed progress, as did a Greyhound bus that accidentally sideswiped the larger box. CIA security had prepared for unexpected encounters, of course. Officers took the driver aside to offer cash for his repairs, plus a bonus, if he didn't report the accident.

California Highway Patrol led the way from Lockheed's lot to the

Nevada border, where the convoy pulled over for a sack lunch, plus lemonade and coffee, prepared by the cooks at Groom Lake. Nevada cops took over from there, taking Highway 95 to Mercury, where the convoy turned off the public roads and into the Nevada Test Site.

At Groom Lake, the two carriers were driven straight into a hangar, where Article 121 was unloaded and reassembled, while the boxes were taken apart and packed up for transport back to Burbank, where they would be used again to move Article 2.

The only thing left to do now was actually fly this thing.

Bob Gilliland had flown pretty much every plane in the Air Force inventory, but his favorite was Kelly Johnson's F-104. When Gilliland first left the Air Force, he began work as a test pilot and before long was hired by Lockheed proper, where he became one of the primary test pilots on the Shooting Star. Gilliland was famous among pilots for having made five dead-stick landings in the F-104—that being when a plane's engines quit and the pilot decides to land the crippled craft instead of bailing out. It was stories like this that caused Kelly to want to woo Gilliland to join the Skunk Works.

He sent his chief test pilot, Lou Schalk, to do the recruiting.

"How would you like to come over here with me?" Schalk asked Gilliland as they drove around Burbank one day.

"Okay," Gilliland replied. "Tell me about it."

"I can't."

With very little information about what he'd be flying, Gilliland took the job. "I guess I knew they were doing something secret and good so that's why I assumed that would be the case," he later said.

Soon thereafter, Kelly called Gilliland into his office and pointed to a photo of the F-104. "I've got something that flies higher than this, flies further than this," he told Gilliland. "Now let's go and take a look at it."

He led his new test pilot out of the engineering suite and over to manufacturing, where Gilliland set his eyes upon an airplane that was unlike anything he'd ever seen before. The A-12 before him had yet to be painted. It was this shiny silver wonder.

A few weeks later, Gilliland caught a morning flight to Area 51 in preparation for flying that wonder.

As the lead test pilot, Lou Schalk got the first test flight—in fact, he was contracted to make the first thirteen flights—but Gilliland wouldn't have to wait too long for his chance.

Every A-12 driver was given a call sign starting with Dutch. Gilliland was Dutch 51.

Lockheed pilots like Schalk and Gilliland would conduct all the test flights, but by the time the A-12 was operational, it needed a deep bench of professional pilots flying for the CIA.

To be eligible for the program, pilots had to be active-duty and qualified with at least 1,000 hours in "latest high performance fighters," as well as "emotionally stable, and well motivated." They should also be between twenty-five and forty years old, under 6 feet tall, weigh no more than 175 pounds, and have wives who weren't too nosy. One prospective pilot was rejected when he told security personnel that he wasn't sure he'd be able to keep this job secret from his wife, who was summarized in a debrief report as "very inquisitive, intelligent, and domineering."

Candidates were called to random hotel rooms, where civilians asked them vague questions and shared virtually no information about the classified program they were being interviewed for. Only that it was an opportunity these patriotic gentlemen wouldn't want to miss.

This was a commitment, because this plane's program had the same parameters as the U-2. The pilots couldn't fly A-12s—spy planes that would overfly enemy territory—and be employed by the Air Force, because should one be shot down and captured, it could be seen as an act of war.

Once they were flying, drivers had to be civilian pilots, which meant that part of volunteering for duty on OXCART required them to transfer, at least temporarily, out of the Air Force. He had to resign his officer commission for the duration of the program.

Publicly, the pilot was resigning, but this was the same protocol they went through on the U-2. Any man who "quit" to fly the A-12 would be reinstated into the Air Force, at the same rank, upon completion of the CIA job. No questions asked.

Eleven pilots—including Ken Collins, the Korean War vet who thought he was signing up to be an astronaut—were chosen, and then trained for this specific plane by Lou Schalk, who was the first-ever A-12 instructor as well as the first man ever to fly the plane.

Pilots selected for OXCART were family men, who were considered more stable and less susceptible to foreign intelligence agents (especially attractive ones), and were asked to move their families to Los Angeles before they even knew what they'd signed up for.

From there, they commuted to Area 51 from Burbank, meaning that they were away and out of touch from Monday morning to Friday night. Their wives didn't know where their husbands were, what they were flying, or who they were with. They had a number to call for emergencies, and messages would be relayed to the pilots as quickly as possible. Otherwise, there was no communication.

Kelly was honest with these pilots. They were part of a truly experimental test program, with systems that were untried and untested. Every flight would raise new problems, forcing engineers to adapt on the fly. And there was no flight simulator, because building one would have taken too long and cost too much money. Instead, Kelly designed a single two-seat trainer version of the A-12 with an elevated rear seat—which allowed a trainee to sit in back and watch his instructor fly—and duplicate controls front and back so the instructor could have his student take over.

But before any pilot got into the trainer, he had to complete exhaustive ground training at Area 51 taught by Lockheed engineers, and it was common to see Kelly himself at these sessions as pilots were introduced to a host of advanced technologies, like inertial navigation, radar jammers, spy cameras, and on-board telemetry.

This so-called A-12T, with its second, raised cockpit, was a little awkward-looking and was nicknamed the Titanium Goose, but it proved very helpful in speeding up flight training for pilots. It had a second benefit, too, allowing VIPs to take a spin in this remarkable bird.

Like Kelly himself, who took off from Groom Lake on September 6, 1963, with his head pilot, Lou Schalk, at the controls. Schalk steered the bird to a comfortable height, then handed the stick to his boss and reported after that "this was the first time [Kelly] had ever been supersonic, and he was delighted."

21

THE MIRACLE PLANE TAKES FLIGHT

No day on the OXCART program passed uneventfully. When a ground crew at the Ranch fueled the A-12 for the first time, the prototype sprang sixty-eight leaks, because the rubbery sealant didn't stick well to the inside of the titanium fuel tanks. This problem was fixed, sort of. But the A-12 never stopped leaking entirely. Which was more of a feature than a bug. The plane's titanium body expanded in flight, so some give had to be built into its skin, which meant that the A-12 would leak, a little, whenever it was fueled and on the ground. This problem resolved itself once the plane was aloft and at speed. At that point, the skin tightened up, plugging those leaks.

The fuel that leaked, like everything else on the bird, was exotic. Kelly's team had considered all kinds of unusual fuels to power the A-12's massive engines, including coal slurries, boron slurries, even liquid hydrogen.

Ben Rich's work on the SUNTAN hydrogen plane project demonstrated that liquid hydrogen could work, but it was not feasible. You'd need special ships liquefying the gas in harbors around the world, in proximity to wherever you were basing your planes. Coal slurries—finely ground coal mixed with oil and water—weren't practical, either; they were hell on turbine blades. And boron slurries clogged injector nozzles.

So, in the end, they went back to basics. The plane would have to use more traditional liquid petroleum—albeit a very special kind—a kind that would need to work in temperatures ranging from −90 degrees during midair refueling at high altitude to 650 degrees when in supersonic flight.

There was no fuel on the market that met these specs, so Kelly called in a specialist again, his old friend Jimmy Doolittle, at Shell Oil.

Doolittle had come through for the U-2, with LF-1A (for Lockheed Fuel 1), and his oil company delivered again for the A-12. It joined forces with Ashland, Monsanto, and Pratt & Whitney to develop a new and very expensive variant it called LF-2A.

This fuel had an extremely high vapor pressure; it did not ignite easily. You could throw a match into a puddle of it and there'd be no fire; in fact, the match would go out. Instead, LF-2A had to be chemically ignited by combining it with triethylborane, or TEB.

TEB is pyrophoric, meaning it ignites in air, when exposed to oxygen. At first, this clever system was named the Pyrophoric Ignition System, or P.I.S. That lasted as long as it took for someone to see the acronym in writing. From that point on, it was known as Chemical Ignition System, or C.I.S.

Just starting the A-12 up was a process. There was no ignition—no button or knob that the pilot could push to activate those giant engines. Instead, mechanics would push the plane out of its hangar, then roll over the ground cart—on which sat two Buick V8 racing engines without mufflers.

These two huge engines would rumble to life and power a shaft that would spin one of the A-12's Pratt & Whitney J58 engines up to operating speed. The pilot would then run that one engine up to full power, to adjust the engine trim, but always just one, because if both engines were fired up, the brakes and chocks couldn't keep the airplane in place.

The power of these new J58 engines was awesome even at idle. During tests, the plane would be chained to the ground as techs moved its throttle into afterburner, causing great jets of blue flame to shoot out both exhausts.

The J58s absolutely roared; anyone nearby had to wear earplugs or risk severe hearing damage, and the acoustic energy they generated was so intense that observers couldn't possibly clench their jaws tightly enough to keep their teeth from rattling, which is why crew members and observers in proximity were instructed to stand with their jaws open.

As Steve Justice, who led the Skunk Works in the early 2000s, would later put it, "You could feel the gooey part—this bag of water we're made of—vibrating, and your ribs moving around from the roar of this jet engine, and see this flame about 20 feet long coming out of the aircraft. It's one of the most incredible sights that you can imagine. It is just the absolute definition of power."

On April 25, 1962, the Groom Lake ground crew pulled the A-12 prototype out of its hangar and onto a taxiway, where that process of firing up the engines was carried out in preparation for a critical moment in the program: the first taxi test.

The idea was for Chief Test Pilot Lou Schalk to reach takeoff speed, lift the nose and nose wheel, then put it back down and bring the plane to a stop. For some reason, though, Schalk lifted all the way off, in a plane that wasn't quite ready for flight.

Kelly looked on in shock as his star pilot was barely in control of the precious craft—its nose was swinging left and right and Schalk wrestled with the plane to try to get it back on the ground as quickly as possible without damaging it, or himself.

He flew about a mile and a half at an altitude of 20 feet before touching back down in a cloud of dust on the emergency dirt runway that had been built on the dry lake bed to accommodate contingencies just like this.

"The trouble," Kelly later wrote, was a stunning failure of preflight inspection. Specifically, "an improper hookup between the rudder pedals and nose wheel steering, as the rudder and nose wheel turned in opposite directions from those desired."

Some members of the flight-test group argued that this should be labeled as a first flight and not a taxi test. Informally, the test engineers called it "the Broad Jump."

A day later, it was time for the real thing.

This was a nervous moment for Kelly. Every first flight was. A million last-minute thoughts pinged around his head as he watched Schalk taxi out to the runway: *Would the engines work? Would they keep working the whole flight? What happens if one quits at a critical time? Will Lou be able to put that thing down on the lake and stay within the test area? What if he has to make an emergency landing outside the fence?*

The biggest question, though—the one he always asked himself last as a pilot was about to alight in his brand-new invention—was: *What did we forget?*

As had been his habit since the early days in Burbank, Kelly spent the last few days before a maiden flight with his design engineers and the pilots, going over every facet of the plane, asking that question about every part and system.

He would instruct his team to retract the landing gear five hundred times, and then inspect it all over again to make sure they hadn't done any damage during those five hundred tests. They would test the ejector seat, the engine plumbing, the fuel plumbing, and the radios, as well as the tires and hydraulics. Everything. And then everything again.

What, he'd keep asking, did we forget?

But there were no apparent issues on the A-12; the test went smoothly. "Lou made a beautiful takeoff," Kelly reported. Then, at around 300 feet, the plane began to shed small pieces from the chines, which looked dramatic but wasn't actually that serious.

Schalk landed beautifully, without incident, and engineers identified, then fixed, the problem before the next flight.

That first plane—Article 121—flew naked, with no paint on the exterior, so the prototype had a silvery metallic sheen from its titanium skin. But it was quickly pointed out that this wasn't going to work once the plane was cruising at Mach 3 plus. That metal skin would suffer from the adverse effects of aerodynamic heating.

The solution? To paint the A-12 with a special matte black paint, an idea derived from Kirchhoff's law of thermal radiation that a very dark object both absorbs heat and emits it, creating a thermodynamic equilibrium. Engineers first suggested painting only the areas most likely to superheat (like the nose, engine inlets, and the leading edges of the wings), but ultimately decided to paint the entire plane black, which is how the plane got the nickname that's now shorthand for the A-12 and all its variants: the Blackbird.

By the end of July, Lockheed test pilots had completed twenty-eight flights, at increasing speeds and altitudes. And Kelly was there to watch every one of them.

The program was pretty much humming, with only occasional interruptions from nearby underground nuclear tests at Yucca Flat, such as the so-called Siren shot, when a 104-kiloton thermonuclear device was detonated, creating a crater 320 feet deep and a giant cloud of radioactive dust that blew northeast over the desert toward the Ranch.

There was an accidental test case in 1962, when a ground radar array in the San Francisco Sector—which monitored the West Coast for possible incoming targets and sent all the data to nearby Beale Air Force Base for analysis—kicked up an unknown target flying at

200 knots, at 60,000 feet. That was too high for the base's interceptor jets to make a visual ID. But one general at Beale had been cleared into OXCART and knew that the A-12 was the only plane in the U.S. arsenal capable of getting a visual on this bogey. He requested help through Strategic Air Command, and got it.

The Ranch by this time was a booming oasis in the desert, with more than 1,100 people living on-site and regular flights from Lockheed in Burbank. These flights had begun in late 1960 on the original Starliner Constellation prototype that Kelly pulled out of mothballs and sold to the CIA for $305,000, less than half its value, provided they didn't ask for more than some basic updates, including seats for passengers and a galley to provide drinks and snacks en route. Program employees nicknamed that slightly outdated plane both "the Hog" and "the Cattle Car," and eventually the flights got so crowded that two Super Constellations were requisitioned, and often all three planes made multiple flights a day.

By May 1968, the three planes had made 11,495 flights, carrying 492,205 passengers.

During the peak times, Kelly himself flew back and forth from Burbank so much that he may as well have been commuting, but he made most of those trips in a bit more comfort, using his own JetStar corporate jet prototype.

He juggled problems on the line with problems on his sprawling ranch and made time for problems at home, too, when they popped up.

All in all, 1962 had been a good year. After all those delays, the A-12 was in the air, pushing its limits on every flight. Its new engines were on the way. And Kelly was very upbeat during his annual year-end Christmas speech.

"Our last year has been one of great confusion, baffling problems, adaptation to new things, but overall a successful year," he said. "I did want to prove to you that the things you're making here do fly."

And perhaps because his staff had exploded in size—swelling from 3,200 employees to 5,400 in a single year—he did something unusual. He defined the Skunk Works for the assembled.

"I've never tried to do that before, but I'll try to define it as I see it," Kelly said. "The Skunk Works is a concentration of a few good people solving problems far in advance and at a fraction of the cost of other groups in the aircraft industry by applying the simple, most straightforward methods possible to develop and produce new projects."

That was a mouthful.

Kelly paused and took another crack at it. "Really what it is is the application of the most common sense to some pretty tough problems; that's a better definition."

He noted proudly that his entire staff—which included engineering, tooling, purchasing, inspection, and the shop personnel, all of whom had worked on "several different projects" at once—numbered fewer than "the engineers alone on one of our nation's comparable bomber projects. Yet we have taken on similar projects for less time and considerably less cost."

The U.S. government, he said, was paying attention. The revolutionary products emerging from the Skunk Works had impressed important people. "They see in our approach the possible means for producing other things such as missiles, satellites, and possibly a supersonic transport cheaper, sooner, and better than with the very complex methods used elsewhere in eluding our own other branches of the corporation."

And most of the world had yet to see their masterpiece.

22

DAMN THAT'S FAST

The Lockheed A-12 was ridiculous on paper, as a model, and especially as a real plane—in every possible way. The leading edge of its wing was twice as hot as a soldering iron, the average temperature of the skin at cruise was hotter than the maximum temperature of an oven, and its engine inlets just gulped air—100,000 cubic feet of it per second, which is the equivalent of 2 million people breathing at once.

And what comes out the other end of those engines was just as ridiculous. The J58 generated twice the horsepower of all the engines on the *Queen Mary* cruise ship, which weighs 81,000 tons. The plane's top speed, at Mach 3, was twice the speed of a bullet, and the plane covered a football field's distance every one-tenth of a second. And sitting inside the cockpit was like working a high-stress job in a moderately hot oven. Pilots had to wear a kind of space suit, with its own cooling apparatus, pressure control, and oxygen supply.

Test-flying the A-12 introduced all sorts of complications. It flew so fast that the first approved test flight areas were just too small; pilots had to fly in a constant bank just to stay inside the boundaries, which meant they were always feeling G-forces, and the problem wasn't solved until the plane's approved flying area was extended to the entire continental United States.

Pilots had to be very wary of accidentally crossing international borders, too. You need 120 miles to make a 180-degree turn in the A-12, so even a slight misjudgment, and you're violating another country's airspace by accident. Also, the A-12 covered so much terrain that air traffic controllers had to pass the planes off every five minutes.

Pilots had other novel concerns, too. Like weather balloons, which flew so high that they'd never been a danger before. Pilots were also surprised to find . . . something splattering on their windshields at 80,000 feet. That turned out to be insects. How'd they get up there? They'd been carried aloft by atomic blasts, and just got stuck, floating around in the thin air high above the Earth.

Probably the most important safety measure on the plane was the pilot's pressure suit. As long as the A-12 was functioning properly, its driver flew in relative comfort, with a cockpit pressurized to sea level and pure oxygen to breathe. But if those systems were to fail, or a pilot had to bail out, he would be killed instantly at these cruising altitudes and speeds. His blood would boil in seconds. The pressure suit, then, was his backup survival system. In the event of depressurization or a bailout, it would save his life. And if the problem was simply a depressurized cabin, the suit would keep the pilot safe long enough for him to complete the mission.

Kelly picked the same company that designed the U-2's flight suits—the David Clark Company, in Worcester, Massachusetts—to tackle the A-12 suit.

A human body is fine up to about 10,000 feet above sea level. Above that, both pressure and oxygen become issues. Pilots were sent to Clark's headquarters, where several women conducted fittings of the so-called S-901 pressure suit, which included a base layer of snug cotton long johns, an inner waterproof pressure garment that fit like a wet suit, a layer of support netting, and a heavy canvas outer garment with attachment points for things like regulators and a flotation bladder.

The pilot's head was then measured for a custom helmet that would be mated to the suit, plus pressure gloves and large combat-style boots that went over the suit's built-in booties.

Once a suit was made, it and the pilot went to a second facility, where a different contractor had a special chamber that could simulate high-altitude conditions. This was to make sure the suit pressurized when needed and, as important, to show the pilot that it actually did what it was supposed to.

The suit was incredibly awkward and required a lot of getting used to. Pilots had to jump off a high dive into the base pool wearing a suit, as well as travel to Lake Mead and go parasailing while wearing one. All in service of learning to be comfortable with this very specific kind of discomfort.

The parasailing exercise really terrified the pilots; it was reported by several of them to have been among the most harrowing moments of their OXCART experience. Pilots were so heavy in their suits that the Boston Whaler boats used to pull them struggled to lift the men and their suits off the ground. So instead of lifting off from the lake's rocky shore, a few pilots were dragged through the rocks and sand and into the water.

One thing the suits did not have: a waste-disposal system. A-12 missions, unlike the endurance marathons flown by U-2 pilots, were relatively short; typically, five hours or less. So pilots simply controlled their food and drink intake prior to flying, with an emphasis on foods that didn't pass quickly through the body or produce gas.

Still, accidents did happen.

The margin of error for a flying machine as experimental as the A-12 was virtually nonexistent. It felt like only a matter of time until there was a catastrophic accident, and on May 24, 1963, that inevitability occurred.

Sheep-dipped Air Force pilot Ken Collins—Dutch 21—was returning to base on a subsonic J58 engine test in Article 123 when he encountered some heavy clouds and rain at around 25,000 feet. Collins climbed to 30,000 feet to try to get above the weather, but the clouds were just as thick at that altitude.

His instruments all looked fine—no alarms—but Collins noticed that his flight controls were slow to respond. He was rechecking the instruments when he saw his altimeter and airspeed indicators both winding down.

Seconds later, according to Collins, "all hell broke loose." Something was terribly wrong, but he was in thick cloud cover, with no visual references, so it was impossible to have any clear sense of where the plane was or what it was doing. Then, suddenly, the A-12 stalled, pitched up, and went into "a flat inverted spin."

Lacking controls or concrete information about his speed or altitude, Collins chose his only good option: He flipped his visor shut, grabbed the ejector ring between his legs, pushed himself back in his seat, and pulled up, triggering a series of near-instantaneous events.

The canopy flew off and a set of small rockets fired, shooting the seat, and Collins, straight out of the plane. Considering that he could see almost nothing, and that he'd just ejected from an upside-down plane, Collins didn't feel great about his chances. And this being his first-ever ejection, in any plane, he wasn't even sure he was going to survive until he'd separated from the seat and his chute opened.

Still, Collins wasn't out of trouble. His heart jumped when he looked up to see his chute break away from his harness.

What fresh hell was this? In fact, it was normal procedure. A-12 ejection systems had two chutes, the first being a drogue to slowly decelerate a pilot who was likely to be ejecting at ludicrous speed.

The main chute, which followed later, after a pilot had descended to a lower altitude, was so big that, if it were to deploy at high speed and high altitude, it would almost certainly kill the pilot instantly from extremely rapid deceleration.

Finally, Collins broke free of the clouds and could assess his situation. He was still 15,000 feet up, and there was his plane, in the distance, spiraling toward the ground, where it exploded. He had no idea where he was when he hit the ground, rolled, and then stood up, as trained.

Collins checked his limbs. Everything seemed in order.

He was scrambling around collecting classified papers from the scrubby desert hillside when he spotted three civilians in a pickup truck headed his way, with the plane's canopy in the bed.

Thinking fast, he scared the men away from his wreck by telling them that it was a fighter with a nuclear weapon on board.

But Collins gladly accepted their help, and a ride. He asked them to drive him to the nearest police station so he could make a phone call, and the program security staff fed the press a predetermined cover story for an A-12 crash—that an F-105 had crashed during maneuvers. It's still listed that way in official records.

Article 123's wreckage was recovered within forty-eight hours and any witness who might have possibly seen something was identified, located, and asked to sign a secrecy agreement by security staff.

Meanwhile, a recovery team was sent to the crash site to cut up any portions of the wreck that were recognizable and truck away all debris for disposal. The only thing of any importance they couldn't locate was an emergency pack lost by Collins during his fall from the skies. That pack had one thousand dollars in cash and several important and secret letters written specifically for a pilot who crashed and was found (alive or dead) by civilians. A team of nine working first from three trucks, and then on foot, searched for three days and failed to locate the pack.

A second search party was then dispatched to the crash site to work on horseback—using calculations provided by Kelly himself—and finally found the pack a little over a mile from where Collins had touched down.

Kelly had arrived at the Ranch within hours of the crash, on a Lockheed Constellation that flew in from Burbank carrying security staff and engineers, plus the boss to lead the investigation. The plane then picked up Collins and a flight surgeon and flew the two of them, plus Kelly, to the Lovelace Clinic in Albuquerque for a checkup.

During a debrief, Collins recalled the flight as best he could, then agreed to a dose of sodium pentothal—aka truth serum—in order to verify his statements and perhaps to unlock hidden details.

The first assumption of any crash is that the cause was pilot error. That's how it is, until proven otherwise. But Collins, it turned out, had done nothing wrong. His plane's pitot tube and air data computer both had failed him. The cause, Kelly later ascertained, was ice in the plane's pitot tube, which measures air speed.

And Collins walked away, feeling very lucky to be alive.

Ten years later, he got a package in the mail from a Lockheed engineer named Keith Beswick. In it was the D-ring that Collins had pulled to eject from the crash.

23

MOVIN' ON UP

Down days were hard to come by—they had been hard to come by for at least a decade at that point—but in the middle of 1963, Kelly Johnson threw a party in the most exclusive possible location: inside the Skunk Works.

This was to celebrate the new and improved Skunk Works, not yet open for business—the fifth and largest iteration of Kelly's shop, which had evolved from a circus tent filled with desks made from engine boxes to a vast stand-alone compound every bit as nice as Lockheed proper.

Kelly inaugurated the shiny new digs one special night when, for the first and perhaps only time, wives were allowed to visit this hallowed place where their husbands worked under the strictest secrecy.

Althea Johnson—Kelly's chief of staff both at home and on the Ranch—had helped her husband organize the event, and as Kelly took the podium to greet the crowd, he explained how this rare open house at the impossible factory had come about.

Apparently, he'd been mumbling aloud one night about who should break the ceremonial bottle of champagne to christen the new base when Althea perked up and said, according to Kelly, "I've been hearing about this guld-durned Skunk Works for twenty years, and if anyone is gonna bust something on it, I'm gonna be the one!"

Some of the shop's metalworkers had fixed up a piece of titanium extrusion from the A-12 production as the target for said bottle, and then shrouded the bottle in a protective casing so it didn't shatter.

"So, none of you are gonna get hurt by flying glass," Kelly said. "But in all likelihood the bottle won't break. So, first we'll have to find the center of percussion. Then if Mrs. J, who is very accident prone, doesn't break her knee with this, she might hit that titanium angle."

Not that she needed additional motivation, but Mrs. J did, indeed, hit the target. The bottle broke, and the crowd cheered.

Kelly ran quickly through a sanitized version of the Skunk Works history, starting with the circus tent next to the wind tunnel, where the F-80 was built, moving on to the first permanent office, "in the southwest corner of the field," and then on to Building 82, where ADP had two separate homes.

"Skunk Works Four we left today," Kelly said, having finished "seventeen major projects and parts in two others" inside its walls.

Kelly being Kelly, he couldn't help but point out the care and precision with which he had built this larger, more lavish home. The cost, he noted for Important Audience Member and Lockheed president Dan Haughton, was just $13 a square foot, not including furniture. "We'd made careful analyses of how many offices should have paneling and carpet, and for a net cost of $3,774.30, we came to the conclusion that we could do it in practically all of them—particularly if we used old engine boxes for paneling, as we have done in many offices."

In true Skunk Works fashion, the job was completed almost exactly on time; it was just four days behind schedule and Kelly noted—seriously, it seemed—that he had vowed to pay for the building costs within two years by creating a special "building fund" that would be funded by improvements in efficiency.

Finally, there was a warning. "I do want to say to you we don't want you to advertise that this is the new Skunk Works," he said. "And, you wives here tonight, we'd like you to forget that this is the Skunk Works. Security clamps on at twelve tonight."

Moving to a larger space more befitting the vaunted machine his Skunk Works had become was important to Kelly, and to Lockheed. It was inevitable as soon as the A-12 program expanded beyond its limited purview as an extremely classified CIA spy plane project to a suite of supersonic prototypes built for the Air Force.

The Air Force, however, was nothing like the CIA. It was a vast, sprawling bureaucracy with many masters, some if not most of whom Kelly didn't like or respect.

One of his most important beliefs—that the Skunk Works should control its own developmental flight-test program—was a nonstarter with the Air Force, which argued, not wrongly, that Lockheed knew plenty about building planes but a lot less than the Air Force about flying them.

Kelly's frustration with the Pentagon, and his inability to bend the defense establishment to his will, was hardly a secret. It is all over his project logs from this period and showed up in surprising ways, too, like the very funny memo he wrote to "all concerned" on January 29, 1963.

The title was "Idiot Charts and Related Subjects." And it's an all-time classic of the Kelly Johnson canon.

The memo begins: "In the last few days, I have seen several reports prepared in ADP which reflect a method of preparation of data which I insist shall not take place in the Skunk Works. Some of our newer people are drawing up 'Idiot charts' of a type which I would never use for any presentation nor allow in our reports."

Idiot charts, he went on to explain, typically take two forms:

1. One that states by written sentence a subject which is perfectly clear to the audience and which should be to the presentor [*sic*]. It is generally designed to allow the presentor to spend 4 or 5 minutes on things having nothing to do with the chart.

2. A systems diagram which looks very much like all other systems diagrams, in that it consists of square boxes starting with an input from some source and ending up with an output. A great deal of time can be wasted with such artwork and very seldom does it make any point with anybody reading the report.

Kelly was noticing the insidious mission creep of federal bureaucracy in other ways, too: "There is also an increasing indication that our new proposal reports are far too lengthy and detailed. I even saw one which contained those forbidden phrases 'value engineering' and 'PERT—Cost.' For the benefit of those new to us, I expect all engineering done in our projects to have value. I expect the maximum effort to be made in the first place to make a thing producible and we will have nothing to do with PERT—Cost studies in any operation with which I am connected."

Finally, he wrote, "We will continue to make charts in the old-fashioned manner, which means that they will be few in number and present things in a clear, sharp manner and, if we can't do this, we won't make a chart, for we obviously don't understand the subject ourselves."

MOVIN' ON UP

Even within the Skunk Works, Kelly's portfolio of projects was large and constantly changing. There were the various OXCART-related projects—the A-12, the SR-71, and the YF-12—as well as the updated U-2s. He couldn't ever focus on just the present and immediate future, either, because once a specific model was built and flying, he needed new work for the engineers and designers who'd created it.

Ideas came to him, and were proposed, such as the one he offered to both the Air Force and the CIA in June 1963 that they "put a long focal-length camera" on top of the A-12's fuselage, pointing upward, "to evaluate satellites."

That same month, the Skunk Works began work on a new version of the U-2 for the CIA under the code name Project WHALE TALE.

Even with its incredible range of 3,000 miles, the U-2 was unable to reach certain areas of interest for U.S. intelligence by taking off and landing from land bases. To cover the whole world, the CIA needed a version that could take off and land on aircraft carriers, and asked Kelly to cook one up. He did so, and proposed building twenty-five of them.

Bean counters within Lockheed management proposed that the planes could be delivered for $765,000 each, but Kelly pushed back. There was no way he could do it for less than $1 million a plane, and he ordered them to "do the quote over."

By late summer, he was ready for a test. On August 4, under cover of darkness, a Lockheed crew working with a small team from the Office of Naval Research (ONR) managed to sneak a U-2 onto the USS *Kitty Hawk*, docked in San Diego, in the wee hours as sailors from the ship stumbled home from a nearby carnival.

The next day, the aircraft carrier "steamed out of San Diego to a point south of San Clemente Island," Kelly noted in the U-2 log. He wrote the flight card himself and briefed a small group of Navy personnel on the plan for a takeoff and landing test. "They were all extremely cooperative, but it seems that the captain on the bridge didn't get word of what we wanted to do, so we had several instances of going too fast or too slow or in the wrong direction."

Despite the confusion, test pilot Bob Schumacher took off, Kelly wrote, "with no difficulty whatsoever" from a carrier moving at 24 knots with a 6-knot wind. The landing, on the other hand, didn't go so smoothly.

Schumacher made three approaches, flying through turbulence that tossed his featherweight plane around, and on the third pass he tried to set the bird down on the carrier's deck.

Kelly watched his pilot bounce, "hit hard on the right wing tip," and then pull back on the stick and abort "just before coming to the end of the angled deck." Schumacher ascended, circled around, and made one final pass so that observers could check "to see that all the parts were on" and then flew off to the mainland to find a more stable place to land.

The Skunk Works, along with the ONR, modified the plane, renamed it the U-2G, and trained Schumacher along with four other pilots to operate this very specific and challenging craft.

On March 2, 1964, Schumacher made the first successful carrier landing in a U-2G, but the new carrier-based variant would fly only one operational mission, that May, when a U-2G was dispatched to observe and photograph French nuclear tests on Mururoa Atoll in French Polynesia.

It was a productive flight. Those photos confirmed that the French were close; in fact, they'd be producing their own nukes within a year.

24

ESCAPE TO STAR LANE

Kelly Johnson loved the pace and importance of his projects at the Skunk Works, but the place nearly broke him more than once. Despite his obsession with the job and the work, he was still a human who needed at least occasional escapes, and that's what his beloved Lindero Ranch was for.

But the value of the ranch he and Althea bought in the 1950s, when the San Fernando Valley was still the country, skyrocketed in the early 1960s, when the Colorado River aqueduct diverted water into the area, enabling large-scale development.

Developers leapt into action. Property taxes jumped 1,000 percent, and it was no longer viable for the Johnsons to keep such an enormous parcel of land in the soon-to-be-booming valley, where new neighborhoods began to sprawl east and west. So in 1962, the couple sold the ranch to a developer who dug a lake, subdivided the land into 746 lots, and called the place—halfway between Calabasas and Thousand Oaks—Lake Lindero.

This was a windfall, but Kelly couldn't survive without an escape. He was immediately on the hunt for a new ranch, and in November

1963, he found it—a giant plot in the Santa Ynez Valley, about thirty miles north of Santa Barbara.

This was an area that the Johnsons knew well and loved. Kelly and Althea never had kids. It's unclear whether that was by choice or because they were unable to. Neither of them ever commented on it publicly. But it meant that any time Kelly wasn't working was free for recreation, and Althea, not one to sit idly at home, was always up for an adventure.

The couple had been visiting the nearby Alisal Ranch for years, to play golf and ride horses, and this property was massive. It had 2,000 acres and came with three hundred head of cattle, a sizable amount of oat hay crop, and a Spanish-style house with a pool that was bigger than the one the Johnsons had at their regular place in Encino. "We weren't exactly roughing it there," he later wrote.

The previous owner, a movie distributor, had named the ranch Star Lane, and Kelly never thought of changing it. Considering Lockheed's tradition of naming its planes for "stellar bodies"—Vega, Orion, Constellation, Shooting Star, Starfighter, and so on—this felt like kismet.

Star Lane was to be his new escape, and he'd need it in the coming years, but this wasn't a place Kelly went to relax, at least not in the traditional sense. This we know because he left behind extensive logs of his time there—as detailed as the ones he kept at the Skunk Works—and even reading them is exhausting.

He took his first vacation at Star Lane a month after purchasing the place, to get things "sorted out," which for Kelly meant hauling in all the farm equipment himself, on a rented flatbed—including a plow he bought for sixty-five dollars and spent three days rebuilding to a nearly new state—and seeding acres of land. In subsequent months, he would improve roads, design a new barn, and repair a plugged sewer at the house of one of the ranch hands.

Throughout the spring, as tests on his various A-12 spin-offs continued and were plagued with issues, Kelly hit the 101 Freeway and escaped to Star Lane every chance he could.

If something broke, and Kelly was there, he fixed it. He built a workshop himself, from the ground up—designed to withstand earthquakes and 140-mile-per-hour winds—and kept a meticulous list of his hardware inventory—every nut, bolt, and screw—in a ledger.

When he needed one of those screws, in any size, it was easy to find because he was meticulous with the storage, too. Each type of nut, bolt, and screw was sorted out and kept in a separate baby food jar, with a label. "Having a shop is one of my joys, and has been since boyhood," he wrote. His Swedish-born father, Per Jonsson, would have loved it.

Farm engineering flexed the same muscles as airplane engineering, but it came without the stress. It wasn't a chore; it was pleasure. When it was time to replace an old bridge over a small creek that would sometimes run wild in the spring, Kelly designed its replacement to be so strong that it could accommodate military vehicles, including tanks and mobile missile launchers, in case Uncle Sam needed to use his property in the event of a war—and then sent a letter, complete with stress analysis, to the Pentagon to let the generals know about it, too.

The Johnsons had ranch hands, and even built a house for the foreman and his family, but Kelly was hardly a gentleman farmer. He himself was out there, in the fields, whenever he was on the property, working the land or tending to injured animals.

One day, he was driving with his foreman in the ranch Jeep when they encountered a sick or hurt calf in the road and, because the mother was nowhere to be seen, Kelly was wary. He approached carefully to check on the calf's status.

The small animal seemed to be okay, and Kelly was walking back to his tractor when he heard "a tremendous noise" behind him. He swung around and froze, because "charging at what must have been at least 140 knots was what looked like the biggest cow I'd ever seen."

Kelly had no time to run; instead, he dived for cover into one of the furrows he'd just dug with his tractor and "the cow went right over me, udders dragging over my face, but hooves not hitting with any full force fatally, I hope."

When the cow chose not to make a second pass, Kelly got up, finished his work, and went back to the house to have lunch with Althea, who was slack-jawed at the sight of her husband showing up in torn clothes, all bruised and scratched.

"What an ignominious end it would have been if I'd been killed by a cow after all those thousands of hours of flying in experimental airplanes!" he wrote. "I've always said it was safer in the air than on the ground."

Near miss aside, Star Lane was a paradise for both Johnsons. Althea became the ranch manager and found as much joy in those hills as her husband did.

During those early years at Star Lane, Kelly later wrote, he and the missus "lived an almost idyllic life, not without its pains and difficulties, but overall a very happy time."

There are few reports of Kelly Johnson sightseeing or golfing, though he did play the latter somewhat regularly, typically with Althea. Most of what we know about his leisure time is from those voluminous reports of the things he did at the ranch, compiled in the logs. A biographical brief that Lockheed PR man Warren Hughes prepared for a curious AP reporter named Ralph Dighton in 1963 also included tales of Kelly's tinkering with agricultural machinery, which indicates that this was a personal detail he was happy to make public.

Hughes's brief reports that, when the Star Lane hay baler didn't bale alfalfa fast enough for Kelly's liking, he "changed the gear mechanism" and increased its output threefold.

The document also celebrates Althea Johnson's ability to operate a tractor, noting that she'd picked up this skill by taking her husband's advice that anyone can master a machine by studying its manual. "Under [Kelly's] aegis, she has mastered the tracked vehicle by patient study of the instruction manual," Hughes wrote. In fact, he reports, the couple had his-and-her tractors, "and operate them on weekends."

You can hear Kelly's voice between the lines of this document, sharing stories that amused him and burnished his legend as a creative engineer who had never met a problem he couldn't solve himself. Like the story of how he found his first ranch, Lindero, by obtaining his pilot's license and flying around the area scouting locations until he found his ideal spot—a spot that he considered living on, and commuting from by plane, until he calculated the probability of weather problems and, according to this press release, "ruled out the idea because 17.0639 percent of the time he would have a problem with weather at one end or the other . . . based on an analysis of weather reports covering 30 years."

25

PROBLEMS, PROBLEMS

Expecting the unexpected was part of being a test pilot, and A-12 drivers were accustomed to adjusting on the fly. Every flight in this audacious new plane was an adventure, especially because the models powered by the new Pratt & Whitney J58 turbo ramjet engines were plagued by an issue known as "unstarts."

Unstarts were a malfunction of one or another of the plane's two inlet spikes—those sharp pointed cones that extended from the engine on each wing—and in the early days of A-12 testing, unstarts occurred on nearly every flight, often a few times.

These spikes were functional, not decorative; and they had nothing to do with aerodynamics. Rather, they were a specific design fix to control airflow into the engine—compressing it for combustion—during supersonic flight, and were one of the A-12's cleverest functions.

The A-12 engines were turbocharged for takeoff and for the first part of every flight as the plane climbed and accelerated, and the spikes remained stationary until it reached Mach 1.6, at which point they began to retract automatically, closing up the air inlets. In this sense, the working inlets are "starting." And from that point on, compressed air would be diverted around the core engine via a series of pipes into the afterburner as the engines shifted to a new method of propulsion.

They began to function, at this point, as ramjets, which means that the high pressure of the air required for combustion is created by forward motion of the plane instead of mechanical compression.

If there was one thing every OXCART pilot could count on in those early days, it was an unstart, typically in the climb out, between Mach 2.5 and Mach 2.9. An A-12 pilot recognized one of these startling events by a sudden and enormous *bang*, followed immediately by his plane yawing violently toward the unstarted inlet.

An A-12 at cruise had about 15,000 pounds of drag, but an unstart on one inlet added 7,000 pounds of drag to that side, all at once. A pilot didn't need to guess, then, which side had the unstart. He could just note which side of the cockpit his helmet slammed into as the plane jerked suddenly; the unstart was on the other side.

Adjusting the electronic inlet controls would sometimes clear the unstart and get the engine back in working order, but more often a pilot had to fly himself to safety. Drivers were trained to descend and slow back to subsonic speed in order to fix the problem.

If the pilot couldn't catch this in time, his plane would decelerate rapidly and begin a "shuddering dive toward the ground," said Blackbird pilot Ken Collins, during which the pilot's helmet would be bouncing repeatedly against the canopy. It was a fight, then, to start at least one engine before things got so dire that he had to bail out.

Back at the Skunk Works, engineers attacked the unstart problem and found that the inlets were not handling shock waves at certain speeds due to the tiniest of design errors that somehow was not caught in wind tunnel tests.

Kelly himself had one fix in mind: "mice"—small fairings specially shaped to smooth the airflow—that would be placed inside the inlets. These mice, plus laborious work by his engineers adjusting the schedule of deployment of the spikes until they could keep the shock waves outside the engine, fixed the problem on subsequent versions of the plane.

Unstarts aside, the A-12 was probably safer and more successful than anyone hoped it would be, considering its advanced technology. And its pilots, chosen from the best of the Air Force talent pool, showed remarkable composure under the most stressful circumstances.

Until July 9, 1964, when the plane's streak of good fortune ran out as CIA driver Bill Park was returning from a Mach 3 check flight in Article 133, one of the newest planes in the fleet. This particular A-12 had flown just ten times, for a total of 8.5 hours in the air, when its controls froze up at 500 feet as Park made his final approach for a landing at Groom Lake.

The plane banked hard left and Park could not regain control, so he made the difficult decision to eject from a plane that was flying at 200 knots and tilted 45 degrees, just 200 feet over the ground.

Somehow, Park survived "what must have been one of the narrower escapes in the perilous history of test piloting," according to the official report of the crash. The primary cause of the accident was a servo failure, which was—much to Kelly's chagrin—a failure of testing. "It seemed that, while we tested for high temperatures and low temperatures and normal temperatures, we didn't test for cold hydraulic oil with a hot servo valve, or vice versa. We did from then on."

Regardless, the test program marched on, as the population of Area 51 reached its peak, with more than 1,800 people on-site to run and monitor A-12 testing. On October 23, an A-12 flew 500 nautical miles at Mach 3.0 at altitudes up to 80,000 feet, with two aerial refuelings. Kelly's baby was getting closer to operational parameters.

But problems with the inlet controls continued to plague the engineering team. And well into 1965, Kelly and his engineers were struggling for a reliable solution. They ran more than 10,000 wind tunnel tests, collecting 1.25 million data points, but just couldn't get the inlet control to work consistently.

Patience at the program office back east was fraying, too. On August 4, 1965, OXCART program manager John Parangosky—a key player in several audacious CIA black ops—flew to Burbank to make the Agency's frustration very clear, and a memo of the summary of events reports "delays in the program to make the O operational" (O being shorthand for the OXCART planes). Parangosky got the best possible assurance that the Skunk Works could get things back on track: "Kelly Johnson is now going to devote full-time to the project."

Kelly acknowledged in meetings that the problem lay in the hydraulic servo system that drove the spike in and out. It was clear to the CIA's young and fast-rising deputy director for science and technology, Bud Wheelon, that this was a fixable design flaw. The hydraulic system simply could not respond fast enough to the changing aerodynamics of this audacious airplane during high-speed flight. He urged Kelly to change it out with an electronic version.

But Kelly did not like or trust electronics, and he kept dragging his feet, pointing out that the program had $35 million invested in the hydraulic solution.

Wheelon's reply? It cost at least that much every time an A-12 went down, and this was no longer a debate. He would cancel the program if Kelly refused to switch. That was the end of the argument.

Agency concerns weren't lessened by the loss of a third OXCART plane, Article 126, which went down shortly after takeoff on December 28, 1965.

That plane's driver, Mele Vojvodich Jr., ejected safely—in this case, just 100 feet above the runway—and said in his debrief that something was just off about the plane. It was as if the Article did the opposite of what he wanted it to. (He also claimed to have had a premonition about the crash the night before, and credited his survival to that vision.)

Vojvodich's parachute barely had time to open before he touched down on the ground, and he feared at first that he'd been badly injured. He was paralyzed, temporarily, from the waist down and had to crawl away from his plane's wreckage in order to escape its inferno, but he recovered quickly and was cleared to resume flying. By the time he got home, he was just limping a bit and simply told his wife that he'd sprained his ankle playing tennis.

An extremely frustrated Kelly blamed Vojvodich at first. He "complained bitterly" to Bud Wheelon and General Jack Ledford (director of the Air Force's Office of Special Activities) about the quality of the CIA pilots, further raising tensions between the Skunk Works and the Agency. This tension peaked when Ledford and Kelly got into it on a flight to Area 51 in the crash's aftermath. According to witnesses, they nearly came to blows.

And in the end, Kelly was wrong. Vojvodich and Ledford were vindicated. Lockheed mechanics had screwed up and reversed the pitch and yaw gyros in the control augmentation system, even though they were color coded for precisely this reason.

Kelly swallowed his pride and admitted as much in a letter to Ledford. He assured the general that improvements would be made, including color coding and labels, and promised changes to inspection procedures as well. "We are finding as our aircraft become more and more complex that we have an increasing problem training even experienced people to follow good practice," he wrote. Which . . . wasn't *actually* the problem here, Kelly admitted. The mechanic who'd reversed the wires? He "had seven years' experience, and much of it on aircraft with complicated electrical systems."

But the most disheartening news of the year, by far, awaited Kelly at home. Althea—the woman he'd met in his first months at Lockheed, when it was a tiny start-up in the early days of aviation, and who'd been at his side through a half dozen evolutionary leaps in airplane design—had been feeling run down and went in for tests. She was diagnosed with cancer.

This would be devastating news at any time, but it felt especially cruel after those two "idyllic" years on the ranch where the couple planned to watch sunsets and ride horses well into retirement. And Kelly, for whom no question was unanswerable, couldn't help but wonder if Althea's health problems were somehow a result of head injuries caused by a car accident—when, a few years prior, a drunk driver had T-boned the Johnsons.

But that's not how cancer works. And it wasn't necessarily a death sentence, either. But a pair of operations did not cure her. Althea's prognosis looked bleak—so bleak that she fell into a depressive funk and one day, while Kelly was out in the fields, she attempted to kill herself with sleeping pills. Fortunately, Kelly came back to the house earlier than expected and was able to get Althea to the hospital in time to save her.

Whether she had one year to live, or five, wasn't clear. No one knew.

26

BLOWING COVER, AND MINDS

On February 29, 1964, President Lyndon Baines Johnson blew America's mind. This was a historic day for LBJ, his first general news conference as president, precisely one hundred days after the shocking assassination of his boss, John F. Kennedy, which set America into a prolonged period of mourning.

Johnson began with some routine personnel announcements—a new assistant secretary of state for Asia, a new assistant secretary of the Army, and a new addition to the Civil Rights Commission. Then came the big news, which had just been approved for release by the National Security Council that morning: "The United States has successfully developed an advanced experimental jet aircraft, the A-11, which has been tested in sustained flight at more than two thousand miles an hour, and at altitudes in excess of seventy thousand feet," Johnson said. The performance of this plane "far exceeds that of any other aircraft in the world today."

The president had not erred in calling this new plane the A-11. That was done at Kelly Johnson's suggestion, as a crumb of disinformation. And the photo shown to the media that day was not the A-12 spy plane, either. It was the YF-12A interceptor.

Kelly and the CIA had worried that revealing the A-12 as a surveillance plane was giving away too much to the Soviets. Instead, the president should show the interceptor, leaving the impression that this was a tactical aircraft and not a spy plane.

To further protect the cover, and the plane's secret Nevada test base, Kelly decided to say that this new interceptor plane would be stationed at Edwards Air Force Base in California—and two YF-12As were flown from Area 51 to Edwards to back his story up. (The planes arrived the day of LBJ's speech and flew supersonic. Their titanium skins were so hot when they landed, Kelly later said, that they caused the hangar's fire-extinguishing system to kick on "and gave us a free wash.")

President Johnson credited Lockheed with this incredible new airplane, but not the Skunk Works or Kelly. "I do not expect to discuss this important matter further with you today but certain additional information will be made available to all of you after this meeting," he said.

Johnson told reporters that he was disclosing the country's most secret military program "to permit the orderly exploitation of this advanced technology in our military and commercial programs." Specifically, the A-11 was being tested as a "long-range interceptor" (a veiled warning to the Soviets about their long-range bomber plans), and what the designers and engineers learned in making such a leap forward would aid development of a supersonic transport—a passenger plane that could break the sound barrier and slash travel times around the planet.

That was the public story. In reality, the new president never explained why he'd chosen to blow the program's cover. But his sharp-elbowed defense secretary, Robert McNamara, was said to have pushed for it, arguing that a program this big—more than $1.5 billion for twenty-five planes—would eventually be exposed.

Most likely it was a mix of reasons—for instance, that a growing number of commercial airplane pilots had seen the Blackbird in flight, so its existence couldn't be kept secret for much longer. It was also possibly a blocking move, especially as over $40 million had been allocated for an Improved Manned Interceptor (IMI) plane. And then there was the political reason—Republican nominee Barry Goldwater had criticized LBJ throughout the campaign for being weak on the Soviet Union. A Mach 3 fighter that flies at 70,000 feet sure projected strength.

An argument was even made that the A-12 and its variants could lower, rather than raise, tensions in the arms race. "As a multipurpose weapon, it seems likely to delay the time when the military must turn irrevocably to the missile—a weapon which, once fired, cannot be recalled," one of LBJ's special assistants, Douglass Cater, wrote in the national newsmagazine *The Reporter.*

Plenty of politicians were shocked by the news. Certain senators, for instance, were outraged that a program this audacious could be funded and kept secret, like Illinois Republican Gordon Allott, who said: "A decision was made. It was made in our name, using hidden funds, and other methods to which I do not have access. But, somehow, it was accomplished," as if the existence of extremely secret programs hidden from even senior politicians were something new in Washington.

In fact, the program *had* been congressionally approved by a select five-member subcommittee of a secret House appropriations committee led by a Missouri Democrat named Clarence Cannon.

Cannon—described in 1964 as a "spry man of eighty-five"—led a tiny, powerful committee that provided oversight for the country's most secret programs. It was formed during World War II, when the senator discovered $800 million in expenditures "tucked away" in dark corners of the defense budget. Cannon feared that some terrifying boondoggle was afoot, and he raised such a fuss that the Pentagon arranged for him and four colleagues to secretly board a midnight flight to Oak Ridge, Tennessee, where they learned that America was building an atomic bomb.

All five men kept this secret until the bomb was deployed. This same subcommittee, which had only one other original member remaining, now approved and watched over funding for the U-2 and A-12.

Reading between the lines of the limited technical data that was made available, J. S. Butz Jr., a writer for *Air Force Magazine*, knew exactly who'd done this: Kelly Johnson. "Under any conceivable set of circumstances, designing, fabricating, flight testing, and bringing a pioneering, first-generation Mach 3 cruise plane to operational status in 3 years would be an almost miraculous achievement," Butz wrote. "True, the CIA-type management system is conducive to rapid development. In effect, the CIA simply says to a contractor, 'bring us one of "these."' We are making you responsible for performing all tests and making all technical decisions."

But everything about this new plane left Butz, a scholar of airplane design, dumbstruck, especially the legacy that its technology should create. "The A-11 will play a key research role in building the technology of Mach 3 plus cruise airplanes of all types—transports, fighters, and bombers," Butz wrote that April. "In this role, its ultimate importance to aviation and the nation may be as great as any aircraft ever built."

Once the supersonic plane was publicly known, complaints about sonic booms breaking windows began to arrive at Lockheed. One man claimed that a boom spooked the mules on his packtrain so badly that they tried to leap off a cliff. Another guy complained about the effect of Blackbird overflights on fishing in Yellowstone Park, saying that fish darted to the bottom of lakes and hid there for hours after a boom.

Kelly was skeptical that many of the complaints were legitimate and, according to Norm Nelson—the CIA babysitter who later ran the Skunk Works—they tested this theory by announcing bogus flight plans in order to generate bogus complaints.

One time, a plane pranked Kelly himself, by causing a boom over his beloved Star Lane Ranch, but the joke wasn't very well received because it shattered the boss's $450 picture window.

It didn't take long for people who followed defense and aircraft design to identify the Blackbird's creator. To those who followed the industry, Kelly Johnson was already the titan of aerospace design and a cult figure among engineers. But he was also a bit of a cipher to the public at large because he disliked attention and couldn't really talk anyway, considering that so much of his work was secret. So when a journalist at the Springer News Service reached out in 1964 for comment about rumors that he had led development of the top secret "A-11" spy plane, Lockheed comms chief Warren Hughes drafted a memo stating that Lockheed could not confirm this involvement, and would not.

However, Hughes was happy to share a few personal details about Lockheed's mysterious chief engineer that he had collected the previous year, for that "AP personality feature" by Ralph Dighton. For instance, that he was married with no children and liked to spend weekends riding horses and farming on his ranch. He was also a "burly man who sometimes moves rocks and builds fences for exercise." He owns a police dog. His "style of talk is fast clipped sentences" and he is "direct and to point but always courteous and cooperative within restrictions of his work." Also, a "quick thinker." For more color, he suggested that Springer's "people" just go ahead and read Ralph Dighton's feature, which had run in December 1963.

But with the reveal of the SR-71, Kelly's profile was higher than ever, and Hughes drafted a memo to Bert Holloway, another of the company's PR men, to get ahead of it. The subject heading was "CLJ Posture."

"Here is a reminder that we have to do something about CLJ and talk or don't talk," Hughes wrote. "Certainly we cannot get him to sit still for a one-at-a-time one-shot interview every time someone wants one. He'd shoot me."

You could make a case that 1964 was the biggest year in Kelly's extremely eventful career to date. That April, he was added to Lockheed's board of directors. This meant more money, and more clout at the company. But Kelly probably cared more about the honors from his peers.

Like the Collier Trophy, awarded by the National Aeronautic Association for "greatest achievement in aeronautics or astronautics in America." This was the second time Johnson had been given the most prestigious award in aerospace design—the first was in 1958, for the F-104 Starfighter.

That wasn't it, either. On July 1, a Western Union telegram arrived at the Skunk Works from the president, Lyndon Johnson, notifying Kelly of another great honor. He was one of thirty Americans who would be awarded the Presidential Medal of Freedom. And when Kelly's secretary passed the letter on to her fifty-four-year-old boss, she added this note: "Did you notice that you are among the four youngest receiving this honor?" (Other recipients included Helen Keller, eighty-four; Willem de Kooning, sixty; Edward R. Murrow, fifty-six; John Steinbeck, sixty-two; Carl Sandburg, eighty-six; and Walt Disney, sixty-two.)

On September 24, Kelly was back in D.C. again, this time to receive his Collier Trophy from a slightly more famous Johnson—President Lyndon—in the White House Rose Garden. "I am honored to receive and to welcome the most distinguished citizen of that most unique and distinguished corner of our land, the 'Skunk Works,'" the president said, noting that he was "very fond of" his Lockheed JetStar, used for shorter hops, which the press had nicknamed "Air Force One Half."

"There are those who feel that Kelly Johnson is a one-man team himself," LBJ told the crowd of reporters. "He works by a rule that is wise for us all: Be quick, be quiet, be on time. I have commended that rule, Mr. Johnson, to my staff here at the White House."

These awards, being massively public honors, had one downside. They forced Kelly out of his preferred state as a reclusive engineer. He even granted a very rare one-on-one interview to a mainstream press outlet, to a reporter from *Look* magazine. The resulting article revealed very little but did include some classic Kelly Johnson lines, such as: "You must have a dedicated bastard in charge—that's me." And: "I believe in flying things I work on, and I've been alongside pilots who were as frightened as I was. When the day comes to quit flying, then I'll quit designing."

By mid-1964, Kelly was two years into working on the fighter version of the A-12, which he called the YF-12A fighter concept. The plane was outfitted with Hughes AN/ASG-18 radar and the Hughes GAR-9 rocket system.

A few months later, it was time to test the YF-12's armaments—to fire the AIM-47 missiles while in supersonic flight. A pilot named Jim Eastham was in the cockpit, and the test went well.

For the interceptor to do its job, a pilot would have to fire missiles while flying Mach 3.0 at cruising altitude, or 80,000 feet. The task at hand—to shoot down an incoming supersonic bomber before it was in range of the continental United States—required these missiles to reach Mach 6.

Legitimate concern was raised about what might happen if these missiles flew farther than expected during tests; they might, for instance, take out some barn or a herd of cows on some innocent farmer's ranch. So the tests were conducted over the Pacific Missile Range, off Point Mugu, not far from Kelly Johnson's Star Lane property.

In order to do this, Fred Trost, deputy director of test and deployment for the YF-12, visited, cleared, and then made a handshake deal with the admiral in charge of the range. No paperwork was signed, no funding was requested, and during tests, certain civilian technicians were placed on the range's tracking cameras.

The A-12's fighter/interceptor spin-off became Kelly's pet obsession throughout the mid-1960s. America needed this plane, he believed, and so did the Skunk Works. It was critical to the future of both.

But if there was one thing he could not predict, it was the fickle winds of the Defense Department, which continued to shift. Different departments at the Pentagon gave him different signals and the process drove him a little crazy. Opinion was even mixed within the Air Force, and the various Blackbird logs are filled with changes of direction.

In advance of the 1965 Pentagon fiscal year, Kelly was asked to calculate the effect of cutting \$38 million and \$64 million, respectively, out of the prospective production budget for the F-12, which called for making twenty-five fighter versions of the Blackbird.

"This would really wreck our operation," Kelly wrote in the log. He would have to lay off "as many as 1,000 people" and "it would delay the schedule for an indeterminate period." His counterproposal was to make reductions in "man hours and facilities" and to change the way money went out the door—"paying invoices instead of commitments."

Just the thought exercise was frustrating for Kelly. The Air Force—which, as a buyer of war matériel, was accustomed to bloated budgets—still wasn't grasping that the Skunk Works ran lean anyway. There was no fat to cut. "We do not have any such lush figures in our quotes to them and there is very little to squeeze out by any means," Kelly wrote.

By this point, he was even frustrated with the CIA (a frustration that was increasingly growing mutual). Kelly felt that his favorite client agency, the one that gave him near-total freedom, was changing under Bud Wheelon, the CIA science and technology chief who'd taken over from Dick Bissell.

Anyone who knew Kelly, or did business with him, was aware of his feelings and felt his frustration. It was never something he tried, or cared, to hide. And on August 26, 1964, he was irritated enough to fire off a letter to Leo Geary that's become one of Kelly's most famous documents. The subject line was "Method of Contracting Between the Government and ADP."

"You and I have talked a number of times recently regarding the pressures that are building up to put our ADP operation into the standard systems used by the Air Force with other contractors," Kelly wrote. His team had been "very fortunate" over the past decade to have experienced such trust, without "the necessity for hordes of people to enter into the operation, which is so standard throughout the rest of the industry.

"As you well know," he went on, "the fundamental principle in our Skunk Works operation is that we operate with the minimum number of people and the least possible red tape"—for instance, just ten people in contract administration. "Because most of our contracts are of the 'black' variety, we are very careful to maintain complete records of how we spend the government money and of all our dealings with vendors and subcontractors. Early in the program, we asked for the assignment of government auditors to be present here on the scene so that, should there be any questions about our vendor contracts and internal operations, we could get a ruling on our problems prior to making any commitments."

Kelly noted that the Skunk Works was not without accountability. On the contrary, his group was subject to a "special audit" within Lockheed "because we do not utilize the CALAC purchasing or contracting groups." What's more, the Skunk Works was audited by an outside firm, Arthur Young & Co., every year. Kelly's group might resist oversight, but it wasn't in order to pad profits. "You well know that our net profits have been lower than other sections of our Lockheed divisions."

The point, he said, was efficiency: "This direct method of dealing . . . has provided the government with the best possible service with a minimum of confusion and direction." And then, for some reason, he changed his mind and crossed that section out.

Kelly painted a picture of what the creeping bureaucracy could do if left unchecked; what the Skunk Works would look like if it had to abide by "normal procurement methods used by the military and NASA."

It was a kind of lesson posing as a thought experiment.

"I would like to spend a few moments on what I think would happen if our contracting should be done through the so-called 'normal' Air Force channels. In the first place, I would have to enlarge greatly my contract and finance group, to a size I would estimate to be at least 4 to 5 times the present group.

"I would like to give you some examples of the type of thing that has resulted when normal government methods of procurement were put into effect, comparing them directly with identical projects carried on by ADP with various vendors."

The point was, very clearly, that Kelly's methods produced results faster for less money, whereas "the impact of the so-called 'normal procurement' procedures is appalling."

In one comparison of a project from another contractor, Kelly noted that he had "been told by responsible people that 75% of the engineering effort on the Vehicle B inlet control does not consist of engineering but merely of providing paperwork in various forms, as currently required by usual military contracting systems." A second example "again substantiates the fact that the simple, tight contracting methods which we currently use get results for the government cheaper than normal procurement." The final two examples were letters from subcontractors, such as the Titanium Metals Corporation, "stating the advantages of our methods."

Even Kelly himself might become—horror of horrors—an "administrator" and not a manager who literally got his hands dirty on the line. "I hope not only that means will be found to continue our present contracting and procurement methods, but, also, that the same methods can be used more widely by the government in developing new systems of any type, in the interest of reducing cost and development time."

On October 22, 1965, Colonel Horace Templeton, director of the F-12's Special Projects Office at the Air Force, wrote to Kelly about the results of a management survey completed by Colonel Benjamin Bellis. It stated that the Air Force was satisfied with Kelly's practices but that before an actual F-12 program could begin—as in, a contract for producing operational fighter versions of the Blackbird—the two sides needed to "resolve and strengthen" a "few additional weak points."

Kelly, whose resolve to stand up to the tidal waves of Pentagon bureaucracy was slowly eroding, agreed to accept nineteen recommendations. Basically, though, Templeton reported that the program was going to vouch for Kelly and his peculiar but effective ways.

Bellis and five other reps from the Air Force conducted a survey of operations at Burbank, Palmdale, and Edwards Air Force Base and the newcomers came away impressed. "For those of us that had not previously been associated with the ADP program, this survey activity was indeed a revelation."

In their first meeting with Kelly, Bellis reported, he told the men that "ADP consists of personnel doing jobs and that there are no manpower allocations in the organization for so-called coordinators or 'people to check on people.'"

What seemed to baffle them most was ADP's practice, "unique in the aerospace industry," that production and inspection groups work "directly from engineering drawings." It was also upsetting to one of the auditors to find an engineering note that read, simply, "shim as necessary," without any further detail.

This study was exactly the kind of thing Kelly didn't want in his programs—a group of people checking on his people who would then pass judgment to some other group of people. But enduring such an indignity was necessary to assure the vast Air Force bureaucracy that—should this new fighter be approved—it could be built in Kelly's mysterious shop. In fact, the report states: "It should be fully understood by all Air Force and ADP personnel that all proposals for an F-12 production program are based on the use of the present ADP system. Under no conditions would it be possible to live within the presently proposed time and cost restraints if a standard engineering drawing system is required."

Shorter version: If we want the plane fast and on budget, we do it Kelly's way.

27

MOTHER–DAUGHTER

The Soviet Union was obviously not the only threat to American interests that the Pentagon—and by association, Kelly Johnson's Skunk Works—was worrying about. China in particular had been a rising threat since the late 1950s, and a small U-2 fleet based on Taiwan and flown by so-called Nationalist Chinese pilots known as Detachment H had begun to overfly mainland China to photograph military installations and industrial centers starting in 1962—at least until the Chinese acquired their own SA-2 missile batteries and started to blast U-2s out of the sky.

Between 1963 and 1968, five U-Birds flown by CIA-trained Taiwanese pilots in the so-called Black Cat squadron were shot down, thanks to new and improved SA-2 batteries that had been nicknamed "Oscar-Sierra units"—after the acronym OS, which stood for what U-2 pilots would yell when an SA-2 locked on and fired: "Oh, shit!"

Still, the task of covering China—in particular, studying its nuclear facilities—was too important to just abandon. Especially after a 22-kiloton explosion at Lop Nor on October 16, 1964, revealed that Communist China was, in fact, nuclear capable.

Gathering human intelligence from inside the Chinese nuclear program was impossible. China was as difficult, if not more difficult, to penetrate than the Soviet Union. But at this rate of escalation, the situation was growing increasingly urgent, and the CIA made the need to solve this problem abundantly clear to its top contractors.

Complicating matters even more was the location of the Chinese testing site. Like all nuclear testing sites, Lop Nor was chosen specifically for its remote location and lack of human population. It's a dry lake bed in the high desert of Mongolia, about as far from civilization as you can be in China. If you want to conduct tests way, way outside the view of your enemies, it'd be hard to pick a better site on the planet.

Not surprisingly, Kelly Johnson had an idea in mind for how to penetrate this veil of secrecy: He told the CIA that he could build a drone—an unmanned baby version of the A-12—that would be launched from the plane, in flight, and could fly even faster and higher than its mother ship.

Kelly had been thinking about the baby Blackbird for a few years. He first proposed a supersonic drone known as the Q-12 to the CIA a few years prior, only to be rejected. But his old friend Leo Geary, in the Air Force Special Projects Office, had been more receptive. Geary rooted around in his funding pools and found $500,000 for Kelly to design a prototype of this drone, which he called the D-21.

Kelly's clients at the Pentagon and the CIA had been hammering him about radar cross section for years at this point and he'd absorbed the message. In addition to being fast and high-flying, he wanted the D-21 to be almost invisible to radar, and he put one of his newest brainiacs, a mathematics wunderkind named Denys Overholser, on the project.

Overholser, an Oregon State University grad, came with a whole new skill set: He was Kelly's first significant hire to be trained on those increasingly important computers.

After graduating from college in 1962, Overholser was hired by

Boeing, and couldn't believe his luck when he was recruited over to work at the Skunk Works. "They were just off the edge, the finest minds in aerospace, in each discipline, all in one group," he later said. "I thought 'Holy smoke, if they figure out what I don't know, I'm history.'"

But this young engineer who could write computer code paid immediate dividends, helping Kelly calculate a way to make this theoretical supersonic drone basically invisible to radar.

The CIA heard about the project and reversed course, proposing a top secret program to buy fifty D-21s from Lockheed for $31 million. The program was code-named TAGBOARD and compartmentalized inside the National Reconnaissance Office (NRO), a top secret department jointly staffed by Air Force and CIA officers.

TAGBOARD was so secret, in fact, that the engineers working in the already extremely secret Blackbird assembly building were blocked from seeing the drone by a wall nicknamed "Berlin Wall West."

The D-21 was 40 feet long, looked like a miniature Blackbird, and was powered by a single ramjet. This was a clever design, almost impossibly advanced and sci-fi seeming, especially when you consider that this supersonic drone was imagined in 1964.

It could cruise at Mach 3.3 at upwards of 100,000 feet, using inertial navigation for guidance, and then, upon returning from its mission, would slow to Mach 1.6, release all the valuable parts—the inertial navigation system, avionics, camera, and film—at which point the crew of a nearby JC-130 Hercules would push a button, causing the airframe to self-destruct.

That Hercules would also retrieve the package by capturing it in midair using a technique originally developed to snatch film canisters dropped from orbit by the first spy satellites.

This project had challenges on top of challenges. For instance, that the drone needed to reach Mach 2 just to get the air compressed enough to start the ramjet, which meant that it had to be launched at ludicrous speed.

The obvious but objectively terrifying solution to this was to use a modified two-seat A-12 *as the launch vehicle*, an idea that Kelly, who thought it up, never loved. It was wildly dangerous.

Two A-12s were modified for the tests and redesignated as M-12s (M for "mother" with the D in D-21 standing for "daughter"). Basically, the mother would carry the daughter on her back on a specially designed low-drag pylon that could support the 11,000-pound drone through the air at Mach 3.

The team hoped to fly the first test on Kelly's fifty-fifth birthday, February 27, 1965, but that date came and went. It was finally ready a little more than a year later, on March 5, 1966.

On that day—with only a tiny number of Americans cleared to know about it—a D-21 drone successfully detached from an M-21 over the Pacific Ocean not far from Point Mugu and, for a few nerve-racking moments, flew frighteningly close to the plane, flown by Bill Park. Then the drone pulled away and took off just fine. The plan was for the drone to fly more than 1,500 nautical miles over the ocean, at Mach 3-plus, and while the D-21 did hit Mach 3.3, it flew only 150 miles before crashing into the Pacific.

To Kelly, it was a success. "We demonstrated the launch technique," he wrote in the log, "which is the most dangerous maneuver we have ever been involved in, in any airplane I have worked on."

The next two tests were encouraging, too. In both cases, the drone detached and took off without incident but malfunctioned later. Then, on the fourth test, Kelly Johnson's worst fear was realized.

On July 30, during another test over the Pacific Ocean, a drone launched from the back of an M-21 flown by Bill Park veered immediately to the right, colliding with the mother ship at Mach 3 and destroying much of the plane's right wing, rudder, and nacelle.

The problem was the Blackbird's old nemesis, an unstart—in this case on the drone. And the entire accident was caught on film, by the other M-21, which had been flying chase at Mach 3.2. It was almost certainly the first time two planes had ever flown in formation at Mach 3-plus.

Both Park and his launch control officer, Ray Torick, ejected over the Pacific Ocean and splashed down 150 miles offshore. Park dumped his chute and made his way into a life raft that deployed when the crew ejected. But Torick broke his arm at some point during the ejection and couldn't make it to the raft. His suit filled with water and he drowned.

Kelly was gutted. He killed the program and offered to return to the NRO what remained of the development money. The Pentagon wasn't ready to give up, however. And Deputy Defense Secretary Cy Vance convinced Kelly to try again, by launching drones from a safer alternative platform—a specially modified B-52 bomber.

Two of America's biggest, slowest bombers were drafted into the program, modified in Palmdale, and nicknamed the BUFFs, for "big ugly fat fuckers." Instead of launching from a plane that was already going at Mach 3, the so-called D-21B would launch from the lumbering bomber at approximately 38,000 feet. Then, when clear, a booster rocket designed by Lockheed's rocket division (adapted from a rocket man-rated for use in the Apollo space program) would fire and burn for 87 seconds, accelerating the drone to Mach 3.3-plus at an altitude of 80,000 feet, at which point the drone's onboard ramjet would roar to life and take over.

Then the booster would detach and the drone could cruise at Mach 3-plus for 3,000 nautical miles, along a preprogrammed course, to overfly and photograph whatever the target was. When the mission was complete, it would descend to 60,000 feet so that the camera, with film and the important electronics—inertial navigation, telemetry system, automatic flight control system, and recovery beacons—could be ejected. That package would slowly float down under a parachute until a Lockheed JC-130B arrived to snatch it out of the air.

The first BUFF was test flown on May 30, 1967, and by August—just over a year after the accident that killed Torick—the planes were flying with a D-21 mounted under the wing. They called this version SENIOR BOWL.

Shortly thereafter, in June, the Chinese carried out yet another nuclear test, raising the stakes even further—because this time it was a hydrogen bomb.

But the revamped D-21 that piggybacked on a B-52 wasn't ready, and a CIA memo from longtime Kelly ally slash foil John Parangosky said as much, while pointing out that Kelly "exudes his usual confidence forecasting the satisfactory demonstration" of the redesigned D-21B in upcoming test flights. "It is a rather optimistic feeling for such a complex reoriented program," Parangosky wrote, his sarcasm seeping through.

In fact, the D-21 wouldn't be deployed on a mission—"to photograph denied areas of impending crisis"—until November 1969, when President Richard Nixon approved the world's first-ever drone surveillance flight.

Nixon craved photos of a new Chinese ICBM base and, while the drone did survive the incredibly complicated launch—which involved a B-52 with two aerial refuelings, Navy ships on station, and the JC-130 plane used for recovery—it failed in its mission when the drone's guidance system malfunctioned. Instead of coming home with important photos captured over remote China, the D-21 just kept flying east at Mach 3-plus until it ran out of fuel and crashed in Siberia.

The next two missions went a little better. In October 1970, a drone flew over China, took pictures, and made it back to the drop zone over the Pacific, but the parachute on the package failed to deploy and the goods plummeted into the sea, never to be found. In

March 1971, the steps were repeated, and this time the parachute worked—but a Navy frigate ran over the package, destroying it.

The final attempt was a step backward. That drone took off just fine but never returned, most likely because the Chinese shot it down.

And that was it. An Air Force "secret history" of drones that was later declassified summed up the program this way: "SENIOR BOWL closed its operational career with a record of 0–4."

28

BIG BROTHER TAKES FLIGHT

Kelly Johnson promised the Air Force that he'd have the new and improved two-seat version of the Blackbird that President Johnson had already publicly bragged about—the SR-71—flying by January 1, 1965, and if he made that date, the Pentagon would reward him with a "milestone" bonus. Extra revenue was always welcome but hitting the deadline was more important to him because Kelly needed all the goodwill he could gather from his fickle benefactors at the Pentagon.

In early December, Kelly approached Bob Gilliland, the test pilot he'd chosen to handle the first flight of the big Blackbird. At this point, the prototype SR-71—known as Article 2001—was far from completed. It had been shipped from Burbank to Palmdale, but at that moment it was still in pieces inside a hangar, awaiting assembly.

"How do you feel about going wheels up on the first flight?" Kelly asked. He wanted to know if Gilliland was comfortable committing to takeoff and flight, from the outset; the pilot would lift off, retract his gear, and go for it, instead of leaving open the possibility of aborting his mission and circling around with the gear down to make an emergency landing.

Like most good test pilots, Gilliland knew almost as much about the planes he would be flying as the people who designed and built them. He spent the time leading up to tests talking to engineers and mechanics. And in this case, he was confident that the SR-71 was ready. Or at least he had confidence in the SR-71's ejection system, because if the landing gear failed, he'd have to bail out, considering that the plane's body was essentially a giant fuel tank. You simply could not land the plane on its belly.

"I'm fine either way," he replied.

Kelly seemed surprised by this. The thought of Gilliland bailing out and letting his prized prototype crash, costing millions of dollars and months of work, may have been racing through his mind.

"Let me think about it and get back to you," Kelly said finally.

A week before the agreed-upon date for that first flight, he called for Gilliland to come see him again. "We're going to go wheels up," Kelly said, and not only that, but if things were going okay, he wanted Gilliland to go supersonic on the maiden voyage.

On December 22, Kelly hopped over the mountains from Burbank to Palmdale in his twin-engine JetStar to watch the SR-71 fly for the first time, with Gilliland at the helm.

Gilliland raced down the runway and took off north, toward Mammoth Lakes, with three F-104s flying chase. Once he was away from civilization, over the Sierras, Gilliland hit the afterburners. As the plane accelerated toward Mach 1.8, a red light began to flash, indicating that the canopy was loose. Gilliland looked over one shoulder, then the other. Things looked good. Hopefully it was just a malfunction in the warning light.

So Gilliand pushed on. He roared over the mountains, swung around, and headed back toward the base, slowing the plane as he neared Palmdale.

"How's your fuel?" Kelly asked by radio from his perch among some VIPs atop one of the airport buildings.

"Fuel's fine, sir," Gilliland replied.

"How about a flyby?"

Gilliland pushed forward on the throttle and ripped overhead, making a low-altitude gear-up flyby as a cluster of generals watched with mouths agape.

"How's your fuel now?" Kelly asked.

"Fuel's still good, sir."

"How about another flyby?"

Gilliland complied again, and those generals were agitating for a third when someone noticed a white stream pouring out of the plane's tail; Gilliland had a fuel leak, and a bad one.

He banked around, squared the plane with the runway, and touched down, deploying his drag chute to help slow the plane.

In the first-ever flight of the larger, two-seat SR-71 Blackbird, which lasted about an hour, Bob Gilliland broke the speed of sound and set a first-flight record by reaching 1,000 miles per hour.

Not long after, Kelly was back in his JetStar for a far more secret test. That same day, he flew to Area 51 for another, very different special event—to watch the modified MD-21 mother ship fly for the first time with the D-21 mounted—and it, too, broke the speed of sound on its maiden flight. "The second first flight of a new type in one day," he wrote in the log.

The same night, Kelly delivered his annual Christmas speech to the U-2 staff, at their new base in Van Nuys, and made the same announcement as part of the good news portion. The bad news included word that the president was planning to cut the federal defense budget, an act that would have, Johnson said, an "unknown effect on our programs." The Skunk Works needed "follow-on work" and "long lead items" in order to live and grow. So it was on him to make the "hard try for new business."

This prospect of down years ahead troubled Kelly, even a day after his latest triumph—the SR-71 breaking the speed of sound and setting a test record on its first-ever flight. But there were other things on his mind, too. Althea—"Cutie" in his notes and logs—couldn't seem to get healthy. She'd been hospitalized again.

On Christmas Day, Kelly drove up to the ranch in the evening, alone, to get some space and air. He rode his horse Wildfire, helped pour fittings on the equipment barn he'd designed, and finished a bridge. "It's nice and green," he wrote in the ranch log. "We need cows—I don't have time to get them. Hope CP"—his ranch log shorthand for "Cutie Pie"—"gets well soon."

29

GO TIME. AGAIN.

Tensions in Asia rose precipitously in 1965, intensifying pressure on the Skunk Works to get one version of its Blackbird ready for operational deployment. China continued to accelerate its nuclear program, and there was much worry in Washington about what seemed to be increasingly imminent in Vietnam—a Communist takeover of the entire country.

A plan was drawn up and $4 million in funds allocated to base a detachment of A-12s at Kadena Air Base, on the Pacific island of Okinawa. Blackbirds flying from this base could, with aerial refuelings, make round-trip overflights of both China and Vietnam.

In August 1965, the CIA's new director, William Raborn, sent a memo to President Lyndon Johnson that the A-12 was almost ready to serve its purpose, even though—back at Groom Lake—the inlet controls (and resulting unstarts) continued to haunt Kelly and his engineers.

The single-seat Blackbird passed an important test that month, when an A-12 simulated an actual mission by flying 7,480 miles with two refuelings and three legs at cruise altitude, during which the plane flew at Mach 3.1 at altitudes over 80,000 feet.

Four A-12s were selected for this operational deployment and—despite some "as yet unexplainable phenomenon at cruise conditions" that would shorten the plane's maximum distance—the first batch of pilots was checked out for Mach 3 flying.

On November 14, Kelly Johnson gave the CIA a thumbs-up. "The time has come when the bird should leave its nest," he wrote to his CIA boss, John Parangosky.

This was the dawn of a new era, and to mark that occasion, the A-12's operational phase got a new code name: Black Shield. Because of the geopolitical risk inherent in missions that flew over hostile territory, A-12 missions would be governed by the same rules as the U-2 flights. They must be approved by the National Security Council and authorized by the president, in this case LBJ.

By January 1, 1966, all systems were go.

Then inertia set in.

The problem was primarily political. Debate raged in Washington among the small group of people who were aware of this incredible new weapon. In particular, there was disagreement over whether the intelligence needs in Vietnam and China justified the risk of putting the plane into service and revealing its secrets and capabilities to America's most important foe, the Soviets.

The CIA and Joint Chiefs of Staff wanted to put the plane into use; Defense Secretary Robert McNamara and his allies did not. And President Johnson sided with the latter group.

The wait was frustrating, especially for Kelly and the Skunks. "Still no deployment," he wrote in the log on October 10. But the downtime was not wasted. His engineers continued to identify and fix bugs in the A-12's propulsion—"duct leakage and basic engine performance"—while CIA pilots based at Groom Lake made forty flights a month, carrying out a series of increasingly impressive tests of the plane's mettle, culminating in a record-setting flight by Bill Park (the pilot who'd survived the first-ever A-12 crash) in October.

In December, Bill Park flew 10,198 miles in six hours—crisscrossing America several times—at an average speed of 1,659 miles per hour, which factored in the extreme decelerations required for aerial refueling. "It was a record unapproachable by any other aircraft in the world," Lockheed historian Jay Miller later wrote. The flight, the CIA's project lead John Parangosky noted, "began at about the same time a typical government employee starts his workday, and ended two hours before his quitting time."

This waiting period also included the OXCART program's darkest moment when, on January 5, 1967, Walter Ray became the first CIA pilot to be killed in the line of duty when his plane, Article 125, crashed near Groom Lake.

Ray, aka Dutch 45, had been with the program since 1962 and was one of OXCART's most seasoned pilots, with 358 flight hours in the A-12. He was near the end of a routine training mission when he radioed to base that something was wrong with his fuel.

Ray was burning it—or more likely, losing it—at an excessive rate and finally, less than ten minutes from the strip, announced that he was ejecting. Both engines had flamed out.

The A-12's ejection was unlike the system in any other plane, because designers had to assume that pilots might have to use it at extreme speed and altitudes—and that they may be injured or unconscious in the process. The whole system, then, was automatic.

When a pilot ejected, he stayed attached to his seat, which then fired a drogue chute to slow and stabilize the seat for descent. Only when the pilot reached a lower, safer distance did he detach from the seat by the so-called butt-snapper straps, which forced his body up and away from the seat. Finally, when the pilot was safely away, his main parachute would fire, and he could float down to earth.

In Ray's case, things went horribly wrong. His parachute backpack got stuck under his seat's headrest, so Ray was unable to detach from his seat, and the main chute never deployed. He was most likely killed instantly when he and the chair slammed into a mountainside.

Shortly after the crash, an investigation revealed that Ray's death was the result of an improper modification to the ejection system. Basically, Ray—who was the shortest of the program pilots—had been uncomfortable with the position of his head on the seat back and someone in flight-test engineering had glued a "balsa wood block" to the headrest. But when Ray's butt snapper fired to separate him from the seat, his parachute pack was wedged against that block, trapping him in his seat.

According to pilot Dennis Sullivan, that tragic error was revealed when engineers hung a similarly modified seat in a hangar to test what would happen in an ejection. Sullivan arrived at the test just as it had concluded but heard the results. "It failed seven times in a row," he said. "This unauthorized, untested mod killed Walt."

In early May, President Johnson finally approved use of the A-12 to overfly Vietnam, and on May 17, Black Shield was put in motion. The first A-12 arrived in Okinawa on May 22, having flown nonstop from Groom Lake, followed by a second on May 24 and a third on May 26. By May 29, Detachment 1 of the 1129th Air Force Special Activities Squadron—which had 6 pilots and 250 total personnel—was ready.

Two days later, on May 31, Kelly Johnson's supersonic spy plane overflew enemy territory for the first time with Mele Vojvodich, the CIA driver Kelly had once blamed for a crash—when mechanics accidentally reversed the pitch/yaw sensors—at the controls.

Vojvodich, thirty-six, had made his name on a daring mission behind enemy lines as part of the U.S. Air Force's 15th Tactical Reconnaissance Squadron. Vojvodich flew 125 combat missions during the Korean War, but this one—in an RF-86 Sabre—was the most daring.

Vojvodich took off with two fighter escorts for what would become one of the longest, if not the longest, photo recon missions of the war. He departed from Kimpo Air Base, near Seoul, and crossed the Yalu River into enemy territory—Communist China—while his escorts peeled off and stayed south of the border. He then flew a remarkable 350 miles with no backup, photographing ten bases and capturing, for the first time, proof that Soviet-built nuclear-capable Ilyushin bombers were at Chinese bases.

Low on fuel, Vojvodich turned for home and noticed two MiG interceptors on his tail—those two enemy fighters increased to four, he later claimed, to an estimated twenty-four MiGs, in the end, that chased him all the way back to the river. Vojvodich then ran out of fuel and dead-sticked his landing into Kimbo, having flown for 3 hours and 15 minutes over enemy territory.

Instead of a celebration, Vojvodich returned to an irate commander. The wing's colonel was sure this reckless mission would get them both court-martialed for risking an escalation in the war, and according to one account of the exchange, Vojvodich inflamed the situation, literally, when embers from his pipe spilled and caught the colonel's car seat on fire.

In the end, neither of them was punished. When intelligence officials saw Vojvodich's photos, they promoted the colonel to general and gave Vojvodich a medal. And in all likelihood, the mission was what got this particular pilot into the OXCART program, where he was ultimately selected to make this first-ever operational flight from Kadena.

Vojvodich (aka Dutch 30) took off in a torrential downpour—almost certainly the first time any driver had ever launched a Blackbird in the rain. Heavy rain was a problem the plane had never once encountered in testing, but the mission went ahead because the weather on the flight path and, most important, over the target area was clear.

As he approached enemy territory, Vojvodich ran a final series of tests of his jammers and other critical systems. One of these tests triggered an alert that told the ground command to activate a telemetry radio link known as Birdwatcher, which would track the plane during its mission.

If Birdwatcher failed, the mission would be aborted. If it worked, the pilot would click on his camera systems and switch to mute mode, which cut off all transmissions from the plane except for Birdwatcher.

Shortly after 10:00 a.m. local time, Vojvodich crossed the North Vietnamese border, made a single pass that had him over enemy territory for only nine minutes, refueled in the air over Thailand, and was back in Kadena by 12:33 p.m.

As soon as the plane was on the ground, its film was taken out of the camera, boxed up, and put on a plane back to the United States, for delivery to a special facility at the Eastman Kodak plant in Rochester, New York.

Art Lundahl, director of the National Photographic Interpretation Center, sent an analyst named Tom Ferrell to Rochester on a CIA plane to pick up the photos as soon as they were processed. That's how urgently the president wanted to know about the state of Soviet surface-to-surface missiles in North Vietnam.

Thus Tom Ferrell became the first person ever to see an image captured from Kelly Johnson's supersonic plane. He studied the film, called the White House to deliver his analysis, and then hand-carried the photos back to Washington on a commercial plane.

Those photos captured clear imagery of 70 of the 190 known surface-to-air missile sites around Hanoi, as well as shots of nine high-value sites, including an airfield, an army barracks, and Haiphong's port.

Even better, no radar stations tracked the plane, and no missiles were fired.

Operation Black Shield missions were planned and directed by John Parangosky's OXCART program office in Washington, D.C. If the weather was clear, Washington alerted Kadena at least twenty-four hours in advance, and a primary and backup plane and pilot were chosen. At several points over the ensuing day, the planes would be inspected and weather forecasts checked. Finally, two hours before the designated launch time, Washington would issue a "go/no-go" order.

Pilots had no idea when, or how often, they'd be asked to fly. So-called Mission Alerts came via secret wire from the CIA, twenty-four hours in advance of an overflight. Whichever man was on duty as the primary pilot would prepare to fly, with his backup also attending briefings and generally making himself ready to step in.

Three hours before takeoff, pilots were given a final mission briefing to go over the flight plan, the camera targets, the latest intelligence on missile defenses, and the escape and evasion plan should the plane be shot down.

Support tankers took off first, three hours before the Blackbirds, to get in position for refueling over central Thailand, and there was always a backup tanker in the air as a fail-safe.

One hour before takeoff, pilots got a final physical, put on their pressure suits, and began pre-breathing pure oxygen to purge the blood of nitrogen. Both pilots then went to their planes and, fifteen minutes before launch, ground crews fired up the primary plane's thunderous engines and set them to full power.

Finally, when all checks were completed, the pilot taxied out and blasted off, on normal afterburner power. He'd run an acceleration test to assess how well the inlet spikes were working. If all looked good, he'd fly on to the tanker rendezvous—150 miles downrange—and only then would the backup pilot stand down.

Mating the world's fastest plane with a lumbering tanker plane was no simple thing, either. In order to match speeds, the A-12 pilot had to slow down almost to his stall speed, while the tanker pilot flew at basically his plane's max. Neither plane had any margin for error.

These Black Shield A-12s were painted black and had no markings other than a red tail number that was changed for every mission. Operational sorties were typically four hours or less and required two or three aerial refuelings—once after takeoff from Okinawa and once over Thailand for each pass the plane would make over Vietnam.

For the first three months of Black Shield, not one A-12 was picked up by enemy radar. But on October 30, 1967, that streak of good luck ran out, when two antiaircraft missile batteries in North Vietnam locked onto pilot Dennis Sullivan's plane.

However, neither site launched a missile, so Sullivan made a clean pass over Hanoi, his cameras firing away, then circled around for a second pass. This time, the batteries did fire six missiles at the supersonic plane streaking overhead.

Sullivan peered into his rearview periscope and saw six white vapor trails rising fast toward his location, at 83,000 feet. Four of the missiles reached his altitude and detonated, but behind his plane—one of them just 100 to 200 yards off his tail.

Sullivan looked at his instruments. Everything seemed fine. So he flew back to base and landed safely, only to learn, after the postflight inspection, that debris from the missiles had struck his plane. Two small metal pieces—debris, not warhead pellets—had penetrated the underside of his right wing.

It was a near miss, and the incident caused program leadership to suspend the flights over North Vietnam for sixty days. But the mission had been a success. Sullivan's cameras captured high-resolution images of Hanoi's air defenses, as well as a surprise: clear photos of the six missiles fired at him, winging upward with their long gray vapor trails toward the A-12.

In January 1968, Black Shield was given a new target, North Korea, after North Korean naval forces fired upon and then seized the USS *Pueblo*, a Navy spy ship, in international waters off the coast of Pyongyang on January 23. This was a bizarre, unsettling move by the North Koreans, and the Pentagon worried that it might be the first step in a series of provocations.

So, on January 26, CIA pilot Jack Weeks (Dutch 29) took off from Kadena in Article 131, bound for North Korea, to capture photographs that could be analyzed to see if the North Korean military was preparing further actions.

Weeks was to make three passes at 80,000 feet, and the mission required what was, for the A-12, fairly technical flying, because at Mach 3.2 the passes went fast, and Weeks would need to turn quickly and bank around for another flyover.

Unfortunately, Weeks ran into some familiar trouble—an unstart—when the right inlet spike failed to retract all the way. He managed the failure, but his plane ran hot and he was burning more fuel than usual, so Weeks aborted the mission after two passes and headed back to Kadena with plenty of valuable intelligence in the camera.

A wire to CIA headquarters from the base commander summed up Weeks's challenging mission succinctly: "Jack had to use some skill and cunning to make this mission good, and he did."

Analysis of the photos Weeks captured revealed no signs of a North Korean mobilization. It also showed that the *Pueblo* was hidden in a small bay near Wonsan harbor, guarded by two patrol boats.

Those photographs helped calm tensions in Washington, where some hawks were agitating for a retaliatory strike, and provided hard proof that the U.S. ship and its men were being held. The Koreans couldn't lie about either thing, and U.S. officials began negotiations to get them back.

Three weeks later, Frank Murray overflew North Korea a second time in an A-12, photographing more than 160 sites and identifying at least one new SA-2 missile battery site, but unfortunately clouds obscured the skies over the location where the *Pueblo* was believed to be, so analysts got no further information on the ship or its crew.

For months after the *Pueblo* crisis, the State Department resisted CIA efforts to mount further Black Shield missions over North Korea because of political fears over what might happen if an A-12 were to be shot down over North Korean or Chinese territory. Finally, the Pentagon prevailed, by assuring Secretary of State Dean Rusk that the plane was so fast it would be over Korean territory for only seven minutes and that, even if it were to be hit—an extremely unlikely possibility—the plane would be flying so high and so fast that it would very likely make it back over the border to safety before crashing.

So on May 8, Jack Layton flew the third-ever mission of the A-12 over North Korea and, once again, cloud cover was an issue. Layton managed to photograph numerous important installations and confirmed that there was no unusual military buildup along the DMZ (demilitarized zone), but he did not come back with any interpretable photograph of the *Pueblo*.

His flight was still historic, however. It would be the last operational A-12 sortie in history.

Seeing the A-12 operational, finally, nine years after OXCART was born, was a massive triumph—a vindication, even—for Kelly Johnson. But he also knew even before it flew that the days were numbered for the A-12, the original Blackbird. Johnson knew this before Black Shield even began—in fact, back in the same month when the A-12 was declared in good enough shape for operational deployment, in November 1965.

A memo circulated at the Bureau of the Budget put into print an idea that was already being expressed at the Pentagon—that with the new, two-seat SR-71 version coming to the Air Force, there was no need for a separate spy plane fleet operated by the CIA.

A study group that included OXCART boss John Parangosky took up the question and offered three scenarios: Keep both fleets; mothball the A-12 but maintain its capability by letting the Air Force and CIA share the SR-71; or terminate the A-12 fleet in January 1968, at which point the SR-71 would take over and become the only Blackbird in the air.

That last option prevailed, and despite efforts by CIA director Richard Helms to overturn the decision, President Johnson formally accepted this recommendation in December 1966. Provided the SR-71 was fully ready, the A-12 would be removed from service on January 1, 1968.

In early March 1968, the first three SR-71s landed at Kadena, in preparation for taking over the Black Shield mission. A month later, Kelly was still doubting this decision and wrote in the A-12 log: "The photographic take of the A-12 is considerably better than the SR-71s because the Hycon camera in the latter airplane isn't doing its job."

Two months later, the A-12 detachment was packing up for home, and Kelly wrote on May 29 that "plans were put into effect for storing the A-12 aircraft at Palmdale," home of Edwards Air Force Base.

There was one final, dark moment in the A-12's very short and otherwise bright history as an operational spy plane. On June 4, Alabama native Jack Weeks left Kadena in Article 129 for a routine check flight, to test out a newly installed J58 engine that had replaced an engine that another pilot had complained about. Provided the engine was okay, the Article would be flown back to Palmdale a few days later.

For this short hop, Weeks had only a half tank of fuel, enough to get him up to altitude, refuel in midair with a KC-135T tanker, and carry out the predetermined engine checks.

Weeks flew off, checked in to say that all was well from his current location over the South China Sea, 520 miles east of Manila, in the Philippines, and then . . . vanished.

A massive search and rescue turned up nothing. No signs of a crash. No wreckage. No clues at all. Neither Weeks nor his plane was ever found and his fate remains a mystery today.

This is how Sharlene Weeks, Jack's wife, found out what her husband had really been doing since 1962. When Jack left the Air Force, he'd told his wife he was going to work for Hughes Aircraft, on a secret project he couldn't discuss. She'd assumed he was maybe being vetted for NASA to be an astronaut.

"I was surprised that I had never figured it out," Sharlene told *The Tuscaloosa News* years later. "I think I was busy raising the children, doing my job and I knew he couldn't talk about his. We shared other things in common but we didn't share that."

Less than three weeks later, Sharlene and the wives of all the other original OXCART drivers were invited to Palmdale for a special ceremony: retirement of the last operational A-12, Article 131.

Driver Frank Murray (Dutch 20) had the honors of making that last flight. He took off from Groom Lake, soared over the Sierra Nevada Mountains into California, and landed at Edwards, where Article 131 would remain, in storage, for twenty years. Kelly was there to watch, too.

Later, he noted this final flight glumly in the log: "While the intelligence community very much wanted to keep the A-12 program going, the present financial situation cannot stand the strain. It's a bleak end for a program this successful."

On June 26, 1968, the five surviving Black Shield pilots—Ken Collins, Ronald Layton, Frank Murray, Dennis Sullivan, and Mele Vojvodich—were awarded the CIA Intelligence Star for valor in a quiet ceremony at Area 51. Jack Weeks's star was accepted by Sharlene, who then had to return it and couldn't tell anyone the truth about her husband for decades, until the OXCART program was finally declassified in 1998.

30

A NEW ERA DAWNS

The CIA's A-12 Blackbirds flew largely unmarked because the idea was that no one should know, if a plane were to crash, what outfit had been flying it. But with the SR-71 replacing the A-12, command of the craft would shift to the U.S. Air Force, and the Air Force required official identification on all its planes.

That meant that before the SR-71 could begin flying for Strategic Air Command, it had to bear the proper markings. Specifically, Air Force brass wanted the branch's stars-and-bars insignia plus "US Air Force" in paint.

Kelly and Ben Rich, his chief thermodynamicist, pushed back. This was a ridiculous ask. The surface of the Blackbird at Mach 3 reached 600 degrees, which, as they pointed out, is roughly the temperature of an oven broiler. Paint a piece of metal and put it in the broiler, Rich said. See what happens to the paint. (Hint: It doesn't last long.)

Even special paints burned off the plane at these temperatures, and those that didn't burn off changed color. White became gray; blue and red both turned purple. Ultimately, the Skunk Works did find a paint that didn't burn off or change color under intense heat, but it cost more than $1 million to manufacture and use it.

Unnecessary paint wasn't the only seemingly pointless requirement the Air Force, with its commitment to bureaucracy, imposed upon the program. Another was that the plane, like all planes, had to meet certain legal requirements—such as the Arizona road dust requirement, which felt a little silly for a plane that flew at 85,000 feet.

The SR-71 would become, by far, the most widely known variation of the Blackbird. So famous that its name is now also shorthand for Kelly Johnson's supersonic masterpiece. There were also far more of them built than the single-seat CIA version, even if most aerospace nerds—and Kelly himself—prefer the original.

The Skunk Works manufactured only ten A-12s, and two of them were converted to YF-12A fighter prototypes. Kelly's shop built three times as many SR-71s. By the end of 1967, Lockheed had built and delivered all thirty-one SR-71s commissioned by the Air Force. The Skunk Works' role then shifted to preparing the planes for deployment to Black Shield and service of the Blackbirds in the field.

Six Air Force SR-71s arrived in Kadena in March 1968 and gradually replaced the A-12s on Black Shield missions. For a short time, A-12s continued to fly, and then remained on Okinawa, on standby, as backups for the SR-71s. But by June, the A-12s had been completely phased out and the Air Force took over all flights in Asia.

The Air Force pilots did adopt a few habits from their CIA predecessors, like the use of the nickname "Habu," which had first been given to the A-12 by locals. The habu is a deadly black pit viper that lives on the island, and when locals began to see this strong plane taking off and landing from Kadena, it apparently reminded them of the snake because they would point at the bizarre jet and yell, "Habu!" The nickname stuck.

Pilots who flew these hairy missions celebrated each one by painting the silhouette of a habu on the side of the plane's fuselage—at least until a *Time* magazine correspondent spotted one on display near Washington, D.C., with forty-two habus on the side and reported it. After that, the Pentagon ordered the planes to be wiped clean of cartoon snakes.

31

DARK DAYS

Highs and lows. The business cycle of a defense contractor portrayed as a line graph looks basically like an EKG, even when you're a hero of the Cold War and arguably the greatest airplane designer of the twentieth century.

As Kelly's A-12 continued to push the human flight envelope and its big brother, the SR-71, was entering service, Kelly couldn't even enjoy his success. He had to think about the future. His business, and the Skunk Works' existence, depended on it.

No project meant more to him than the YF-12A interceptor version of the Blackbird. And the fact that he couldn't get Washington to commit to this plane really loomed over Kelly like a dark cloud.

In these times, Kelly was as diligent about his Star Lane ranch logs as he was about the logs for his various projects. The place clearly centered him as much as anything could, and unlike the Skunk Works, he never had to worry about running out of work. On the contrary, the logs are action-packed, almost exhausting to read. There were always bridges to fix, fields to tend, and steers to brand.

"Cutie and I worked 13hrs today putting braces in corner of hay barn," he wrote in May 1966 after a long weekend with Althea, who, he seemed happy to note, was enjoying a period of improved health.

That trip had a morose moment, too, one that revealed Kelly's soft spot for animals—in this case, a bull that he was preparing to send to slaughter. This "bull with the sore foot" died one day before the Johnsons planned to sell him. "I'm glad God let him die on this beautiful ranch without going to the stock yard," Kelly wrote. "He has suffered so much!"

It's hard to know from Kelly's logs if he went to the ranch more in 1966 than in other years, or if it just seemed that way, because the YF-12 project was causing him so much angst and the ranch was a release valve for his frustrations.

Kelly was ordered to stop flying the planes because the Air Force wanted to oversee testing, the Pentagon rejected his latest Hail Mary idea—to convert two of the YF-12As into prototypes of the F-12B for $31.6 million—and one of his three prototypes crashed and was badly damaged while landing at Edwards Air Force Base.

In August, Johnson laid off half the program's testing staff, "maintaining only enough people to store the airplane or send it to Burbank," he wrote. "We are very near the end of this program."

The ranch log over these months is filled with pessimism. Althea was healthy and making trips with him again, but also, if Kelly's entries are to be believed, increasingly difficult for him and the ranch staff to deal with. "I can't see her wreck lives—she is violent & unstable," Kelly wrote after witnessing a fight between Althea and his ranch manager. "Could be beautiful life if she didn't *have* to *have* some*one to hate!*"

At the Skunk Works, Kelly was boss, but on the ranch, his power had a check: Althea. The couple was regularly at odds over the ranch and how it should be run, especially as it pertained to employees. "I've had bad skipping and racing all day. Don't want Cutie to be mad at me and everyone—she still persists in trying to run ranch. Says I'm a softy with our help. Not true—only fair to them."

Weekends were short, and busy. There was rarely even time to ride the horses, because increasingly bad traffic on the 101 meant that the Johnsons had to start their trip home from the highlands on Sunday afternoon.

And Kelly went from stress at the ranch to stress at the plant. Every month seemed to bring a new crisis.

On October 8, a U-2C was lost over Vietnam, bringing the total number of operational aircraft down to eighteen out of the original fleet of fifty-five. Nine days later, on October 17, another U-2 crashed in California, this time because of pilot error, reducing the fleet to just seventeen planes.

Somehow, perhaps because he had no choice, Kelly was bullish about chances that the F-12B fighter version of his YF-12A would actually go into production. But Defense Secretary Robert McNamara continued to block funding for the ninety-three planes the Air Force had ordered in 1965, citing the rising costs of the Vietnam War.

At the same time he was lobbying for the supersonic fighter, Kelly was thinking about how to adapt the SR-71 fleet, too. "Even I could not deny that 40 reconnaissance airplanes are more than required under the present political situation, unless we have an actual war." As he put it: "In a desperate effort to save the OXCART program, suggested to Cy Vance, and then to John Parangosky, that the Skunk Works convert half of the Blackbird fleet to bombers to 'counter the ABM threat.'"

In March, he shared some of his worry with the ADP's engineers in a speech at a favorite watering hole, the Sportsman's Lodge. He called this "an unfortunate time" in which the Skunk Works was, for the first time in its history, "between projects" and facing an uncertain future.

"At this point, many of you are thinking . . . where was our forward planning?" Kelly assured his engineers that he was pursuing every possible lead, from new versions of the U-2 and Blackbird to entirely new projects. "In conclusion—tough & dark but I see some daylight showing through." The company's entire fate could turn in a day, he told them, if the Pentagon decides to put the F-12B into production.

But the grind was wearing on him. Already, Kelly had begun talking to Lockheed's leadership about retirement, or at least phasing himself out as boss of the Skunk Works.

In March 1967, he invited his key deputies—including Dick Boehme, Rus Daniell, and Al Viereck—and their wives to Star Lane for the first time to "work out Skunk Works plans." The primary purpose was to tell his most important managers that he planned to retire in one year's time. It was time to plan a line of succession.

And yet, a few months later, he was clearly still torn over it. "When or if or why do I retire??" he scrawled at the end of one log entry. It's impossible to know how Kelly felt on any given day, but his log entries from this tumultuous period veer from eager and optimistic—practically childlike at times—to dreary and downtrodden, bereft of hope. His wife's health, and the health of the Skunk Works, weighed heavily on him.

Also, his own health was not good. He'd begun to have fainting spells again, and when he went to see his longtime doctor, Kelly learned that he also had diabetes, bad ulcers, esophagitis, and mental fatigue.

His doctor put him on a strict diet of 1,800 calories a day, which cut his eating and drinking, as well as his weight, which at 206 pounds was the lowest it had been in decades.

Still, he could feel the stress. "Mental probs and worry are bad," he wrote in the ranch log, noting that one close colleague had cancer.

Another had dropped dead on a ship on his way back from England. "Is it later than I think, too?"

To make matters worse, Kelly's relationship with Althea was as low as it had ever been. Althea had been, for some time, running the ranch because she had more time and enjoyed the responsibility, but this wasn't always the easiest thing for the ranch or for Kelly. She, like he, could run hot and didn't always get on well with the staff.

Kelly couldn't abide that tension; it grated on him. So, on Thanksgiving of all days, Kelly told Althea there was going to be a change in management. He was taking over Star Lane. And Althea did not take the news well.

"I don't want this ranch run from a position of fear," is how he explained it in the log. "I think it's jealousy about our possessions again. It is totally miserable. I wrote her three letters saying how we all need her but *she* doesn't respond. I'd settle on any terms except she *must not* be so suspicious of all people trying to cheat us."

This, as you might expect, made for a desultory holiday season. On December 16, Kelly was again at the ranch alone, because Althea was now boycotting Star Lane. "Cutie has another spell. Came up & took her things to Encino! Left the ranch cold!"

This, he fretted, was likely to be the fourth straight unhappy Christmas. He'd spent three in a row alone on the ranch, and you can hear in the writing just how miserable this man was on his own. The entries are emotional, bordering on mopey, as Kelly dwelled on the negatives in his life.

He was feeling the weight, more than ever, of being responsible for the Skunk Works and its large staff. "It isn't right for me to come up here alone at this stage of life and look back & forward to what I see. All we have to do to have marvelous health, a purpose in life & *fun* is to be reasonable—not to hurt others—accept their faults as they must accept ours & look at the mountains!"

The year ended with perhaps the darkest moment of Kelly's career. After seven years of development, more than one production order for F-12Bs, and just days after telling his engineers at the annual Christmas speech that should the program go "we will be swamped," the Special Projects Office of the Air Force notified Kelly that it was terminating the project and canceling all air defense programs at the Skunk Works.

McNamara told the Senate Armed Services Committee that the decision was part of a strategic defense strategy that would combine Airborne Warning and Control System (AWACS), Federal Aviation Administration radars, and F-106X interceptors, and save $1.4 billion over ten years over a version that included the F-12B, while keeping the mainland United States just as safe.

The secretary of defense told senators that he just didn't consider the Russian bomber threat to be a real danger, especially compared to ICBMs. "We have no evidence that the Soviets are developing a new advanced intercontinental bomber," he added.

That's not how Kelly saw it. This wasn't about one program. McNamara's crew of eggheads were out to destroy him and his shop. And General Benjamin Bellis, a key Pentagon deputy, "is the hatchet man put in to take the Skunk Works apart. . . . He was either directed to do it or, more likely, took it upon himself."

At stake was a fundamental difference of opinion for how things should work. The Pentagon under McNamara simply couldn't abide by Kelly Johnson's way of doing things.

On December 30, after downing four Scotch and sodas, he sat down to write one of his glummest ranch log entries yet. "Wrote C.P. a letter," he began. "Don't know whether it will work—doubt it. It's too bad she can't come back to the world we both love. At long last I've learned—give lots—give everything—it comes back to you. Have had terrible business & *more important* national news—*F-12B is out*. A disaster—first nationally then to *my* or *your* very best people. Now (after four Scotch & sodas)—I wonder for the first time—what purpose in life. I have no *personal* ambitions. I can contribute to our defense—our real life . . ."

And then the entry ends, abruptly, with just this short final note, underlined: "*out of ink*."

The Pentagon's rejection of the YF-12A and the F-12 fighter demoralized Kelly. He had built the world's fastest plane and then proved it could be armed—making it essentially unstoppable as an interceptor jet—and yet the DOD wouldn't buy in. Meanwhile, the Soviets were pouring money into airplane development, and the United States was in danger of losing air supremacy. It was, Kelly thought, a disgrace.

It's telling that the vast majority of interviews and public statements made by Lockheed's reluctant icon came in the late '60s and the early '70s, when he was feeling so frustrated.

Kelly distrusted, even disliked, the press. But using the media as a bullhorn was one way to pressure Washington.

Lockheed's resident genius was a tough get and a tougher interview. He was a colorful man, as an engineer and boss, but that was rarely conveyed to reporters. Kelly would answer questions, when the subject wasn't classified, quickly and deliberately. As one reporter noted, Kelly Johnson "prefers building aircraft over granting interviews."

Reporters who visited Kelly's office, on the rare occasion they were granted such an audience, inevitably noted how it was mostly devoid of decoration, and how his desk was often completely bare—"not even a single sheet of paper," wrote one reporter in *Air Force Magazine* in January 1966.

One issue did animate him on that day, however—and it's not surprising when you consider the context of his ongoing frustrations with Robert McNamara and a Department of Defense that acted more and more like a business.

There was, Kelly warned, a growing gap between the United States and the USSR in the development of manned aircraft systems. The Soviets were not only ahead; they were innovating more quickly. Thanks to bureaucracy and competitive bidding, the lead time for U.S. weapons systems was now 30 to 100 percent longer than for the Soviets. Kelly suspected that the Russians were well on their way to a Mach 3 bomber that could destroy a U.S. city before any American interceptor or coastal defense could shoot it down.

Considering the audience—this was *Air Force Magazine*—it's hard not to read this part of the interview as a warning to generals and policymakers.

Some challenges of going head-to-head with the Soviets were just inherent; they couldn't be overcome. As the reporter was leaving, Kelly pointed out a copy of an aerospace magazine on his desk. Any issue of this publication was loaded with detail about U.S. defense systems. "I would give my right arm if I could get anything half as detailed about the comparable Soviet systems," Kelly said. "Here they can buy the whole thing at the newsstand for fifty cents or so."

In February 1967, he told *Space Digest* that the United States had fallen behind the Soviet Union in advanced R & D projects. "Five or ten years from today the Soviets may well be ahead of us in military technology and science," he said. Also problematic was the fact that the Pentagon had learned nothing from his Skunk Works model. Current projects were taking longer, and costing more, than ever before. And that gap was only increasing.

Among the mistakes he observed was a lack of vision. The Pentagon, under Robert McNamara, had shifted its entire national defense posture from airplanes to missile systems. "No good carpenter would tackle his job with just one kind of tool in his toolbox, and there is no reason why our national defense should do so," Kelly said.

Kelly's criticisms of the Pentagon and its services sound like a critique of all bloated bureaucracies—and are, in retrospect, an early warning about a problem that would only get worse over time. Development times had grown so long that weapons systems were sometimes obsolete by the time they were ready to deploy, and much of the fault for that, Kelly said, lay in too much focus on particulars instead of a holistic, whole-system approach.

Obsessing over every single piece of a machine will inevitably cause a project to take forever. The better way was to make that machine fast, figure out what wasn't working, and fix those things on the fly.

Specifically, Kelly told the reporter, the United States should be focusing its R & D on the two things that most affect your ability to make quantum leaps in progress: materials and propulsion. It was only after shifting to titanium—and completely innovating his tooling processes—that Kelly was able to make the Blackbirds. Titanium was a massive leap, and even years later, in 1967, the United States wasn't milling this important metal in any large quantity.

During the interview, he handed two nearly identical-looking bolts to a reporter. One was much lighter than the other.

The heavy one, Kelly said, was titanium. It cost about $1.50 a bolt. The other was beryllium, which was both lighter and stronger. It cost $100 a bolt. "We've got a lot of work to do before we can bring the price down to where we can afford to use this material the way we should," he said.

It's rare, and refreshing, to read these stories about Kelly speaking freely. Locked away at the Skunk Works, he developed a reputation of being a grumpy taskmaster. But in print, on the rare occasion when he felt free, Kelly made for fun copy.

One of the earliest mass-market profiles of Kelly Johnson had appeared in the Sunday, December 22, 1963, edition of *The Arizona Republic*, under the headline "U2's Silent Father." Before the byline, the paper appended an editor's note stating that "one of the real geniuses of modern aviation is a man little known outside the profession."

Reporter Ralph Dighton then quoted Johnson explaining his aversion for the media: "I learned a long time ago [that] you can't put your foot in your mouth if you keep your mouth shut." Dighton reported that Kelly had been in seclusion, working on improvements to the U-2 and other secret projects since the downing of Francis Gary Powers over Russia in 1960. And that he had only now "stepped back into the spotlight" on the occasion of an award from the Air Force for "providing the free world with one of its most valuable instruments in defense of freedom." (That being the U-2.) And also because the California Museum of Science and Industry had recently opened an exhibit on the history of aviation featuring sixty different Lockheed planes, the design of thirty of which Kelly had overseen.

His was a career of hits, even if the public knew very little about him. Dighton mentioned two designs from the 1940s that had been too advanced to entice buyers. One, from 1949, was, according to Kelly, rejected by airlines for being "too radical for public acceptance." A boast, Dighton pointed out, that had merit: "On paper it looks remarkably like today's latest swept-wing transports with aft-mounted engines."

"I have several letters of apology from airlines people in my files," Kelly told him.

For a media recluse, Kelly Johnson seemed to enjoy cultivating his eccentricity when given the opportunity. Dighton wrote that fifty-three-year-old Kelly pursued efficiency in every way; he took the stairs two at a time. The reason, he explained, was mechanical. "Actually it's easier that way. There's less internal friction because you bend your knees only half as many times."

32

RAY OF HOPE

Even after the F-12 was officially killed, Kelly wasn't completely giving up on the plane he felt he needed to save his operation; there remained some tiny sliver of hope. And mostly it had to do with this: Robert McNamara was leaving his post as secretary of defense.

In February, having already laid off 130 engineers, Kelly flew east to explain to some of his favorite generals why the SecDef's decision to kill the Mach 3 fighter was a mistake—one that left America vulnerable to attack from the air. "Perhaps the battle is not yet lost," he wrote in the project log. "Can't do anything about it, however, until McNamara leaves on 29 February."

Kelly really did see this as a failure of vision by the secretary of defense and his key deputies—"McNamara and his band," he called them—especially when you put his plane against the F-106, the program the Pentagon favored. Among Kelly's objections to the F-106 was that the Air Force was about to commence a new "armament system and missile to put in an airplane that will be 15 years old when it rolls out in 1974"—an airplane that was inferior to the F-12 in most every way. "Its cruising speed will be ¼ of ours, its range ½ as much, and even the new armament system will be substantially poorer" than what the F-12B could offer.

The big worry, the thing Kelly did not want to see, was that he would be ordered to destroy the Blackbird tooling—all the custom tools and molds created in order to produce the planes.

Some of the top generals in the Air Force, including Chief of Staff General John P. McConnell, sided with Kelly on this. "The F-12 is by far a much better aircraft than the F-106 against improved bombers or bombers which have a long-range air breathing missile which they can launch against this country," he told the Senate Armed Services Committee in April 1968.

McConnell was among the harshest critics of McNamara and openly doubted the accuracy of the most recent National Intelligence Estimate, which predicted that the Soviets wouldn't have this bomber capability within the next eight years. "I think they will develop that type of a bomber and have it in operation," he said, and recommended that SAC use a combination of improved F-106s and "a few F-12s" to have adequate cover. This combo, McConnell stated, "would give us all that we would require against that particular threat."

Congress, or at least the Senate committee, supported him. It rebuked the outgoing secretary's opposition to the project, which had been at times in open defiance of Air Force desires. Specifically, McNamara had refused to release $55 million in funds that had been allocated for two straight fiscal years. "No matter what steps are taken to modernize our current interceptor forces, the F-12 is an interceptor of the truly superiority necessary for the defense of the U.S. as well as being the only interceptor available to us which can provide effective defense against likely future developments in the Soviet bomber threat," the committee said in a report.

August 31 was another milestone—this one, the anniversary of his marriage to Althea, which was not the joyous day it should have been.

"Cutie depressed today," Kelly wrote in the ranch log. She was angry, again, about the missteps of their ranch hand, Frankie, who'd gone out, gotten "stoned" (according to Althea), and "ran his car up on a divider."

Mrs. Johnson wanted to fire him, but Kelly was trying to be understanding. Frankie had been with the Johnsons for years; he was practically family. Cutting him loose would only make things worse. "I want to keep him and try to put him on the right path," Kelly wrote. "A thing we owe people! He is a nice, hardworking, intelligent guy. But wilder than a March hare."

That Christmas, just a few weeks after his latest promotion—from vice president of advanced development programs to senior vice president, Kelly decided not to deliver his annual series of speeches to the various departments that reported to him. Instead, he gave just a single talk to his core staff.

The prognosis for the F-12B was grim, he told the team. And while a new drone program showed promise, the Skunk Works was mostly making do with small updates and design studies, and by pitching in wherever those extra brains and capable hands could be useful.

"Nineteen sixty-nine has been a rough year," was how Kelly opened. "In spite of our best efforts, we have been unable to sell additional Mach 3 aircraft. The disarmament actions our country is taking make it quite improbable that we will be getting any further programs on the SR-71 or F-12B." This, he said, "in spite of a brilliant combat performance" by the operational SRs flying over Asia. "Only yesterday I received a call from the Air Force saying how pleased they were with the overall performance of the SR-71."

With no new advanced development to do, Kelly had saved two-thirds of his engineers by making them very useful across the company on Lockheed commercial projects. Skunk Workers had provided 25 percent of the group engineers for the L-1011 TriStar wide-body jet (designing and building the cockpit and part of the forward fuselage) and were "doing a great deal of important static testing" on components for the enormous C-5A military transport.

There was, he noted, a "very promising aircraft in preliminary design which conceivably could be given a start in April." And if so, "we will have to show that we haven't forgotten our basic Skunk Works procedures as the delivery date will be early as usual." But those in attendance could almost feel his lack of confidence in those lines.

In these public settings, Kelly was trying to keep a brave face, to boost morale. But in private he worried, a lot. Things were getting bleak. The Skunk Works had been bleeding staff for two years, and he was down to 3,200 employees, with "little possibility of holding at that level."

"We are holding on desperately now that the elections are over in the hope that President Elect Nixon will meet his pre-election campaign speech promises of rearming the United States," he wrote in the log on November 7, 1968. "Do not know how long it will take before something is done along these lines. The F-12B is still far and away the best interceptor in the world and will be in spite of the recent Russian developments."

In February, Kelly flew to New York City to receive yet another award, this time the 1969 Billy Mitchell Award from American Legion Post 743—the Aviators' Post—presented by Mayor John Lindsay.

On the afternoon of the event, Kelly was still fretting about the state of U.S. fighter tech. "There hasn't been a true air-superiority fighter developed by this country since 1958," he said in an interview at Lockheed's New York City offices, in the iconic Chrysler Building. "I think we need to build a new fighter for air superiority, beyond the F-15, that can go at least Mach 3 and be able to operate up to 80,000 to 90,000 feet. . . . I'm sure the other side will have a plane that can operate up this high, and I don't think we should let the enemy have this airspace all to himself."

Basically, what Kelly was saying was that he still wanted to make the F-12. And that slim hope he clung to, fueled by his worry about America's defense against newer, faster Soviet bombers—lasted until February 5, 1970, when he received a telegram at the Skunk Works ordering the thing he'd been worrying about for years at that point: Lockheed was to destroy all tooling for the SR-71 and its variants.

This wouldn't necessarily prevent Lockheed from ever making Blackbirds again—the plans existed, and everything could in theory be rebuilt—but to start over with the tooling on a plane that was wildly expensive to build on the best of days would raise the cost exponentially. To say building more of the planes, someday, was technically still possible felt like an asterisk.

This is what Robert McNamara had wanted and even pushed for. He left office before he was able to accomplish the feat. But finally, two years after Kelly Johnson's foil left the Pentagon, it happened anyway.

Kelly had stored all those tools for three years as he pushed for the F-12, but he'd always known this could happen. That it probably *would* happen. Until the tooling was gone, his fighter was a threat to funding that other defense officials—and contractors—needed for their own next-generation projects, like the B-70 or F-15.

Ben Rich would later refer to February 5, 1970, as "one of the most depressing days in the history of the Skunk Works." That's when "all the molds, jigs, and 40,000 detail tools were cut up for scrap and sold off at seven cents a pound." They left only the bare minimum that Lockheed's field crews would need to produce spare parts for the remaining planes.

There's a security justification for doing such a thing. Technically, an adversary could put certain tools to work—or make larger leaps in their own tech by using them. But really, the destruction of the Blackbird's tools was a political move: to kill, once and for all, a program that certain DOD factions viewed as a boondoggle, one that threatened funds they wanted for other projects.

"Ten years from now the country will be very sorry for making this decision of stopping production on the whole Mach 3 series of aircraft in the USA," Kelly noted in one of his final entries in the SR log.

Whether it was just the rumor mill, sparked by the decision to kill the Blackbird program once and for all, or backstabbing by someone at another company or the Pentagon, Kelly was forced to deny reports that he was retiring and that his famed Skunk Works might be shuttered. "We are not going to close Skunk Works and I am not going to retire," he told *Newsweek*.

The truth was less rosy. Privately, Kelly worried a lot about the future of the Skunk Works, which, he admitted in a letter to his old friend Leo Geary, was "now down to a service organization" in the wake of the F-12's demise. All his key Skunks had either retired or moved to other projects. "Won't be too long before I'm all alone or moving up to the ranch," he wrote.

On top of that, Kelly's ulcers were back, and there were, as he wrote in his ranch log, "bad troubles" at Lockheed proper, where a series of mishaps had led to a "desperate" situation.

This had nothing to do with Kelly or the death of the F-12. He was referring to a crisis raging at the larger corporation that threatened to destroy the entire company. The company announced nearly $33 million in losses for 1969 and was teetering on the verge of bankruptcy—so close that CEO Dan Haughton had written to the deputy secretary of defense, David Packard, who was a defense contractor himself before taking the job at the Pentagon, for financial assistance.

How Lockheed got into this mess could fill a chapter or more, but basically came down to two things. First, the company had staked its future on the chronically delayed L-1011 TriStar airliner, forcing Lockheed to take on a $400 million line of credit in 1969.

The other problem, the one that caused Haughton to prostrate himself before Packard, was the enormous C-5A cargo plane. This program was also plagued with delays and, worse, a runaway budget that made the C-5A's development the first program in U.S. history to overrun its projected cost by $1 billion.

The Air Force was furious about all this and had stopped paying bills. But Lockheed couldn't stop production.

Haughton's honesty—his display of vulnerability in trying to save the venerable institution he had inherited—was, to Kelly, "a brave, brilliant thing to do." The boss needed support and Kelly was happy to give it. He told *Newsweek* that he "had to help Lockheed" and make his next concept work.

Earlier in 1969, Kelly had tasked his staff with developing the concept that could fit the Air Force's International Fighter Aircraft (IFA) competition. Winning an order like that would be huge for a company, transformative. This would be a volume buy, not a specialty item.

His idea had been to modify the existing Starfighter—specifically, the F-104G, which many foreign air forces were still using—to have more wing and a larger tail, plus a host of technological improvements that would make it faster, more nimble, and easier to fly.

Internally, he called the lightweight fighter project the CL-1200. The plan, he told journalists, was to develop "an advanced fighter designed to meet the threat, not only as it exists today, but as we think it will exist in the next five to ten years." This jet would be roughly the same size as Soviet fighters but would be able to outmaneuver them *and* the generation that would replace them. And because he was talking to a newer, more frugal Pentagon, Kelly accentuated the value here: Seventy-five to 90 percent of the "expensive parts"—radar, cannons, missiles, cockpit, forward fuselage—would be shared with the existing F-104G design.

Unfortunately for Kelly, there was no momentum, or money, for a prototype in 1969, so he'd been forced to abandon the plans. "A couple of years ago, I would have said 'the hell with it' and got the money," he wrote. "But now we have a cash flow problem."

Still, this wasn't wasted effort. And he wasn't totally giving up. He noted that the Skunk Works had built the first Starfighters in 357 days and "I don't think we're so decrepit that we couldn't build a new one in less time." If only the company could back him.

Which, for now, it didn't want to do.

Anxiety was a constant for Kelly, even in good times. But there was a big difference between the stress of meeting a deadline, or even the stress of solving some confounding technological problem, and the stress that came with chasing new work. Because that stress was all about his ability to keep his key staff employed—really, to keep the Skunk Works running.

And the situation at home wasn't helping. Althea was hospitalized for yet another cancer surgery, this time to remove part of her lower bowel. She "came through very well," Kelly wrote in the ranch log, "but it's been a rough month."

Kelly's own health was hardly better. His annual physical revealed that his chronic ulcers had damaged his stomach so severely that the gastric outlet was "down to [the] size of a pencil," resulting in "constant pain." His doctor and friend Lowell Ford urged him to have surgery because he feared "it will burst on a trip" and kill him.

Kelly agreed to have that surgery in April, after he'd finished his proposal for the CL-1200. "It's been a long time coming but pain is too constant & too great."

On April 6, he gave in and submitted to the surgery to open his duodenum, which was badly scarred from decades of chronic ulcers dating back to his time at Michigan.

Even in the midst of one of the worst months of arguably the worst stretch of his career, Kelly was able to find some humor in the whole mess. Trapped in a hospital for nine days, with no engineering—or ranch—problems to tackle, he created some work for himself and wrote up a "trip report" for his bosses. This was, he said, in compliance "with Lockheed Management Directive 801-13-70, which requires complete reports on lengthy trips where important people are met or which require large inroads into Corporate medical funds."

The report, he noted, would be handled in "typical Skunk Works engineering style" and would include discussion of any "new product lines" inspired by his stay.

Kelly cracked jokes about the various indignities—having his stomach pumped, his belly shaved—and described equipment as if it was all meant for airplanes and not humans, but did allow himself (and his readers) a rare moment of earnest, sober thought toward the end of the day: "Most wonderful care on the part of the whole operational staff. Visits all hours of the day and night. In spite of what we think is hard work in the airplane business—we're a bunch of lazy, undedicated bums compared to our brothers in the medical profession."

Kelly was unable to eat solids for a week. On Sunday, April 12, he woke early "to try out my new stomach." He even decided to "follow Pentagon practice" and "develop an acronym" for this new design. He called it "New Stomach Revision—A Reverse Flow, or NEWSARF, and then, after his first meal of a soft-boiled egg, red Jell-O, and tea, noted that "NEWSARF stalled on the first turn! Decided it wanted to be a gas factory not a stomach."

The operation was a success. It eased Kelly's chronic belly pain and had an unplanned benefit, too; it improved his health in the process, by causing him to lose weight, which he noted with a joke about outsized costs on the job: "This method of weight saving seems quite direct. Will probably cost $500/pound, same as on the C-5."

Prior to the surgery, Kelly weighed over 220 pounds, and he'd been around that number for a while. His ideal weight, he told doctors, was closer to 195—which is what he weighed in college, as an athlete—and that's where he landed, and stayed, after the surgery.

Unfortunately, the surgery didn't fix everything. It couldn't cure the stress Kelly felt at work, or the long hours he felt compelled to put in. So in August—after Lockheed officially unveiled the CL-1200 Lancer for international buyers—the company's doctors ordered Kelly to take a month off, and then shared that order with his boss Dan Haughton, to make sure that the order didn't just end up in a trash bin.

Haughton called a meeting with his top engineer to discuss the matter, and Kelly agreed, reluctantly, to take nearly two months off—from August 10 to October 1. To make sure he wasn't secretly working, Haughton banned anyone at Lockheed from calling or sending paperwork to Kelly's house. Which, at least intellectually, Kelly understood. "I must stop these periods of irregular beats—particularly, I can't relax," he wrote.

Even with a lighter workload, someone had to step in and lead the Skunk Works in Kelly's absence, and the job fell to Rus Daniell, one of his longest-serving engineers. The other obvious candidate for stand-in boss was Ben Rich, but a few months earlier, Kelly had loaned Ben and a small team over to the regular Lockheed engineering operation to help design a new Navy sub hunter. So he wasn't available.

Daniell was a lifer, one of the old hands, and he had Kelly's respect. But his stint running ADP was, at least according to Rich, both "brief" and "a personal disaster" for an otherwise outstanding engineer. Daniell had become the odds-on favorite to replace Kelly upon his retirement, which was clearly looming, but Daniell approved a proposal for the Air Force during his short time in charge that included (according to Rich) "a glaring mathematical error," and this rare mistake from Kelly's vaunted design shop was a gift to his enemies at the Pentagon.

One general who had probably "waited for years to stick it to Kelly and his know-it-alls" called Kelly in the hospital and, according to Rich, "raised hell about our sloppy work." This call did not make the Trip Report, but it infuriated Kelly, who ordered Rich to return to the Skunk Works immediately and take over. And this rare misstep by Daniell, at the worst possible time, very likely changed the order of succession.

The matter of whether, let alone when, he should retire nagged at Kelly throughout the summer. "Having difficult time about deciding should I retire," he wrote in his ranch log on August 5, noting that he was "trying to stay connected" to Lockheed's commercial operation, to help the struggling company land some new projects—like the CL-1200 lightweight fighter.

Kelly could afford retirement. He was by objective measures a wealthy man. He'd summarized his assets in July in the log: $4.45 million in land, houses, and ranches; $399,086 in stocks; $1,006,000 in bonds, and $292,554 in cash. His $350,000 life insurance policy, he wrote, should—in the event he died first—cover Althea's inheritance taxes, without requiring her to sell any of the property.

But that was an unlikely outcome. Althea has been battling cancer for five years by this point, enduring three surgeries, countless miserable days, and at least one suicide attempt. In the hardest moments, and especially after that last surgery, the family's longtime physician, Dr. Lowell Ford, would sometimes sleep at the house to provide comfort and to help fill in for Kelly, who, in addition to his ulcers, was suffering from angina attacks almost every night.

Kelly and Althea spent those weeks together at the ranch, "neither feeling well," with Dr. Ford making regular checks on them. Kelly couldn't help but fix things, against everyone's advice, and Althea's energy was flagging. In late September, she was rushed to the hospital back in Encino for urgent surgery after diverticulitis turned into a bad infection.

These close encounters with mortality had Kelly thinking about the future, about his legacy and the family's. He started to talk to schools and hospitals that might, one day, want to take over the sprawling Star Lane property. "Must decide what to do with ranch on our joint demise!" he wrote. "Hope we can be intelligent."

On December 1, 1970, Althea died. She was sixty-six. Kelly's wife of thirty-two years—who'd met him when Lockheed was just a handful of people working in a ranch house and been at his side as it grew into arguably the world's preeminent advanced projects machine—had two dying wishes.

The first was to endow a chair in her husband's name at Cal Tech, and Kelly arranged for just that; the chair would be funded upon his death, with $650,000 from the family estate. She also asked not to be buried. Althea wanted to be cremated and to have her ashes spread on those lush yellow hillsides at Star Lane, her favorite place.

Unfortunately, this wasn't legal, as Kelly learned when he looked into the logistics. So, being a law abider, he hatched a backup plan. He borrowed a small plane and flew, with Dr. Ford and his longtime test pilot Tony LeVier, across the valley and west, toward the ranch—"over the mountains where we had ridden so frequently in happier times."

The group continued on to the coast, and out over the beach and the waves into Santa Barbara Bay, until the plane was "far enough out to sea where legally we could fulfill her last wish." In international waters, there is no such ban on the disposal of remains.

And then, for the first time since his early years in Burbank, when he'd arrived as a young college grad with his pal Don Palmer at his side, Kelly Johnson was alone at home and at the ranch.

It didn't last long.

Althea knew that her death was coming, and she worried about what might happen to Kelly as a bachelor. Ben Rich claims that she flat-out ordered her husband to remarry, "because he was not the kind of man who could live alone." She worried especially about his diet and his drinking.

Althea even had a good idea of *whom* Kelly should marry: his pretty, petite administrative assistant, Maryellen Meade—"a vivacious redhead twenty-five years younger" who was newly single herself, having divorced her first husband in late 1970.

Meade had been Rich's secretary before moving over to work for Kelly. And about six weeks after Althea's death, Kelly summoned Rich for a talk.

The boss seemed "embarrassed and troubled." He wasn't the kind of man who talked much about his personal life, but Rich could tell that he wanted to. He needed to. So Rich slowly coaxed it out.

Kelly told Ben what Althea had told him in her final days, a thing that sounded crazy and self-serving but was honest-to-God true: that she wanted him to remarry, and to remarry his own secretary. Kelly wanted Ben's opinion on what people might think of this, and of the age difference. He was even worried about how to react if Maryellen said no.

"Since when do you worry about what people think?" Rich replied. "All that matters is what you think. No one around here will think you're a dirty old man, if that's what you're worrying about."

So Kelly asked, and Maryellen said yes. They married on May 20, 1971, in a small Lutheran church in Solvang, a quirky Danish-inspired town a short drive from Star Lane, and then went to Hawaii for a very short honeymoon.

From the perspective of someone who spent quite a few years digging around in the ruins of Kelly's largely buried life, his obsessive tendency to document even mundane things is possibly the most useful tool to understanding who he was.

On May 25, 1971, Kelly started *Star Lane Book #2*—the second and final ranch log—and signed his name in the upper right-hand corner. Like all his logs, it is sparse and filled with abbreviations and shorthand, but these books are also filled with personal color. At times, they seem almost like a safe space for real emotions.

The second Star Lane log begins: "Book #1 has information covering period Oct 1963 to Jan 1, 1971. This book carries on from Jan 1, 71. Dec 1, 1970, A passed on. May 20, 1971, I married Maryellen at Solvang. We went to Hawaiian Islands for a short honeymoon. Returned to ranch Tuesday May 25th."

The log continues with brief reports of his new life, on the ranch. The day-to-day with Maryellen seemed not so different from the old one, with Althea. The couple planted 300 acres of Kanota oats and got a decent crop, despite a lack of rainfall. The machinery was "in good shape," the pool "filled but not yet ready to swim in," and Maryellen—often just "ME" in the log—had the "main house spic and span," with a new bedroom setup and "several new pictures we bought at art sale in Conejo."

Kelly was, at least at home, happier than he'd been in years.

33

HAIL MARY

In February 1972, Kelly Johnson gave a "seminar on prototyping" to the National Defense Industrial Association that was subtitled "Prerequisites for a Successful Skunk Works." By this point in his career, Kelly's methods were legendary, and he was highly sought after as a speaker by industry groups, whether or not they had anything to do with defense contracting or aerospace. For this particular group, Kelly emphasized the role of prototyping.

Something Kelly learned very early about prototyping was that if the production was to be done by a different group than the one who built the prototype, you're almost certain to encounter what he called the NIH ("not invented here") factor.

Handing a Skunk Works prototype to a Lockheed production group tended to take longer, cost more, and result in too many changes. The solution, he decided, was to take half the workers who would be engineering the production models and embed them in the Skunk Works, to help invent the prototype. At the end, they're relocated back to production—hopefully to help stop anything from being reinvented unnecessarily.

Then there was paperwork. Few things in Kelly Johnson's world were more infuriating to him than pointless paper, especially progress reports that mostly existed to justify layers of management. He presented, for example, a series of slides showing the number of 8½ x 11 progress report pages issued by a single electronic flight control system subcontractor. One project, "Dinosaur," generated 6,000 pages a month. The B-58 generated 15,000. And the Apollo program—200,000!

The SR-71, he said, got off to a bad start, for him. There were 300 report pages the first month, but 30 of those were on Bernoulli's theorem, and when Kelly saw this, he scolded his vendor: "Don't impress me with that. Where are we on the program?"

From that point on, the Blackbird program averaged 35 pages a month.

Even within the Blackbird family of programs, there were stark differences. Problems with the air inlet control on the SR-71 required the Skunk Works to redesign portions of the system, and it requested a certain electronic part from a vendor. The same change was to be made on a different airplane, which was being built on a regular Lockheed production line and thus subject to what Kelly called "normal system development procedures."

In other words, *not my system.*

The Skunk Works purchase order was 3 pages, which received a 4-page proposal from the vendor in response, resulting in a swift approval and order. Changing the part on the version running on the regular Lockheed production line required a 185-page purchase order and a 1,200-page proposal.

Bureaucracy blossomed across the industry. Throughout the Blackbird program, Kelly had used his own security measures. One of them was that no documents should be stamped as classified; this only invited nosy people to pick them up. "Everybody wants to read secrets," Ben Rich later said. "So we stamped nothing."

But in the 1970s, the Pentagon cracked down. It required that all classified documents be stamped as such, top and bottom, so Skunk

Works security personnel had to open up every file and add stamps by hand—including on more than 90,000 U-2 documents that were two decades old and not really secret anymore. By that time, the Soviets and Chinese had shot down five U-2s and picked apart the wreckage. There was nothing else to learn from those papers.

Kelly was alarmed by what he was seeing around the industry. Demoralized even. Pentagon regulations and oversight were overwhelming programs and causing contractors to swell with bloat, resulting in what he called "paperwork empires" on both the military and industrial sides. He wasn't sure that the companies of this new era even knew how to get back to simpler systems if they wanted to, which he doubted. "Great big organizations don't want to give up and shrink to 10 percent their current size," he said, and if you know anything about business, it's hard to argue with that.

Prototyping isn't always a solution, either, Kelly said. You should only make a prototype for a plane you intend to build; otherwise it's just fun work for engineers and a waste of taxpayer money. Also wasteful was building a prototype without the weapons or avionics because that's a sure way to build a plane that isn't practical. "The minute the guns are ready, you realize the plane's too small, and then it's back to the drawing board," he said. "Every line we draw, and every report we write, we write it with the idea being that we're making something useful, and we intend to produce it."

It's hard to object to any of what Kelly told that room. But it's also hard not to see him as, increasingly, a relic. His methods were almost inarguably better at working faster and cheaper, as well as at driving innovation. You have to remove hurdles to do those things.

But the military-industrial complex of the 1970s was very different from the one that Kelly thrived in. Wars—and threats—had caused federal budgets to swell, which caused money to pour into the Pentagon, which created massive opportunities for companies to thrive, and bloat.

In retrospect, Kelly Johnson's fall from the highs of 1966—when both the A-12 and SR-71 were flying and a full order for the F-12 fighter seemed imminent—to where he and the Skunk Works were in 1972 was precipitous. It's almost as if the process of fighting so long for the F-12, only to lose it and all the Blackbird program's tooling, broke him—or at least snuffed out most of what remained of the fire that had driven so much innovation for so long.

History will show that Kelly did pursue the Pentagon's next big combat aircraft program—the so-called lightweight fighter—but it sure doesn't seem, given a few decades of distance, that his heart was really in it.

And Lockheed didn't seem to have much heart in it, either, at least not in 1969. But that failed work did prove useful in 1972, when the Air Force issued a new RFP (request for proposal) for a plane that was, roughly, in the same niche—a "lightweight fighter." The specs, as laid out, were for a plane that weighed 17,000 pounds, carried 5,000 pounds of fuel, and had a 275-square-foot wing.

When Ben Rich offered to assemble a team and set to work designing a plane with these specs, Kelly scoffed at the idea.

It's not enough fuel, he told Rich, pointing out that a fighter on afterburner uses a thousand pounds a minute. And every fighter that carried only 5,000 pounds in Vietnam combat "ran out in tight spots and the pilot wound up at the Hanoi Hilton," he said, according to Rich's recollection. "I won't submit a proposal for something this wrong."

Rich pushed back, saying that this was only a game. They could give the Air Force a design according to its specs, knowing that of course the plane would change in the actual construction and testing. Just tell them what they want to hear and win the job. *Then* fix it.

It made sense. But Kelly wouldn't have it. He "got sore," Rich says.

"Ben, if I teach you anything, it's this," he snapped. "Don't build an airplane you don't believe in. Don't prostitute yourself for bucks."

Some part of Kelly's irritation over this bid probably stemmed from a belief that *he* had planted the seed for this contest, or whatever it was, in the first place—back in the middle of 1970, when he sent an unsolicited letter to Air Force secretary Robert Seamans and SecDef David Packard, proposing his CL-1200—what he called an "air superiority fighter" that was roughly half the weight of the F-14 or F-15.

Seamans and Packard did not reply. Seven months later, he wrote again, but this time he hand-delivered the letter to Packard after hearing that there was activity around a so-called lightweight fighter. During that seven-month period since his first letter, Kelly told him, "three separate Russian fighters had appeared in prototype form."

The Air Force was not a monolith. It had competing factions, with favorite programs and contractors. And what Kelly already knew was that the factions within the branch who had brought the F-15 into service feared the arrival of a lightweight fighter that might be more useful. As he explained in a log entry: "There has been a desperate struggle in the Air Force, which continues, in an effort to prevent a lightweight fighter from competing with the F-15."

Making the F-15 obsolete wasn't Kelly's goal. In fact, when he first proposed the CL-1200 to Packard, he described it as "complementary to the F-15." And yet, on paper, the new lightweight fighter was a threat to those who ordered the F-15, and especially to those who built it.

Kelly's proposed fighter had higher performance in practically all categories, at half the weight. So while certain factions in the Air Force may have successfully stopped a perceived rival to the F-15 the first time, a different faction ultimately prevailed. As he wrote to Packard, "There were enough Air Force people . . . who fought the battle to get a simple, cheaper, high performance fighter which the country desperately needs and which has led to the RFP referred to above."

He had an economic case to make, too. The F-15 was estimated to cost between $9 million and $11 million per plane, if the Air Force built at least seven hundred of them. Kelly told his Pentagon friendlies that he could build the CL-1200 for $4 million a plane, and they only had to buy three hundred.

Within days of the RFP, Kelly met with Lockheed CEO Dan Haughton and other key company executives to lay out the situation. He didn't like the Air Force specs and felt that Lockheed had little chance of winning, especially given the vast field of bidders, which included powerful rivals like Northrop, Boeing, General Dynamics, and North American.

Haughton was annoyed. This didn't exactly seem like an attractive opportunity Kelly was laying out. But he grudgingly conceded.

Kelly got his crew ready. If Lockheed was to be named one of the two finalists—a decision expected within forty-five days of the February 18, 1972, deadline for proposals—he wanted them to begin tooling immediately on what he had been referring to as the CL-1600, with a target date of June 1973—fourteen months later—for the first flight.

This design was built on the specs he thought the plane required, and not the specs that the Air Force had asked for in the first place. It was classic Kelly, and also—it turns out—one of his greatest miscalculations, because the plane Kelly was proposing wasn't inspirational or game-changing. It was an update of an old design, and his apparent passion for it wasn't driven by a need to innovate; it was driven by a need to win business to save his outfit.

On February 16, two days ahead of the deadline, Kelly flew to Wright Field in the Skunk Works JetStar to deliver his CL-1600 proposal to the Air Force. He'd given it his best shot.

Then he flew back to Burbank and settled into a new home life routine with Maryellen. They played golf together every Friday morning when they were able, and traveled. Maryellen often joined Kelly on his client trips, which were still frequent. "She is a sheer delight and very helpful," he wrote in the ranch log. His health was "quite good," for the first time in a while, but he recognized that his lack of physical exercise was a problem, especially during this "very tense period . . . on new projects."

All along, Kelly knew that Lockheed's bid was almost certainly dead in the water. He knew that when he temporarily sidelined the CL-1200 staff, despite having genuine interest for that fighter from abroad, but he couldn't not submit a bid.

But by mid-April, what he suspected was all but confirmed during yet another trip to Washington. The Skunk Works' CL-1600 would not be the Air Force's new lightweight fighter. In fact, Lockheed was not even a finalist.

"It was perfectly clear that we never had a chance to win the lightweight fighter," Kelly later wrote in a summary of the debrief. "They liked no part of the rapid schedule I proposed, nor the fact that our aircraft was large enough to take sophisticated equipment, which the others could not."

All year, Kelly had been railing on the idea of prototyping as a necessary step in building a new weapons system. This process only invited waste—of time and money. He had proposed to build two lightweight fighters, both of which would fly in just over a year, for a design that exceeded all specs, to meet what the Soviets were flying.

Instead, the Air Force decided to commission one prototype each from General Dynamics (for $37.9 million) and Northrop (for $39.1 million), with a deadline for flight of eighteen to twenty-six months.

Forget the designs. The process itself was broken.

In May, Kelly made this case to the Senate Armed Services Committee, saying that if the Skunk Works had been picked, and funded, on his terms, he would have had his lightweight fighter "flying by last Thanksgiving."

Later, he was asked by an *Aerospace Daily* reporter to comment on that quote, and Kelly amended his statement slightly, saying that he should have said "could" instead of "would."

The irony of this frustrating chapter, according to Ben Rich, is that Kelly was right. He lost out on the contract because he submitted a design that he believed made sense instead of the one the Air Force asked for. The CL-1600 weighed 19,000 pounds, carried 9,000 pounds of fuel, and had a 310-square-foot wing. All were significantly above the Air Force specs.

The contract went to General Dynamics, which had designed to the Air Force specifications exactly. But by the time that design became an operational fighter, it had grown significantly. That plane, which became the F-16, weighed 19,000 pounds, carried 7,400 pounds of fuel, and had a 310-square-foot wing.

In other words, Kelly refused to play the game and instead designed the plane he wanted to make, while General Dynamics did what Rich had suggested. It submitted a design to the Air Force specs, and then—after winning the bid—built the plane it wanted to make.

Losing out on this plane that Kelly hadn't really wanted to design stung most because of what it meant for the Skunk Works. There was no next fighter to shift production to.

Kelly made one final attempt to sell his original lightweight fighter design—the CL-1200—by pitching a revised version to the Navy, where he was told that it was a "hell of an airplane"—but also an unwanted competitor for the F-14.

And so, in June 1972, Kelly was asked to "practically disband our fighter project group." The CL-1600 was officially dead, and what activity remained was transferred back to the CL-1200, for the foreign market.

Kelly Johnson would never again lead the design of a new plane.

34

ANOTHER YEAR, ANOTHER SCARE

Life at home, and especially on the ranch, was far better than what was happening in Burbank or in a changing military-industrial complex that Kelly increasingly felt uncomfortable in. His marriage to Maryellen, while hastily arranged, was sturdy and mostly very happy. The second Mrs. Johnson loved Star Lane as much as the first one did, and was far more able to travel.

In December 1972, she tagged along on a European trip, to Nice and Rome, where Kelly visited air bases to talk to pilots about what they liked and didn't like about the current batch of fighters.

Upon their return, Kelly fired his oft-troubled ranch hand Frankie—who finally ran out of chances—after he failed to show up for work a few times and then drove his Jeep through a fence. A few days later, while Kelly was working in the fields, a cow panicked over some perceived threat to her calf and pancaked him when he stepped off his tractor. He felt lucky he wasn't badly injured or even killed.

This wasn't Kelly's only brush with mortality, either. Death was more and more present, as a real threat, in his mind. He was having chest pains again and booked himself an angiogram in February

1973 to see if he needed heart surgery. "Churchill was & is right—if you are afraid to die—you don't live," he wrote. "I've taken so many risks flying, I don't mind this except for worrying my wonderful sweetheart MEJ!!!"

The angiogram revealed three blocked arteries. Kelly needed open heart surgery, a triple bypass, and soon. "Sweetheart"—as he had taken to calling Maryellen—was at his side, "wonderful and strong."

Kelly made a special trip up to Star Lane on March 10, thinking that—with open heart surgery scheduled for the next day—he wouldn't be back for a long time. If things went sideways, he might never come back.

He was back less than a month later, having recovered enough to handle the car ride. He was, he wrote, "improving well," but grew tired very easily and required a two-hour nap after lunch. By doctor's orders, there'd be no work, at all, until May 1 at the earliest.

Not that Kelly was very good at staying idle, either. On April 28, he oversaw the annual roundup of free-ranging cattle, in a different role. Typically, he loved to be in the thick of this wild day, when a crew of local cowpokes on horseback would help him chase down and lasso the cows that roamed the ranch's remote meadows.

This time, Kelly just observed and gave orders as forty-two men came to Star Lane and finished the roundup by 2:00 p.m., at which time a party commenced. (A budget watcher through and through, Kelly noted this in the log: "Cost of steaks and food exorbitant but all had a good time.")

When he did return to Lockheed, Kelly eased back into work. He did some partial days and attended meetings of the board, but through early June he was at Star Lane as much as he was in Burbank. He returned full-time on June 11 and got immediately on a plane to Washington—with Maryellen in tow—to talk to the Navy.

Kelly had neither the drive nor the stamina to lead the Skunk Works in the way he had since it was born—especially in this new era, with a Pentagon run by practices he couldn't abide. He was gradually phasing himself out, to spend more time on his marriage and his legacy. After staying mostly silent for decades, Kelly started to say yes to appearances.

In September, he and Maryellen took off for Fiji, for a short vacation en route to Sydney, where he was to deliver the prestigious Sir Charles Kingsford Smith Memorial Lecture as part of an eleven-day tour of Australia.

Kelly being Kelly, he couldn't help himself. If there was a way to help the Skunk Works—or Lockheed—on the trip, he was going to try. He pitched the L-1011 passenger jet hard, over lunch in Sydney with Qantas executives, only to be told the company was happy with its fleet of 747s.

He used the Kingsford Smith lecture, on September 19, as another chance to push the company line. The topic was long-range flying over the past forty years, and he made sure to talk up the L-1011 there, too, as well as during his press conference and TV appearance the next day.

Frankly, there wasn't a lot to feel good about back home at the Skunk Works. By this point, the Vietnam War was winding down, which meant that the fleet of SR-71s based at Kadena were coming home to Beale Air Force Base, having flown more than six hundred combat missions, including one very special operation to help buoy American hostages at the Hanoi Hilton.

A message had been sent to POWs at the infamous camp that, given a signal, they should make a break for it and run to a nearby river, where Navy SEALs would be waiting to pick them up. That signal was a special kind of "thunder"—produced by a sonic boom from a Blackbird.

Planners asked the SR-71 detachment if it was possible to arrive at a specific point at a specific time and deliver two sonic booms, as close together as possible. This was possible. So on May 2, 1972, two SR-71s took flight and flew toward the Hanoi Hilton with plans to crisscross their flight paths over that base thirty seconds apart—a very tight and hairy window considering that the two planes would be closing on the same airspace at a combined speed of Mach 6.

Pilots in those two birds didn't know what, exactly, they were doing, but they took off from Kadena, refueled over Thailand, headed north along the China border, and then made their high-speed passes over the preassigned point just thirty-two seconds apart. Two days later, a set of Blackbirds did it again, and shaved a half second off the time between booms.

Later, they looked at the film under that spot and spotted a POW camp, but they didn't learn the actual purpose of that mission until years later, when retired SEAL Kevin Dockery wrote about it in his book *Operation Thunderhead.*

This story has a tragic postscript: The POWs never did escape, and the botched mission to save them—Operation Thunderhead—resulted in the last death of a Navy SEAL in Vietnam. His name was Melvin Spence Dry.

Its impressive missions notwithstanding, the Blackbird was basically an afterthought by this point. It felt, increasingly, like an exotic luxury—a phenomenal piece of technology that no longer justified its cost. As a result, the Blackbird's budget was to be cut by $11.8 million in 1974, which followed $7.2 million in cuts for 1973.

This left enough money for just one squadron of twelve planes.

The Blackbird's fan club wasn't going away quietly. Colonel Kenneth Collins, vice commander of that last SR-71 unit, at Beale, told an *Aerospace Daily* reporter at the annual AIAA (American Institute of Aeronautics and Astronautics) meeting, in St. Louis, that the plane was "far superior" to the satellites that were increasingly relied on to replace it. These satellites flew over a target once per day.

"We can position ourselves over a given coordinate at any given time"—as demonstrated by that incredible POW sonic boom exercise. "We are timely," Collins said. "It is the most complete tactical reconnaissance system around." And the SR-71 also brought its spy photos immediately back to base for handling.

Kelly was at that meeting, too, and when a reporter asked for his thoughts, he unearthed an old skeleton, because the thing he prophesied had come true. "I would like to have seen an interceptor," Kelly told the reporter. "The threat which the YF-12 was designed to counteract is now here, namely the Russian backfire bomber."

Funny thing, too. Kelly *still* wasn't done tinkering with his baby, the most famous airplane he'd ever design—arguably the most iconic aircraft in history. That October, he went to the still fairly nascent Defense Advanced Research Projects Agency (DARPA) with a clever idea—another way to convert the Blackbird into an offensive weapon. He told DARPA's director, Stephen Lukasik, that he could convert six of the planes into a "national crises control force"—basically, a supersonic plane that could, in a pinch, deliver a devastating blow that wasn't nuclear: a "streamlined iron bomb with a guidance system," for instance.

This idea, of a "non-nuclear clean kill," was apparently appealing enough to get him, and the Skunk Works, a small contract to do a study. But the program never went into production.

Much is known about Kelly's big projects, the planes that started on a drafting table and ended up in the air. But smaller jobs—where he served as a sort of guest brain—were always a part of his purview, too.

Kelly's mere presence inside the Lockheed family meant that he never knew when he might be summoned for help, even to pitch in on things where he had little if any expertise.

Like missiles. The Air Force and the CIA asked for Kelly's help with a program at one of Lockheed's newest divisions—Space and Missiles. The program, under Air Force control, was in trouble. And Kelly was asked, he recalled, "to get the program in hand."

He replaced the program's head, built a mini Skunk Works up in the Bay Area, and set about identifying the friction points. Like, that the program had 1,200 people in quality control for a rocket with an average success rate of 13.8 percent. In one case, he found that a subcontractor had 35 employees building the missile's horizon sensor—with 60 Lockheed QC managers overseeing them.

It was the opposite of the Skunk Works "K.I.S.S." dictum. This outpost was slow and bloated. "Fire everybody else, and let them build it," was the thrust of Kelly's advice.

He insisted, instead, that the Air Force adopt a very different model, more like the one they'd used with him on the SR-71. "By using our Skunk Works methods we built the first twelve for one-half the time, one-half the cost, one-tenth of the drawings."

And then there was the infamous Project AZORIAN. In 1968, the Soviet ballistic missile submarine K-129 went missing in the Pacific and the Russians never found it. The United States, however, knew exactly where it was—thanks to the Air Force's highly sensitive, seafloor-mounted AFTAC (Air Force Technical Applications Center) acoustic sensor system—and in 1970, an audacious top secret program was commenced by the CIA to recover that sub and plunder its secrets.

Using a bold cover story—that Kelly's old friend Howard Hughes was planning to mine the ocean floor—a huge ship was commissioned. The so-called *Hughes Glomar Explorer* was outfitted with an enormous claw that, dropped through a set of "sea gates" in the ship's bottom, would descend on a steel pipe-string and grab and lift the sub from the ocean floor back up into the ship, without anyone on the surface having any idea what was happening below the hull.

Lockheed got the job of making that giant steel claw, but again, it was assigned to the Bay Area division, Missiles and Space.

In this case, there were issues with the claw's design. The operation was highly classified; the CIA still won't acknowledge much at all about what happened on the program, and no record of Kelly's small role in AZORIAN has ever been released. But some stories have leaked out.

We know that he sent his best metal guy, Henry Combs, to the Bay Area to help fix what Kelly saw as a disaster in the making—a giant claw that wasn't up to the job. The project gets a brief mention in *Kelly: More Than My Share of It All*, where Kelly notes that "skimping on static testing of the remotely-controlled arms—failure to conduct one last test before the retrieval attempt—resulted in less than 100 percent success."

The only other thing Kelly ever said about his role was a short answer given to a writer who sat him down for an oral history in retirement.

"I spent some time up there drafting it, helping with the controls," Kelly said. He had no further comment.

It was increasingly clear to everyone that this Lockheed legend was a lion in winter. And every month, his exit seemed more imminent. Especially given what was happening at home.

In the early months of 1974, Maryellen was hospitalized for complications of her diabetes, while Kelly got pneumonia for the third time that winter. His bosses, he wrote in his ranch log, were again "worried" about how healthy it was for Kelly to keep working.

Maryellen was too ill to attend the annual cattle roundup (eighty-two calves branded!), and Kelly was too worried about her to drink or play poker at the party that ensued. She was in and out of the hospital all spring, with low blood pressure and other complications, but still a joy when healthy. "She is getting better in golf, but gets disgusted on bum shots. Won't be long before she will lick me! Has beautiful swing—tires easily."

The state of his wife's health made it difficult for Kelly to keep focus. Her problems, he wrote, "make it advisable for me to quit about Jan 15, 75"—though Kelly was "not enthusiastic" about the decision and waffled on it from week to week.

It wasn't just Kelly. A historic era was winding down at Lockheed as well.

On March 11, Willis Hawkins, who'd been with Kelly since his earliest days, announced his retirement from Lockheed, and from his position as senior vice president of science and engineering. Two months later, in May, Tony LeVier—the legendary test pilot, who'd risen to assistant director of flight operations—followed him out the door, having served under Kelly Johnson in some capacity since 1941, on the Hudson bomber.

The SR-71, meanwhile, was on its way to becoming a museum piece, but before Kelly's greatest design was fully retired, the U.S. government had a few final jobs for it.

Arizona senator Barry Goldwater proposed and got Vice President Gerald Ford's support for a special flight—to try to take back the world closed-course speed record from a Soviet plane, the MiG-25 Foxbat—prior to the Blackbird's first-ever appearance at a public air show, that September in Farnborough, UK.

This highly specific tool still occasionally came in handy, too. That spring, SRs flying round-trip missions from the U.S. East Coast (which refueled in midair) played an important role in the Yom Kippur War. Blackbirds covered the battlefield from 80,000 feet and passed intelligence to the Israelis, who used that intel to help hit and stop Egypt's advancing Third Army.

In August, the Air Force surprised everyone by asking for $6.4 million to pull several Blackbirds out of storage and return them to flying status. And Kelly used the occasion to scold the Pentagon, one final time, for killing the program. He could not even restart production of the SR-71 if it was essential to America's survival. "Those teams we built up to build it are all scattered," he said during a seminar. "The suppliers have gone out of production on many of those very special items. . . . We've lost our chance. . . . Now the Russians have come up with the type of bomber the YF-12 was designed to counteract"—the Tupolev Backfire—and we haven't got a real threat against it."

PART V

A LION IN WINTER

35

PASSING THE TORCH

In the fall of 1974, as he approached his sixty-fifth birthday, Kelly Johnson asked his longtime protégé Ben Rich to come see him in his office. Kelly looked tired and discouraged. But also a little . . . relieved. He told Rich, then forty-nine, that he had submitted both his and Rus Daniell's names to Lockheed chairman Carl Kotchian as possible successors, but only as a formality. His pick to take over the Skunk Works was clear.

Ben Rich would carry Kelly's torch—and expand the Skunk Works legacy—into a new era.

"I've given Ben the toughest assignments and he's never let me down," Kelly had told Kotchian, expecting some pushback. "He won't let you down either. He's the future."

Kelly lobbied the other board members, too. Late one night, he called Lockheed's president, Roy Anderson, at home. He was, Anderson thought, a little tipsy, certainly loose from some cocktails.

"Goddamn it, Roy," Kelly told him. "I raised Ben in my own image. He loves the cutting edge as much as I do, but he knows the value of a buck and he's as practical as a goddamn screwdriver."

Message received. Kotchian and Anderson honored Kelly's pick.

They liked Rich, who had a brain like his mentor, but also a very different manner of leading. As Anderson said, Rich "told outrageous jokes and talked faster than a machine gun when he got excited about something, which was most of the time." He could also play politics and, unlike Kelly, knew when *not* to pick a fight.

Rich's elevation was assumed by many at Lockheed for some time, and certainly Kelly had long since made his mind up. Among other things, Rus Daniell was simply too old, he thought—just two years his junior. And Rich had the more visionary mind.

In some ways, Rich had already taken over key duties, especially in dealing with the Pentagon, where a new generation of leaders found the Skunk Works founder's towering ego and intractable nature grating. He was just way too difficult to deal with.

And when Rich went to see Carl Kotchian, the primary takeaway was just that: There will never be another Kelly Johnson, nor should there be. The point, going forward, was to do great work *and* get along with the blue suiters. "I understood, as did Kelly, that he was unique in his power and independence, which was nontransferable to any successor," Rich wrote in his memoir. "The Skunk Works was still expected to produce giant results, but the new guy sitting in the boss's chair would be a lot smaller than the original Gulliver."

Rich was a popular choice. He'd been at the Skunk Works since 1950, save for some brief assignments to other projects at Lockheed, and had been responsible for some of the biggest triumphs. Most important among them was designing those exotic engine inlets on the original A-12. He was also, unlike his longtime boss, a gregarious leader who didn't want to rule by fear alone.

Ben, according to Carl Anderson, was "more collegial" than Kelly and "more willing to compromise." He also tolerated, even respected, management, willingly including his bosses in major decisions.

Whereas Kelly's attitude, Anderson said, could be summed up as: "Goddamn it, if I tell you something that's it."

This much was clear, and had been for a while: The board was going to push Kelly into retirement, whether he was ready or not. But increasingly, he *was* ready. The way Kelly wanted to operate—the only way he knew *how* to operate—was no longer the coin of the realm. Gone were the days when he could just barge into conference rooms and tell the generals what they needed to buy.

And the Air Force's lightweight fighter project was proof.

The Skunk Works didn't just lose that important job. Kelly's open disgust during the process, especially over who would run the flight tests, had burned one of the last bridges still standing between Kelly's desk and the Pentagon.

The way he saw it was that his organization had always done its own flight tests, and he wouldn't be changing that now, after decades of the Skunk Works process working just fine. But the Air Force was no longer willing to hand over the responsibility for this, either.

Its pilots would test its new planes, and Lieutenant General Jim Stewart, who ran Aeronautical Systems at Wright-Patterson Air Force Base, was fed up.

Stewart called Ben Rich at home one night and told him that he was tired of Kelly and his demands. Until things changed, Lockheed would get no fighter contracts.

On January 14, 1975, Lockheed chairman Dan Haughton signed Kelly Johnson's official retirement papers, ending his run just before his sixty-fifth birthday and thirty-two years after Kelly's founding the Skunk Works under a circus tent in a parking lot, across the tracks from a very smelly plastics plant. Haughton lauded Kelly's service—to the company and the country—and noted that "we will certainly miss you, but not completely because you are going to continue to work with us on a part-time basis."

Kelly wasn't going away. He would stay as a "senior adviser," working 120 days a year for $24,000, plus $250 per day for every day he worked after that.

Haughton and the board had tried three times to get Kelly to run all of Lockheed, not just special projects or design. But he just was never interested. "I was much too smart for that," he said later, meaning that such a job would mean operating under someone else's rules. "I've been doing what I wanted to do since I was twelve."

Two weeks later, four hundred of Kelly's friends and co-workers, as well as numerous military officers and VIPs, convened at the Sportsman's Lodge, where Kelly had hosted many parties over the years, to mark his quasi-retirement.

There were eighteen presentations made that night, starting with a letter from President Gerald Ford, read by Carl Kotchian, and a variety of gifts were presented, including "a picture of a U-2 done in elegant needlepoint on cloth by the wife of a pilot," a table lamp made with an engine shaft pillar, and—the gift from his Skunk Works team—"a complete tool kit in a huge rollaway stand."

Generals, admirals, spooks, even competitors all paid tribute. And, according to one report, "as praise piled on praise from those who'd come the distance with him, it overwhelmed Johnson." When Kelly strode to the podium to take his turn at the mic, his voice was "muffled and cracking."

"In the Skunk Works and elsewhere, I've always stood on the shoulders of thousands—those who did the work I'm honored for," Kelly said. "And I've always accepted those honors in that spirit, that I am a symbol for them. That still holds true tonight. And for them, I thank you all. I can't say any more."

The Air Force also held a ceremony in his honor, at Wright-Patterson Air Force Base, in Dayton, Ohio—the same base where

Kelly had received the orders that led to the XP-80, and thus the Skunk Works, way back in 1943.

That event in Dayton drew such a crowd of top officers, according to *Time* magazine, that a bystander was overheard asking, "Who's back at the Pentagon running the shop?"

A general's response: "Even more would have come if they'd been able to get away."

Kelly had designed some of the most important and influential planes in history, but he also retired with more than 2,300 hours as a flight engineer. Kelly flew backseat on nine first flights, verifying the safety of his prototypes—which were often quite radical—in the most responsible way possible.

In the weeks and months that followed Kelly's retirement announcement, accolades rained down upon him. The bio in the program for his 1975 Wright Brothers Memorial Trophy—billed as "aviation's highest award"—practically gushes. It noted that Kelly has been called "the best aircraft designer in the history of aviation" and suggested that "his record is his spokesman." As well as: "Perhaps the late President Lyndon Johnson put it best as he was awarding Johnson the National Medal of Science in 1964—one of the highest awards this nation can give a civilian: 'Kelly Johnson and the products of his famous Skunk Works epitomize the highest and finest goal of our society, the goal of excellence. . . . His record of design achievement in aviation is both incomparable and virtually incredible. Any one of his many airplane designs would have honored any individual's career.'"

Ben Rich took over, officially, three days after Kelly's resignation—on January 17, 1975. That was the day he arrived at work and parked in his boss's old spot, just outside the windowless concrete bunker that had housed the Skunk Works and its staff of engineers, machinists, and shopworkers since 1962.

"Even our rivals would acknowledge that whoever ran the Skunk Works had the most prestigious job in aerospace," Rich later wrote. "Beginning with this mild day in January, that guy was me." He was fifty years old.

Rich was eager to take over but also, like his department heads, anxious about replacing a living legend. Of course, the senior engineers were also excited about what a new boss meant—less fear, more responsibility. "I was not a genius like Kelly," Rich wrote, and that's what he told his key staff. "I have no intention of trying to make all the decisions around here the way that Kelly always did."

Rich's style of leadership was to tell the staff what he wanted to see, and then to stand aside and let them execute.

Whereas Kelly, Willis Hawkins later wrote, "was intolerant of average performance and intellect" and was a "technical one-man show," Rich was "the antithesis of this." He "delegated completely, he was open to suggestion—almost to a fault—and he was quick with praise for any, even small, contribution."

The two men had similarities, too. Rich was also prone to dreaming big and did not want entrenched forces to stand in his way. His résumé, Lockheed chairman Dan Tellep would later note, included "legendary feats in challenging bureaucracies."

The main difference between Kelly and Ben Rich, however, was that you *could* actually change Rich's mind, sometimes.

Back in December 1972, Rich had written a memo pointing out the many strengths and some limitations of Kelly's operation, as observed over sixteen years in the Skunk Works. He ran through the greatest hits: strong leader, delegated to subordinates, paid close attention to costs, encouraged flat management and open communication, promoted collaborative environment, empowered individuals, and made quick decisions.

He also fretted about "two scourges that have hit our industry:

the IBM and the Xerox machines. The Xerox created paper. The IBM—as in the computer—made engineers lazy and blunted expertise.

There were fewer limitations, Rich said, but one was a problem he didn't know how to solve. "Kelly Johnson's great aeronautical genius tends to overshadow everyone else."

Kelly didn't steal credit, but his presence was so enormous that it was hard for others to get noticed. There was another, related problem. The company's relationships and contracts were almost entirely dependent on Kelly. He had the friendships and took the meetings.

"It tends to leave a void when he is not around." One way to fix this, Rich suggested, was for Kelly to "take others with him, even to only carry charts, so they will be given visibility." Kelly had built a historic organization; he could also quite easily tear it down with him, without really knowing it.

That's why Rich had begun to travel with Kelly on trips to Washington in his latter years as boss, serving as his travel buddy, assistant, and volunteer chart holder. He was there, alongside Kelly, when Defense Secretary McNamara basically ignored his final pitch to buy the Blackbird interceptor.

Mostly, the SecDef just ate his soup and salad, according to Rich, while also skimming some report. He "never once looked up until we were finished," Rich recalled. "Then he wiped his lips with a napkin and bid us good day."

Running the Skunk Works was a glamorous job, one of the best in the business, but this wasn't a great time to be assuming leadership of a defense contractor. The Vietnam War had ended, and its hangover was real. "Defense spending," Rich noted, "was about as low as military morale."

In 1969, the Skunk Works employed 6,000 people; on the day Rich took over, in 1975, it was down to just 1,500 employees.

Morale at Lockheed wasn't much better. A bribery scandal, followed by the L-1011 commercial jetliner debacle, nearly destroyed the company, and a few months before Ben Rich took over, Lockheed had barely avoided a hostile takeover by the Textron Corporation.

If the Skunk Works was going to survive, it needed Ben Rich to find new business, and fast. Fortunately, this played to his strengths. Whereas Kelly only grudgingly prostrated himself before the generals who purchased his wares, Rich enjoyed the dance. He was, in his own words, "ebullient, energetic, a perennial schmoozer, and a cheerleader with an endless supply of one-liners and farmer's daughter jokes."

Kelly Johnson's pitch was essentially the same every time: *Of course I can do it. It will be great. Now leave me alone.* And Ben Rich had the same confidence in the Skunk Works' ability to make magic. But he knew that the organization's reputation alone wasn't enough to win over this latest generation of DOD bean counters.

From the jump, Rich was forced to endure more bureaucracy than his mentor. There was simply no way around it, which isn't to say he was happy to have more rules or oversight. The problem, Rich often said, was that if you give someone a job, he will do it. And he'll use all the time you give him, too.

Rich liked to share the example of a "gun mod" contract he'd once presented to Kelly, who asked him how long it would take. Rich said nine weeks. Johnson told him he had six. The job was finished in seven.

Ben Rich's first task as the new boss of ADP was to find business. And he had an idea. He flew east and told the Air Force chief of staff, General David Jones, that it was time to restart production of the U-2, which was still flying—a "mainstay of our airborne reconnaissance efforts," Rich pointed out—more than twenty-five years after its first flight.

The idea wasn't just to build more U-2s; it was to build a newer, better version, with a stronger engine and more advanced avionics.

This wasn't some Hail Mary, either. Rich had been agitating for years to update a plane that still had tremendous value, and which, because of regular updates but no wholesale redesign of the airframe, had swelled from 17,000 to 40,000 pounds.

General Jones liked the idea, but Pentagon bloat was real. It took two years before Lockheed finally had a contract to build the new plane, which the Air Force renamed TR-1 for Tactical Reconnaissance.

Rich had other ideas, too. One in particular had been circulating around Burbank. It, he thought, had the potential to alter the course of the Skunk Works, and aviation: a technology, he later wrote, that was "a gift from the gods assigned to take care of beleaguered executives"—a technology that would produce "the most significant advance in military aviation since jet engines, while rendering null and void the enormous 300-billion-ruble investment the Soviets had made in missile and radar defenses over the years.

"No matter how potent their missiles or powerful their radar," he said, "they could not shoot down what they could not see."

36

LIFE AFTER KELLY

Sometime in the early 1970s, mathematician and radar expert Denys Overholser came across a forgotten science paper that would spark him to reinvent the entire concept of stealth.

Kelly Johnson's shop had been pioneering stealth since the U-2, when it first strung all those beads and wires across the bird in hopes of foiling the Soviet Union's rapidly improving radar installations.

The Skunks made great leaps in stealth on the A-12, with its sharp angles and radar-reflective materials, and then again with the D-21 drone. But war planners in Washington were increasingly panicked about the viability of using even the most advanced fighters in combat.

And the Yom Kippur War proved that they were right to be worried. Over eighteen days in 1973, Israel lost an astounding 109 planes as radar-guided missile batteries and antiaircraft guns—given by Russia to Egypt and Syria—picked the Israeli jets out of the sky.

A huge portion of Israel's air force—the same planes, in most cases, also flown by the United States—were erased in less than three weeks. And watching it happen sent shock waves through the Pentagon.

What the U.S. Air Force wanted, more than anything, was to make its jets invisible to radar. That urgency was widely felt throughout the American aerospace industry, and the Lockheed Skunk Works had a secret weapon in Overholser, a thirty-six-year-old ex–college wrestler with degrees in mathematics and electrical engineering.

Overholser was newly arrived from Oregon State when Kelly had asked him to help lower the radar signature—the RCS—of the D-21 drone, and for years after that, he worked long hours writing code to predict the radar signature of certain shapes in the hopes of finding a way to make flying machines all but invisible to radars.

He scoured the world for knowledge and found, in one obscure Russian paper, the key that unlocked stealth for Lockheed and America.

The paper, published in 1966 by Russian scientist Pyotr Ufimtsev—and, for reasons that remain unclear, never classified by the Soviets—was titled "Method of Edge Waves in a Physical Theory of Diffraction," and it was a dense read for laymen. As Ben Rich later recalled it, "The paper was so obtuse and impenetrable that only a nerd's nerd would have waded through it all."

Rich is sometimes called the father of stealth because of the incredible plane that this Russian math paper would enable, but Overholser, Skunk Works engineer Larry Dilger once said, "was the Wozniak"—referencing the brainy but lesser-known engineer who designed the tech that allowed Steve Jobs to build Apple computers.

The reason Lockheed was making only incremental improvements with stealth, Overholser realized, was that it was trying to improve a plane's design without first considering its shape.

But the solution *was* its shape. The way to make a plane nearly invisible was to make it faceted—to break its shape up into plates and connect them in a way that reflected radar away from its source.

One way to imagine this is to think of a mirror versus a prism.

Both are made of mirrored glass, but when you look in a mirror, your image is reflected straight back to you. If you shine a light at a prism, the facets on that prism scatter the light in many directions; only a small portion of it returns to the source.

If you could build a plane comprised of facets, most of the radar that hits it will be scattered away; only a small portion will reflect back—ideally so little that it's lost in the noise.

Overholser was convinced that the Skunk Works could apply this design philosophy to an aerodynamic shape, but when he presented that idea to the aerodynamicists, they scoffed at the idea. No plane shaped like this would never fly.

A challenge was laid.

Overholser worked on a Cray computer with an octogenarian math whiz named Bill Schroeder, who came out of retirement to work alongside his protégé to develop a computer code, called Echo 1, that used math to evaluate the radar return from different objects.

Their work revealed that a plane's shape, if you wanted it to be invisible to radars, simply couldn't be aerodynamic. It could be reduced, using math, to thousands of two-dimensional triangular shapes that could be analyzed by the computer. And that data could then be used to create an actual plane out of an assemblage of flat panels. They called it "faceting."

The designs Overholser was circling around flat-out scared the aerodynamicists, who looked at some of these shapes and told the brainy science guy that it was impossible to make a jagged box fly.

But Overholser was certain, and he convinced the only person who mattered to believe in him: Ben Rich.

Rich was convinced enough to approach the Air Force, which, in conjunction with DARPA, asked the Skunk Works to study faceting a little more closely.

They code-named the project Have Blue.

Now Overholser had to think a bit more practically. He needed to figure out the best shape for reducing a plane's visibility to radar that could—in theory—*actually be built.*

The Skunk Works could then construct a three-dimensional model, stick it on a pole, and run tests to see how invisible it was (or wasn't).

Ben Rich was still early in his tenure, and he continued to lean on his own mentor for counsel—especially because Kelly Johnson's contract had him in the office twice a week.

And when Kelly caught wind of this wacky new stealth project, he scoffed. For one thing, he told Rich, manned combat planes were on the way out. The future was in missiles and drones. Why not, he wondered, just revive the 40-foot D-21 drone—which had the lowest RCS of any craft ever built by the Skunk Works—and turn it into a pilotless attack vehicle?

Kelly hated everything about Have Blue. This was the man who saw air; it was anathema to him to rely on computers to design a thing he still viewed as art. It was, also, in his mind, a huge gamble for his protégé—to bet the farm on this iconoclastic concept.

According to author Walter Boyne, Kelly warned Rich "that he should not attempt such a high-risk venture as his first project" because "his reputation would be riding on it."

In short, when it came to his protégé, Kelly doubted his own maxim: Assume it can be done.

Less than half a year into his new job, Rich looked up to see Warren Gilmour, the Skunk Works' Soviet weapons specialist, walking into his office with purpose. A friend of his at the Air Force had let on that DARPA was giving $1 million each to five companies to present a stealth fighter concept. Lockheed was not one of those five.

The company hadn't designed a new fighter since the Korean

War. More important, almost no one at the Pentagon knew about Lockheed's long history of working in stealth because it had all occurred on highly classified spy plane projects.

We need to get into this competition, Rich said. But he also understood that the only way to make that pitch was to share Lockheed's prior work, and to do that required the CIA to declassify the results of its radar signature work on the Blackbird. This felt like a long shot.

To improve his chances, Rich asked his former boss to put in the request with the Agency.

So, Kelly wrote up a history of stealth at the Skunk Works and sent a letter to the CIA's deputy director, Carl Duckett, in July 1975, asking if the Agency was okay with Lockheed sharing its prior work for this important new project.

Duckett said yes. Lockheed would need to maintain the proper security classifications but, given that, he said, the CIA could drop "any special Agency controls" and the company could share its prior work.

The CIA declassified all the Blackbird's stealth data, and Rich sent it immediately to the head of DARPA, who replied that this was all great but it was too late. The money allocated for this study had been spent.

Rich knew that if he could get Lockheed into the competition, the chances of winning were very good because only he had Overholser's Echo code, which could compute changes to any design overnight. This was a tremendous advantage.

He begged Lockheed president Larry Kitchen for permission to spend $1 million of his own development money on the project, to allow the Skunk Works to catch up. It was not an ideal time to be pitching radical projects to Lockheed's management, let alone asking for money to do it. But Lockheed also needed some wins, badly. So Kitchen grumbled about the cost but ultimately agreed.

On May 5, 1975, Overholser presented Rich with a sketch of what he had determined to be the most stealthy design—"a diamond beveled in four directions, creating in essence four triangles," is how Rich described it. "Viewed from above, the design closely resembled an Indian arrowhead."

Rich wanted to know what the math said. Just how small was the RCS of this bizarre design? Was it as small as a Piper Cub, a condor, an eagle?

"Try as big as an eagle's eyeball," Overholser replied.

Most engineers, especially the old-timers, were skeptical. Hydraulic specialist Dave Robertson nicknamed the design "a flying engagement ring," while Ed Martin took to calling it "Rich's folly." And Kelly Johnson, not surprisingly, hated it the most. He called the early designs, which looked like rejected versions of Darth Vader's helmet, the Hopeless Diamond, and bet Rich a quarter that this monstrosity of a plane would never beat the RCS of his D-21 drone.

Rich took the bet. He asked his team to build a quick wood model of the new design and tested it, along with the original D-21 model, in the shop's anechoic chamber. This, he later recalled, was September 14, 1975—"a date etched forever in my memory because it was about the only time I ever won a quarter from Kelly Johnson."

In fact, the results weren't even close. The Hopeless Diamond was a thousand times stealthier than Kelly Johnson's D-21 drone.

"Pretty quickly," Overholser recalled, "I went from being seen as the village idiot to the village genius."

Ben Rich put Alan Brown, a propulsion engineer, in charge of stealth and named Norm Nelson—the CIA spy planted at the Skunk Works during OXCART to help stop the delays—as program manager.

Kelly wasn't completely wrong. Have Blue was about as aerodynamic as a paperweight. But there were ways to make even ungainly shapes fly—thanks to advances in computing and, especially, fly-by-wire technology.

Fly-by-wire simply means that an onboard computer takes over and makes the inputs a human would previously have made but can't possibly make quickly enough. Think of skid control or antilock brakes in a modern car.

This was the only way a plane with Have Blue's bizarre shape could ever fly, by handing most controls to the computer, which also meant that if the plane's computer was to fail, well, then the pilot was basically useless. His only option was to hit the eject button and pray his chute worked.

When it came time to test the model in the wild, Lockheed went to the nearest outdoor radar range, in the Mojave Desert. Engineers affixed the awkward wooden shape to the top of a 12-foot pole and had operators fire up the range's radar dish, positioned about 1,500 feet away.

Rich was in the control room, alongside the dish's operator, when the system came on. The operator looked puzzled. He suggested that the model must have fallen off, and Rich assured him it had not. Then—as he tells it in his memoir—a large bird landed on the model, at which point the operator spotted the object on his screen.

"I wasn't about to tell him he was zapping a crow," Rich wrote. "His radar wasn't picking up our model at all."

DARPA selected Lockheed's design, along with one from Northrop, and gave the two companies $1.5 million and four months to prepare a larger and more refined scale model. The two designs would then be tested, head-to-head, using the world's most sensitive radars. May the most invisible plane win.

In March 1976, the two companies commenced testing. They alternated days and never saw each other's designs.

Five different radars, all with different frequencies, targeted the models, and on the first day operators were shocked to see that the pole itself had a bigger RCS than Lockheed's design.

Rich was in Burbank, anxiously following from afar. But his spirits soared when Overholser reported back that his design had the approximate RCS of a golf ball.

On certain days, when the heat was most extreme, it warped the results, making Lockheed's design even smaller, and that data got back to Rich, who asked Overholser to tell him what size ball bearing was closest to that result.

This was just like Rich, to embellish things when it was expedient. And on his next trip to the Pentagon to see the generals who would ultimately approve a new fighter contract, he used those ball bearings as a prop.

"Here's your airplane," Rich told the group, and rolled the bearings across the table.

"In the end, we creamed [Northrop]," Overholser later said. The Hopeless Diamond was ten times less observable and had the lowest radar cross section ever measured.

Just as important, the radar test range results matched the predictions his computer program had made exactly, giving Overholser confidence that he could accurately predict RCS for any shape he punched into the magic box.

Lockheed got a green light to build two prototype planes, and Have Blue moved into the black. It was classified to the greatest extent possible and shifted from DARPA, which is staffed mostly by civilians, to the Air Force's Special Projects Office, to minimize the risk of leaks.

The Have Blue contract very likely saved the Skunk Works, and the program would ultimately define Ben Rich's career more than anything he did under Kelly Johnson, but it came with strings.

For instance, the Air Force insisted on making the program "Top

Secret—Special Access Required," which is the highest possible tier, what the Manhattan Project was.

This status ruthlessly limits the number of humans who can be briefed on a program and requires extensive compartmentalization, which means that the vast majority of people working on some particular piece of the project have only a myopic view of it. They may not even know what the whole plane looks like.

When Kelly heard about this demand, he told Ben to refuse it. Abiding by such restrictions, Kelly told his protégé, would raise costs by 25 percent while lowering efficiency.

Rich didn't disagree, but he also wasn't in a position to overrule the Air Force and ask them to just trust him, as Kelly would have. Not on a project this important.

So Rich jumped through the Pentagon's hoops and submitted a security plan, pointing out that the Skunk Works already *was* a black operation. The engineering office had no windows and thick walls, with a "special bank vault conference room, lined with lead and steel" for the most sensitive conversations. Secrecy was baked in.

Among the rules forced upon Rich by the Air Force was one that no single employee could ever be alone with a blueprint. If two men were in a room, and one needed to pee, the other had to lock the blueprints up in a safe until the one came back from the bathroom.

But the challenge of making a near-invisible plane was far from solved. The Have Blue model was static. It was, as a wooden sculpture, almost invisible, but the actual plane would have a lot of things that also cause reflections on radar—like cockpit glass, engine, exhaust, flaps, and landing gear doors.

Rich put Ed Baldwin, one of his oldest and most experienced engineers, in charge of design. Baldy had been at the Skunk Works for decades. He made the original drawings of the U-2 and was the lead designer on the A-12.

And he was decades older than many of his engineers, who were inevitably surprised to see that this Skunk Works legend they reported to did not have an office. He insisted on being right there in the engineering design room, working on the same drafting tables they used, and though Baldy could be salty, like Kelly, he also let talented people do the work they were hired to do.

Have Blue was Baldwin's greatest challenge yet. He had to take a radical static model—a design that looked to anyone who knew planes like a shape that couldn't possibly fly—and turn it into a plane that *could* get into the air and fly around.

37

OLD HABITS DIE HARD

Of course, even after the announcement and the farewell tour of events where peers showered praise upon him, Kelly Johnson didn't fully retire. He just couldn't. Kelly still showed up in Burbank, at least a couple of times a week, ensuring that Ben Rich had a skeptical voice in his ear. But Kelly had more time for other worries, too.

There was the ranch, with its ever-growing list of repairs and chores—oats to sow, cows to brand, roads to patch. "They sure go to hell in a hurry if I'm not here," Kelly wrote. Plus, his favorite horse, Wildfire, was on his last legs. "Today Wildfire came to me and said 'goodbye' as clearly as a horse can," he wrote in 1975, in a long passage that reveals Kelly's sensitive side as much as anything you'll find in the world.

The horse was, he figured, twenty-two or twenty-three years old, which is a pretty good run for a horse, but Wildfire had grown very thin in his dotage. "Followed me to corral gate, put his head on my shoulder & nickered," Kelly wrote, in a passage that's almost hard to imagine, given his gruff, fiery character at work. "I petted him, kissed him and thanked him for all my rides. . . . Petted him some more & *cried.*"

Kelly's biggest worry, by far, was Maryellen's health, which drained his time and spirit. The second Mrs. Johnson rarely made it to the ranch anymore, so Kelly's trips were shorter and lonelier. He was there, alone, on Christmas, to see some friends, ride Wildfire one last time, and take the ranch foreman's family to church.

"I have come to the place where I really understand people—it only took me 50 years," he wrote in the log, in an entry marked as 8:50 p.m. that Christmas night, by which time he surely would have had a few cocktails. "Maybe being good God will help MEJ with her health & sight!"

This entry stands out. It is one of the few, over all his decades of writing, where Kelly mentions religion. He was, it seems, seeking out a little extra strength, or hope, or something. "I will try to be better & talk to Him as I have for years," he wrote. "Please help my Sweetheart & our friends."

Maryellen's diabetes simply wasn't improving; she'd begun a pretty clear downward slide. Kelly had taken his wife to the Scripps Clinic, near San Diego, multiple times, including for laser treatments on her eyes, in the hopes of saving her vision.

But things had gotten bad enough that Kelly wasn't comfortable traveling outside California and leaving her behind. He canceled a trip to Israel to stay home, and then noted in the ranch log that this was probably a good thing, since the region was "too unsettled politically."

Most worrying were Maryellen's kidneys, which were failing. For five months starting in June 1975, she spent four to five hours, every other day, on a dialysis machine. Which was only a temporary solution. Ultimately, she'd need a new kidney.

That October, surgeons at USC's Renal Transplant Unit replaced the failing kidney with a healthy one donated by her sister, Irene. But Maryellen's vision had deteriorated so much that Kelly had to design some quality-of-life aids, like the "closed circuit TV machine" he installed at the Encino house so she could read.

This was no off-the-shelf item. When some engineers at the Skunk Works who'd worked on the spy camera for the Blackbird heard about Mrs. Johnson's problem, they "rigged up a special magnifying lens and light to help her to read in bed and presented it to Kelly as a gift, just to show their affection for him," said Ben Rich.

Kelly's writing in the log during this period was alternately dejected and upbeat. "MEJ improving but has problems—sight, stiff leg, uncertain balance of insulin etc. Irene also still sick but improving. I spend most of my time in hospitals or on freeways."

The next April, Maryellen was admitted to St. Joseph's for surgery, after a bad fall in the bathroom that also broke her nose. The operation went fine, but on April 21, she had a stroke and was back in the hospital, spending a week in the ICU.

When he could find time, Kelly still drove out to the ranch, for short solo trips. Maryellen missed the 1976 roundup, held on May 16—a typically festive event that Ben Rich and his wife, Faye, also attended, as did friends Billy and Nancy Horrigan. (Nancy would soon become a very important figure in Kelly's life.)

On that day—May 16—the future of Star Lane—"this beautiful place"—was on his mind. Kelly had talked often of moving to the ranch full-time, and if Maryellen had been healthy, the couple may well have done that already. But it was just too remote for their situation—too far from doctors and hospitals.

He even broached the issue of selling, again. "When I mentioned this to Sweetheart she said 'No, you love it too much—after all—it's only 7 months since the operation. I'm going to be better!!!' Such a wonderful, brave, darling—The Lord has blessed me two of the best, most courageous people in the world—I will try to equal them."

Whether or not she believed, there was little hope of actual improvement. By August, Maryellen was in "desperate trouble." Her body was rejecting the donor kidney and Kelly was forced to take her to USC's renal care unit for dialysis every day. At home, she couldn't even brush her own teeth anymore. "Poor sweetheart," Kelly wrote. "I help all I can. . . . She is greatly depressed!!! Who wouldn't be?"

Alone at the ranch for a day on August 13, Kelly wrote, for the first time, about what seemed increasingly possible, even likely. "I expect to lose her soon. Yet she stays cheerful & unbelievably brave. . . . I willingly spend all my time with S.H. who is in so much pain."

Surprisingly, Kelly seemed to feel okay himself. He battled fatigue and a sporadically recurring irregular heartbeat but assessed his own health as "good to fair," which was a marked improvement from the recent past, when heart issues had caused him to go in for fibrillation on two separate occasions.

A medical checkup affirmed this. Kelly's health, according to his new physician, was "very good." The patient, he reported, "leads a very active life including much physical work on his ranch up until last 4 months"—when Maryellen's health began to really slip.

Kelly's weight, at 202, was up a bit from where the doctors wanted him to be—193 to 195 pounds—but prior to the 1970 stomach surgery Kelly had weighed 220. His eating habits were "quite good" and his drinking was moderate, at least by his own standards. Kelly allowed himself "just" 6 to 10 ounces of vodka before dinner every night and ate immediately after cocktail hour. Also, there was no more day drinking "except at lunch in Washington DC where it is a *forced* operation."

Kelly was proud of this newfound moderation, and told the doctor that his "father, uncle and grandfather drank a similar amount till their deaths at 84 and 94 years of age respectively."

Finally, there was the matter of sleep. Kelly had relied on Seconal for years to help him sleep—to overcome a mix of anxiety, leg cramps, and his irregular heartbeat—but since retirement he'd cut his intake from five tablets a night to two.

This medical report also reveals something about the way that Kelly had managed the often crushing stress of his job all those years. He had apparently "lived on" the prescription anxiety medicine Miltown when working full-time at the Skunk Works—to the tune of a shocking eight tablets per day. In retirement, however, he'd curbed the habit and was down to, at most, one pill a day.

Still, Kelly's respite from the stress at home was work. He was spending three or four half days a week at Lockheed, and his brain remained on loan to the Pentagon.

In August, he flew to Washington to take part in a strategic analysis of the Soviet Union's new Tupolev Tu-22M Backfire, the kind of supersonic long-range bomber that he had been predicting since the '60s. This was the kind of plane Kelly had wanted the Blackbird interceptor to defend against—a supersonic bomber that could get into range of American territory before the Air Force could do anything about it.

Fortunately, the Backfire was not, in practice, the grave threat to the continental United States that Kelly had been fearing. Its performance, he thought, had been overestimated. And he made this clear in a letter to "My dear Mr. Secretary" Henry Kissinger, having read an alarmist letter that Kissinger sent to *Aviation Week*.

In Kelly's professional opinion, the Backfire was a "mediocre aircraft" as an intercontinental bomber, and stories were already leaking out of Russia that pilots were struggling to fly it, especially on longer routes.

So at least in this version, the Tu-22M was not a significant threat to the continental United States. It did, however, pose a risk to NATO countries and to U.S. Navy ships. So the United States should definitely not ignore it or the Soviet Union's capabilities.

That last point, especially, was critical. "The most important thing the Soviets have which poses a threat to the United States," Kelly wrote, "is a total dedication to producing advanced systems rather than just arguing about the cost and usefulness."

In other words, while the United States formed committees, revised budgets, and bounced proposals from department to department, the Soviets just set goals and commenced design. Which is admittedly a lot easier to do when you don't have to worry about how many pilots you kill in the process or how many votes you need to scrounge up to stay in office.

Kelly did seem to miss the action of work, but what appeared to concern him more was maintaining relevance—inside Lockheed and especially outside, in a defense establishment that was evolving and metastasizing into something he barely recognized.

And as much as he tried to contribute to the company and the national security establishment, he was increasingly anchored by his responsibilities at home, because Maryellen's health continued to slip. When Kelly did manage to bring his wife to the ranch, it was a production, requiring them to bring their maid, the three cats, and a wheelchair, because Maryellen was often too weak to stand or walk. This did not seem to dampen Kelly's spirits, even in his private writings, which read at times almost like a pep talk to himself. "MEJ making a tremendous effort to get well! What a courageous, wonderful, beautiful princess!!"

There's one thing of note in this particular entry, though. Kelly mentions, for the second time, a name that would become very important in his life, sooner than later: Nancy Horrigan: "No matter how she feels, every Tuesday I take her to the beauty parlor & Nancy Horrigan (another angel) takes her shopping for clothes."

38

REINVENTING, AGAIN

The Skunk Works may have had a new boss, but the stealth fighter prototype, like all the other planes that had made the place so legendary, would be built the same way, which is to say: fast and cheaply. And the pressure to do so wasn't just internal. The Air Force wanted the Have Blue test planes flying in fourteen months, so despite this being the most radical plane design in history, there was no time for engineers to reinvent things or design entire novel systems.

Instead, Ben Rich's team went shopping for avionics "right off the aviation version of the Kmart shelf," as he put it—some $3 million in parts from the Air Force inventory, including flight control actuators from the Grumman F-11, the flight control computer (and the pilot's seat) from the F-16, and the inertial navigation system from that trusty old workhorse, the B-52 bomber. It even had a cheap, plucked-from-storage landing gear that came with such limited braking capacity that the plane needed a giant drag chute (hidden in a compartment between the two tails) to be deployed in emergencies.

The stealth plane's engines were from the Air Force, too. Sort of. They came via a major named Jack Twigg, who'd been assigned to serve as the Have Blue expediter. Twigg went to see General Electric's jet engine group and convinced some friends to let him very quietly requisition six J85 engines bound for the Navy from the assembly line and then, according to Rich, "had them shipped in roundabout ways, so that nobody knew the Skunk Works was the final destination."

Burbank had grown and sprawled greatly since the early days of the Skunk Works. It was now dense with houses, restaurants, and shopping centers—plus a fast-growing airport that was directly abutting the Lockheed land.

Developing the most secret fighter in U.S. history on a parcel of central Burbank, even a large parcel, right next to a working airport where people arrive and depart daily, presents some challenges.

Engine tests took place outside, near the fence that divided Lockheed's lot from the Hollywood Burbank Airport runway. Security would park two flatbed trucks on either side of the engine, to block the view of anyone who happened to pass by, and draped a tarp over the top.

They'd then fire the engine up and run it at full power, which created quite a din, especially in a densely populated city, at night. Noise complaints from locals were not uncommon. But it also wasn't all that weird to hear engine noise coming from an airport.

The Have Blue program was as secret as anything the Skunk Works had ever done, but the place still followed Kelly Johnson's dictum that making things *look* secret only invited curiosity. So, even during Have Blue, the Skunk Works engineering office, in Building 82, wasn't secured in the way you might expect from watching spy movies. And the informality of this arrangement occasionally caused problems.

Like one day when the phone system crapped out. Someone called Pacific Bell, the local phone provider, and a technician was dispatched. This tech showed up at Lockheed and was sent to Building 82, where he was directed back to the engineering shop to check phone lines. But it hadn't occurred to anyone to sweep the room first. And a wooden mock-up of Have Blue—this incredibly weird-looking plane design—was just sitting out on a table.

The phone tech fixed the lines, and maybe didn't even see the model, but a little while after he left, someone in the shop noticed that it had been out in plain view.

Security was displeased. Brows were furrowed. And a solution was proposed. Call Pacific Bell, tell them the lines are still wonky, and the same tech needs to come back to fix his shoddy work.

When that technician returned, he was pulled aside and put in a room with security officers, who asked him an odd question: *Did you happen to notice anything . . . unusual on your visit?*

The tech considered this. *No, nothing odd—oh, wait. Do you guys mean that weird model thing, for the spaceship?*

Before he could be released, the tech was forced to sign an inadvertent disclosure form and sworn to secrecy. Say a word, and risk serious prison time.

Security concerns would plague Ben Rich throughout the F-117 program. He liked to say later that Kelly could never have tolerated the swollen bureaucracy imposed upon the Skunk Works by the Air Force during Have Blue. OSHA (Occupational Safety and Health Administration), the EPA (Environmental Protection Agency), the EEOC (Equal Employment Opportunity Commission)—an alphabet soup of agencies hounded Rich, and the Skunk Works, to meet an ever-growing list of regulations. But it was security, more than anything, that stressed Rich out.

Things got especially tense when what he referred to as a "disgruntled employee, bypassed for promotion," complained to a staff member at the House of Representatives that the Skunk Works had lost classified documents right around the time that a model toy company unveiled a "stealth bomber" that looked nothing like the actual thing under construction in Burbank.

Congress tried to order Rich to testify, only to have the Air Force intervene and quash the idea, for fear of what he might disclose; instead, Lockheed's Larry Kitchen was sent, and was "browbeaten unmercifully" before a subcommittee on procedures and practices.

That same committee also sent a contingent to Burbank for a spot inspection, as well as an audit of all security documents, dating back to the birth of the Skunk Works.

"I almost had a stroke," Rich later wrote. He raced immediately to Kelly Johnson's house, knowing that his former boss surely had materials at home, and grabbed up all the files he could find—"cartons of documents and blueprints and God knows what else, all stored in Kelly's garage."

Kelly never leaked a thing in his life, but he lived by his own rules, and one of those rules was that he didn't care what the rules were. He took things home all the time and never worried a minute about it.

Three months shy of two years after the first C-5A cargo plane left Burbank under the cover of night with a camo-painted Have Blue test mule in its belly, bound for the radar cross section (RCS) testing, another C-5A departed Burbank for Nevada for the world's first-ever stealth fighter flight.

Bill Park, Lockheed's test pilot, would be on the stick. Park was the Tony LeVier of his day, and his counterpart from the Air Force for flight testing was Ken Dyson.

Dyson had been happily flying F-15s at Edwards when he was selected out for a special assignment and sent to Burbank with only minimal information. He was to show up at Lockheed, enter at the first gate, and ask to see a guy named Norm Nelson on unclassified business. If asked who he worked for, he was to say "self-employed."

The guards who received Dyson directed him to an "old dirty white ramshackle building," as he later recalled it. He rang the bell, and Norm Nelson answered, then escorted Dyson into "his inner sanctums"—the Skunk Works engineering office, which he later remembered as an "old mostly open bay room with no windows" where engineers were working at tables.

Nelson briefed Dyson about Have Blue and the pilot was "completely dazzled" by what he saw and heard, despite learning that this incredible plane was also going to be "quite unstable" on both its pitch and yaw axes.

Thing is, Dyson loved a challenge. He took the job. Then he "slowly stopped showing up for work at Edwards and eventually didn't show up at work at all anymore." No one on base knew what had happened to him except for General Thomas Stafford, a former Apollo astronaut who oversaw the flight-test programs.

Dyson and Park spent hours and hours inside a Have Blue simulator at Lockheed in Burbank, prepping for flight. And when they got to Groom Lake—now a permanent installation with a robust staff—Dyson flew chase on Park's first flight, "tucked in on his wing, watching that thing."

A few flights later, it was Dyson's turn, and he flew the prototype, with Park flying chase.

As Skunk Works stealth ninja Alan Brown tells it, Area 51 had "the best radars and operators" anywhere. So the stealth fighter would be put to the test against America's all-star radar team, and the most dramatic telling of these tests comes in Ben Rich's memoir.

It's so dramatic, in fact, that Rich opens the book with it. The scene described by Rich is of a radar test "on the scorching Nevada desert" at Groom Lake—the so-called Ship Two prototype against a contingent of U.S. Marines manning a Hawk ground-to-air missile battery that would try to track, but not shoot down, the plane.

The Marines were there to try to lock onto the experimental plane as far out as possible—ideally fifty miles or more.

Rich was handing them an advantage. He gave the Marines the plane's flight plan, "which is like pointing my finger at a spot in the empty sky and saying, 'Aim right here,'" he wrote.

Rich was confident in his plane but also nervous. Plenty of people at the Pentagon doubted his promises. And if the plane failed this test here, the program would likely be killed.

These Marines were not cleared into the secrets of Have Blue. What they were told is that the plane had a special "box" in the nose that emitted powerful beams, reflecting radar away from the source.

Rich stood outside, with a single NCO who would provide visual verification that the Marines were not duped and that a plane had in fact flown over. The rest of the operators were in an air-conditioned mobile command van parked nearby. They had no line of sight on the plane's flight path.

Just after 8:00 a.m., Rich spotted the jet incoming. It was just a speck in the distance, closing fast. Neither the radar dish atop the van nor the missiles in their launcher moved at all as the plane closed in on the site and then screamed overhead.

Rich was absolutely tickled. He ran to the van to check on the operators, who didn't know yet that the plane had passed over. They were still staring at their screens, watching intently for the target. Finally, one of them spotted a blip incoming. It was big, and growing in size as the plane approached.

The operator announced that this looked to him like a T-38. His superior, Rich said, was almost smug at the notion that his allegedly stealthy plane had a huge signature.

But this *was* a T-38—the chase plane that followed Have Blue, for safety, on every flight. The stealth plane had long since flown past and was now circling back to land.

Just then, Rich writes, "the van door opens and the young sergeant steps into the dark coolness, still looking as if he hallucinated in the desert heat—seeing with his own eyes a strange diamond apparition that his missiles failed to lock onto.

"'Captain,'" he began, "'you won't believe this . . .'"

It was an auspicious start. But Have Blue was far from flawless. And one of its biggest problems was an aerodynamic flaw that caused it to land way too hard. On May 4, 1978, Bill Park experienced this when his plane slammed down and bent its right landing gear.

Park didn't know if the gear was broken or just stuck, so he pulled up, gained some altitude to give himself some space to work, and tried to free the gear by using a number of tricks.

None worked, and finally, with his plane damaged and nearly out of fuel, Park did what he had to do.

He reached down, grabbed the ejection seat ring between his legs, and yanked. The seat blasted out of the plane with such violence that Park was knocked unconscious when his head smacked the headrest.

His parachute deployed automatically, as designed, and a limp Park floated down and then smacked into the desert floor, breaking his leg and cracking a vertebra. He spent six weeks in intensive care.

This wasn't Park's first harrowing experience in a Skunk Works plane. He was one of Lockheed's greatest-ever test pilots, having flown everything from the Blackbird to Have Blue. And back in July 1964, he'd ejected from an A-12 at just 500 feet when his plane's hydraulics malfunctioned as he prepared to land.

That time, fate had Park's back. His chute deployed fully "just as his feet touched the floor" and he walked away without a scratch.

In July 1966, Park bailed out of an SR-71 again, this time flying

2,000 mph at 80,000 feet over the Pacific during the TAGBOARD drone test.

This was the infamous test when the D-21 drone crashed into the mother ship right after launch. That crash had killed his reconnaissance officer, Ray Torick, and ultimately the program.

The Have Blue prototype that Park crashed, Ship One, flew thirty-six times before disintegrating in the Nevada desert. It was Park's fourth ejection, and his last. It was time to stop pushing his luck. "I smile a lot because I am just happy to be here alive," Park said upon his retirement. "I believe that circumstances can occur that you cannot overcome no matter how good you are."

After Park's accident, it was left to Dyson alone to test Ship Two, the only surviving prototype. He flew sixty-five Have Blue tests over the next year, the first one only a couple months after Park's near-death experience, until Dyson crashed, too, in July 1979, destroying the second and last test plane.

Which was okay, because Have Blue had more than proved its worth by this point. Even the best ground-based radars could detect the plane only after it was too late, and the jet was well inside the minimum range for any surface-to-air missile on the planet. It was all but invisible to airborne radars, too.

The Hopeless Diamond, then, was exactly what Denys Overholser predicted. And it was time to put the world's stealthiest—and, in theory, deadliest—warplane into production.

39

HOLD ON, SWEETHEART

Even if Kelly Johnson had loved Have Blue, or if Ben Rich had desperately needed his help, it's possible he wouldn't have been able to contribute, at least not as he had on the other great planes in Skunk Works history. His attention in the latter part of the 1970s was increasingly devoted to Maryellen and her declining health.

Kelly's second marriage was born of unusual circumstances, but it served him well. He fell for Maryellen and loved to show her off. Having a younger wife was good for Kelly as he aged. Maryellen was active, which forced him to be more active, too.

They golfed often, and Kelly did that his own way, too. He liked to play fast and carry minimal gear. If you believe Ben Rich, Kelly would sometimes play a full round with only a six iron. Willis Hawkins remembered it as a two iron, saying that an efficiency nut like Kelly didn't see the logic in carrying a full bag of clubs.

Maryellen had been a positive influence on Kelly's health in other ways, too. She drank, but not heavily, which caused him to cut down. She disliked the way that Scotch smelled on her husband's breath, so he gave up the brown liquor and replaced it with vodka.

Kelly was much older than his new wife, but it was *her* health, surprisingly, that failed first. Maryellen lost her eyesight, then a kidney. Then, during a visit to Mayo Clinic, in Minneapolis, doctors discovered an infection in her toe, which is a precarious situation in diabetics. That toe was amputated, but it wasn't enough; surgeons subsequently amputated her leg below the knee and fitted her with a prosthetic. In a single year, she had five major operations.

This was life for Maryellen and Kelly. "For most of our marriage, it was one long disheartening struggle against a siege of failing health," he later wrote. Without sight, she was prone to falls. She got angina. And it was nearly impossible for either of them to correctly monitor her glucose. Low blood pressure was a constant issue.

But Maryellen was also clearly a fighter. The ranch logs of the late 1970s are filled with entries written on Kelly's short, solo trips to Star Lane in which he documents her small steps forward and larger stumbles back.

Walking was increasingly a struggle, even with her artificial leg and a cane. "She's very weak. Can hardly walk," Kelly wrote, and then, as was often the case, he punctuated the thought with a tribute—surely to raise his own spirits as much as anything. "Shows magnificent courage as always."

But this ranch log entry is notable for another reason. It's among the most revealing entries Kelly ever wrote, the rarest of looks into a very personal subject. Kelly never had children. Whether he was unwilling, or unable, or was just too busy and ran out of time, he'd reached age sixty-seven and had two wives, but no kids.

And yet, on this day in May 1977, he admitted something surprising to an audience of none. The subject of children, at least in that moment, was still unsettled in his mind.

"Yesterday had a call from adoption service—they have a child for us!!" he wrote, adding a second exclamation mark and noting that this call had come five years after the couple had first applied to the service.

But so much had changed in those five years. Especially Maryellen's health. "MEJ & I must discuss—it's tragic to be so late—the way of God is strange—is he trying to tell us something. Should I adopt the child anyway?"

It's one of only two times the subject appears in his writings, the other being in his memoir, where Kelly breezes by the matter, noting only that he and Althea never had kids, "but it was not from lack of desire." Whether Kelly kept trying for kids—or how seriously he was still considering fatherhood in his retirement—is a thing we'll never know. We just know that the Johnsons didn't adopt that child. Or any children.

Maybe the notion was lost in the daily stress of Maryellen's decline. "It's been just hell for both of us," Kelly wrote, noting that the "traumatic experience" of his wife's ongoing health troubles had caused her to ask for a psychiatrist.

He'd found a good one, who then went to work. "All kinds of probs came up!! I was and am amazed but will do my best to understand & adapt," he wrote in another revealing entry. "This is the time to really examine your life & objectives & see if you are really the one you have wanted & tried to be. As of tonite here at Star Lane—I have the firm belief that I have done everything a loving husband can & should do. I could give me a great big star for what I have done but I *must not*, for I do it out of love for her. MEJ will never know how much I do love her & the desperate battle I fight *for* her!! I accept her tantrums willingly & understand. But it seems she will never understand me!"

In the summer of 1978, Kelly took his wife to Mayo Clinic in Minneapolis for yet another transplant, this time her pancreas. He even rented an apartment so he could stay nearby during her recovery.

She was in poor spirits—understandably irritable at the diminished state of her life. She lashed out at Kelly, too, writing him a letter in which she "gave me hell," and then followed that up with a second letter, apologizing. "I deserved both," Kelly said.

When they got back to California, he rigged the back seat of his Cadillac into a bed so that Maryellen could sleep on the road, and the two resumed the old routine of going to Star Lane every weekend, at least for a while. "Need to be up here," he wrote. But even the improved bed hack didn't work for long.

Reading Kelly's logs, often written while alone on the ranch, you can feel the turmoil brewing in this publicly stoic man. Kelly wanted desperately to be at Star Lane, his retreat, but being there meant he was away from Maryellen and Lockheed. "God loaned us a part of Heaven for a while but Sweetheart can't use it & I don't know what to do without her!" A day later: "I love Star Lane but *must go home. She is very sick*. Can I come back??? Very important. God only loaned me Star Lane!!"

Heading into 1979—as Ben Rich's Have Blue program roared toward production—Kelly was still hanging on at Lockheed, working three days a week while also shuttling Maryellen to appointments and escaping, whenever possible, to Star Lane. "Don't know how much longer I want to but I have lots of drive etc. to contribute," he wrote. "Next year I must retire from Board of Directors. Just as well."

40

THE HOPELESS DIAMOND TAKES FLIGHT

Ben Rich knew that the Skunk Works after Kelly Johnson still had magic to sell. Of his mindset in 1978, he later wrote, "I was bouncing on pink clouds, enjoying the hosannas reserved either for angels or the head of a research and development outfit that produced a technology everyone wanted," that technology being stealth.

In one of Rich's favorite stories, he attributes his sale of the world's first stealth fighter to that set of ball bearings he'd taken to Washington and now carried in his pocket to any important meeting. He would sit across from a group of Air Force blue suiters and tell them that their absurdly expensive planes had the radar cross section of a Greyhound bus. But the RCS of his new plane, the super secret Have Blue? It's this, he'd say, and roll the ball bearings out onto the table.

What Rich meant was that, to a radar operator, the Skunk Works' new stealth plane would show up as a marble—so small that it was likely to be dismissed as a radar anomaly instead of a fighter-bomber with a belly full of smart bombs roaring inbound to obliterate some critical target.

Apparently, the pitch worked. The Air Force gave Lockheed a

contract to go into production on the stealth fighter a month before the Have Blue prototype even flew.

Rich was stunned. No one in aerospace, or defense, he said, could recall an occasion when the Air Force ignored its own cardinal rule of R & D: Fly it before you buy it. And this was in the new era of bloated bureaucracy—a Pentagon built on bricks of copy paper—that had scared the illustrious Kelly Johnson into retirement.

The Air Force expected Lockheed to have its first production test plane in the air by the summer of 1980. Specifically, by July, which is why the Skunk Works gave the first Have Blue test model the serial number 780 (as in 7/80). And it was clear to Rich by the spring of 1980 that this date looked shaky at best.

That summer, Rich would later say, was the low point of his life. Problems mounted, and every good day seemed to be followed by a bad one. But the worst of his troubles weren't at the office. Rich returned home from work one night, long after dark—as he often did—to find his wife, Faye, looking shaken.

Faye was fifty and had gone in for a routine physical, only to learn that she had a spot on her lung. A biopsy confirmed that the spot was cancerous, and shortly thereafter, a surgeon removed that lung.

Faye was out of the hospital and back home on August 1, but less than three weeks later, on August 18, she complained of feeling poorly while watching TV. Faye's breathing became labored, and she collapsed. Rich called 911 but it was too late. Faye, he later wrote, died in his arms from a massive heart attack.

Among those who comforted Rich was Kelly Johnson, whose own wife was perilously ill. And Maryellen—who'd been tight with Ben and Faye for decades—was devastated by the loss of her close friend.

Kelly surely would have advised Rich to do what he would do in such a time: dive back into the work. And that's what Rich did, not that he had a choice. This experimental stealth plane he'd promised the Air Force was behind schedule.

Rich returned to work and found a note from Deputy Program Manager Alan Brown on his desk, with his next birth date—June 18, 1981—scrawled on it. This was the date that the F-117 would fly, Brown told him. "The date is firm. In granite. Count on it."

Brown's prediction seemed, to Rich, unlikely. Ambitious at best. Especially since one particular problem—lowering the RCS on the plane's square tailpipe, which was made of nickel alloy in a special honeycomb design—was driving the engineers nuts.

This was brass tacks time and Rich truly embraced his mentor's philosophy that if you can just remove bureaucracy, miracles become possible and delays can be avoided. Like when engineering needed to whip up a new shield for the exhaust pipe but didn't have any steel handy, they just cut up a steel work cabinet in the shop. Or when Rich realized he needed a way to get pilots into the cockpit. Rather than design some novel ladder, purpose built for this craft using military specs—requiring time and piles of money—he went to a local hardware store and picked up a ladder. It worked fine.

And so, on June 18, 1981, Alan Brown's prediction came true. The first production F-117 took off. And the Have Blue program gave way to Senior Trend, the production program's new code name.

"She flew like a dream," Rich later wrote in a bit of revisionist history that's forgivable in retrospect. The test flight, by Lockheed test pilot Hal Farley, *was* a success, Farley later explained, in the sense that "we took off, and landed."

This was just thirty-one months after the Air Force had given Rich his green light, when Farley hopped into the flight-test model—known as Aircraft 780 or Ship One—and took off from Area 51.

Farley had flown in the Navy, and as a test pilot for Grumman, before Bill Park hand selected him for the program, after the injuries Park suffered in the prototype crash forced him to give up test flying himself.

Ship One was painted in desert camo to disguise its appearance;

this pattern would make it hard for anyone who happened to get a glimpse of the plane to discern its faceting, a precaution necessitated because these early flight tests would be conducted in daylight, for safety.

And it was the only stealth fighter to get that particular paint. Subsequent test models had matte gray paint, until the Air Force decided that production models should be black.

Everyone knew that the YF-117—as it had been designated for production—was unstable in pitch and yaw. It literally could not be flown by a human without the computers making thousands of small corrections per minute. But for Farley's first test, engineers worried about the probes on the nose that collected air data for the computers vibrating on takeoff. They asked Farley to take off without them.

Lead was added to the nose, for ballast, to make the plane more stable, and he was supposed to fly to 10,000 feet, then turn on the probes. But shortly after takeoff, the plane started to yaw—about 6 degrees to the left.

Farley frantically tried to correct, which caused the plane to yaw even more dramatically to the other side, creating a sensation of skidding sideways that Farley later described, in classic understatement, as "very uncomfortable in an airplane."

The plane was not acting as it had in the wind tunnel tests. Feeling panicked, Farley turned on the air data probes earlier than he'd planned, and the plane stabilized. Clearly, the design needed some refining.

The takeaway from this first test was that the YF-117's tail fins were too small. Engineers went back to work and designed fins that were 50 percent larger. That worked. Once outfitted with these larger fins, the plane was stable.

But no amount of modifications could change the fact that this radical new fighter simply wasn't easy to fly. Without the computers, the country's best military pilots would have been virtually useless, but even the earliest models proved very dependable, and there wasn't an F-117 crash until April 1982, ten months after that first flight.

On April 20, 1982, Lieutenant Colonel Bob Riedenauer—one of the principal test pilots, who went by the handle Bandit 103—was only 30 feet off the ground after takeoff when the plane pitched violently to one side, flipped over, and crashed, pinning Riedenauer inside the cockpit. He survived but broke both legs and badly injured his back.

The problem? It was human error and not a computer failure. The plane's roll-rate and pitch-rate gyroscopes had been installed incorrectly. They were reversed, which caused the plane to roll 90 degrees left after takeoff instead of climbing.

The subsequent audit nearly ended program director Sherm Mullin's career. It found one hundred problems, including the way in which instruments were installed on the jet and tested on the ground. (Mullin survived the incident and then some. Eight years later, he would succeed Ben Rich as the third president of the Skunk Works, assuming the reins of one of the world's top engineering groups despite never having finished college.)

Those problems were fixed, quality control was buttoned up, and the program marched on, as the number of planes produced and pilots trained grew every month.

Back in October 1979, Colonel Bob "Burner" Jackson, an ex-Thunderbirds pilot on the requirements staff of the Tactical Air Command, had been told to assemble a team of pilots for a very special, highly secret duty: to become America's first stealth fighter unit. Jackson sought out older, experienced pilots, at least captains and preferably majors, who had flown both air-to-ground and air-to-air fighters. They should have at least 1,000 hours in fighters, and ideally twice that, with experience in and comfort with high-performance jets.

He told the candidates very little about the job, just that they had an opportunity to join a top secret unit flying a brand-new fighter, and that they'd be spending a lot of time away from their families. They had five minutes to decide.

Most said yes, and on October 15, 1979, the 4450th Tactical Group was created. The pilots were housed at first in a remote corner of Nellis Air Force Base known as Lake Mead Base or Nellis Area II. Technicians began to arrive, too, thinking they were supposed to take part in avionics testing for a new weapons system on the A-7, a workhorse Navy fighter.

Eventually, one at a time, the pilots were cleared into the Have Blue program—told that they would be the first pilots to fly the most cutting-edge plane on Earth, arguably the most experimental aircraft ever put into production.

Of course, this plane was still in development at the time, in Burbank, so the pilots trained up in the meantime on the A-7, which was chosen because its cockpit layout and avionics were most similar to what the production F-117 was likely to have.

One of the first pilots chosen, Major Al Whitley, was asked to create a flight training program for a plane he'd never even seen. Whitley began to travel back and forth to Burbank, where he'd spend days inside the mysterious Building 311, grilling Ben Rich's engineers about the F-117 and its characteristics.

Here's how a story about the 4450th that ran decades later in *Air Force Magazine* recalled those conversations between Whitley and the Skunk Works engineers: "When discussing its flight profiles, handling, and stealth characteristics, they often used such phrases as 'ought to,' 'should,' or 'probably.' It was clear that the program was a radical departure for all concerned."

The Air Force needed somewhere remote and secret to locate the unit, and it picked an all-but-abandoned airstrip in the high desert about 140 miles northwest of Las Vegas to build a base around.

The airstrip—known as Tonopah—was originally built by Sandia National Labs to support nuclear weapons tests but came with almost no frills.

Just as Kelly's Skunks had at Area 51, in preparation for the U-2, the Air Force scrambled to create a compound at Tonopah, hastily constructing additional runways, maintenance buildings, fuel storage, water tanks, a fire station, and a fully equipped hangar for each plane.

For the first year, staff lived in sixteen winterized mobile homes previously used at a Canadian mining camp.

The first five Air Force flight-test planes arrived at Tonopah in June 1982, just forty-three months after Lockheed was given the production green light by the Air Force. The stealth fighter had by now gained a nickname—the Night Hawk—and each plane in this first batch of Night Hawks was given a call sign, too: Scorpion 1 through Scorpion 5, so named because someone had found a giant scorpion in the program office, and the creature, which strikes without warning, felt like the perfect symbol. The unit called themselves the Baja Scorpions, and a black-on-red scorpion insignia was painted behind each cockpit.

The wing went live on October 15, when Whitley himself became the first pilot to make an operational flight in the F-117 stealth fighter, earning himself the nickname Bandit 1. The next pilot became known as Bandit 2, and so on until the program's end. Sometime later, Whitley was given a plaque that said only: "In Recognition of a Significant Event. October 15, 1982."

The Air Force improved things at Tonopah gradually. The trailers were later replaced by dorms and slightly nicer hotel-style rooms for pilots. The rooms came with blackout curtains, which allowed the pilots to sleep during the day, since all flying was done at night, when the planes couldn't be seen.

But hiding the F-117s wasn't enough. Having a fully equipped and staffed air base with planes that stayed hidden would only arouse suspicions from Soviet analysts looking at satellite imagery. So the Air Force constructed an entire cover story for the base, complete with faux secret jets. It moved a wing of A-7 attack fighters to Tonopah and then leaked a rumor that the planes were equipped with what Whitley described as "super secret atomic antiradar devices."

Each plane was affixed with a painted canister that looked, plausibly, like a piece of secret experimental technology, and the planes were flown regularly over Tonopah in order to show off for the satellites.

The ruse didn't end there, either. Anytime these A-7s deployed to another base, Air Force police would close down the base and surround the field with heavily armed, ominous-looking security vehicles in order to further the ruse.

Meanwhile, pilots of the 4450th lived by regular hours on weekends, and then flipped day and night from Monday to Friday, maintaining the schedule of vampires or, as Ben Rich liked to say, owls.

And their test flights, especially the first ones, were harrowing. A pilot would suit up in the wee hours and be ready to depart by 4:30 a.m. He'd taxi out in the pitch black, with all the base's lights extinguished, and take off in the dark, disconnected from all controls.

This plane, though, was unlike anything they'd ever flown. And for these first flights, in the dark, they had to trust the engineers. Because in this radical new jet, the computer was in charge. There was also no two-seat trainer. Pilots trained in a simulator until they were deemed proficient on the controls, then hopped into a plane and took off.

41

FAREWELL, SWEETHEART

By early 1980, Maryellen Johnson's blood pressure was dangerously low and her spirit had all but broken. Kelly hired a nurse to spend most of the day at the Encino house at her side. Things felt increasingly tenuous, and this rational empiricist was even considering hypnosis to get his wife out of her "suicide tendency." "I really believe much of her stomach & heart problems might be helped by some relief of mental pressures."

Kelly's part-time employment, which was supposed to end in 1978, kept chugging along. It seemed like it might never stop, which was fine with him. But one era was about to come to a close. Kelly was turning seventy in May and would be forced to retire from Lockheed's board of directors.

"Time marches on," he wrote.

At home, Maryellen was spending almost all her time in bed. "For most of the 10 years of our marriage, it was one long disheartening struggle against a siege of failing health," Kelly would later write in his slim memoir. And he wasn't just referring to his wife.

In the decade that the two were married, Kelly suffered from frequent ulcers and an irregular heartbeat, had triple bypass surgery, had a section of his stomach removed, and had one slightly awkward procedure offered up without explanation—the "uncomfortable and not uncomplicated necessity to remove a piece of bamboo I'd accidentally driven into my lower colon." On the plus side, he noted, "the operations were successful and my health was restored."

Kelly kept most of this saga to himself. He rarely talked about his personal life at work, except to Ben Rich, who'd become a close friend outside Lockheed.

By early 1980, the situation was dire. One single day, Maryellen fainted six times, and it seemed that she wouldn't make it through the summer.

She did, but barely. Maryellen Johnson died on October 13, 1980, at home in Encino. "The end of a splendid, terrible battle," Kelly wrote. "I buried her Oct. 16, 1980 in Forest Lawn."

This was the cemetery in Burbank that Kelly had chosen in advance for his wife and for himself. It's on the backside of the Hollywood Hills, abutting Griffith Park, and he had walked the grounds in search of the perfect spot, which he found near the top of a newer section, just ten yards or so from where the grass switches to brush. He could see all of Burbank from this spot, including the airport runways that were once property of Lockheed.

This was the second wife Kelly buried, and as before, he didn't stay single long, which was thanks in part to Maryellen, who had continued a bizarre tradition begun by Althea: She handpicked her successor.

In this case, it was Nancy Powers Horrigan, a close friend and golf buddy from the Lakeside Golf Club in Toluca Lake who had visited the Johnsons a few times at Star Lane with her ex-husband.

As Kelly would later tell it, Nancy was the last person at Maryellen's funeral. She spotted him, alone, lingering on the hill by the grave and went over to keep him company. Nancy had been, he said, a rock of support for both him and his wife.

The two wasted little time. "A scant month after Maryellen's death, I asked her to marry me," he wrote. When Nancy asked—quite rightly—if they were moving a bit too fast, Kelly replied that "life was too short." He said that he had been mourning Maryellen for years as her health failed. "Let's put the past away," he told Nancy. "I don't have time to wait as a mere matter of form. Let's get on with life."

His argument apparently worked. The two married that month, honeymooned in Hawaii, and returned to California to begin "a marvelous life together," he wrote in his log.

Having a preplanned marriage transition plan is, objectively, quite odd. But I can only imagine that Kelly saw it as . . . efficient. That he engineered his romantic life the way he engineered his airplanes—seeing time and energy as exhaustible resources that must not be wasted.

And so, Nancy Horrigan became the third Mrs. Kelly Johnson and the third wife to be nicknamed "Cutie" or "Cutie Pie," perhaps, her son later joked, as a matter of efficiency—so that Kelly didn't accidentally call her the wrong name.

Nancy, like Maryellen, was sporty. The newlyweds played a lot of golf, adopted two "police dogs" (named Wolf and Prince), and spent as much time as possible at Star Lane, which continued to grow in value every year.

Kelly did not like surprises, and he didn't want to leave any behind for Nancy. He was already thinking about how his wealth—in particular, this ranch—might burden the missus after his death.

His plan was to leave Nancy the Encino house, a much smaller Star Lane (with large chunks sold off to lessen the tax burden), and some "other stuff, so she will be wealthy for life." He also planned to give $600,000 to $700,000 to St. Joseph's Hospital, to expand its ICU—which had saved/repaired him more than once—and was pleased that the new tax bill by recently elected president Ronald Reagan "seems very helpful for me to do this."

But other than estate planning, he wasn't acting like a man approaching his end. Kelly was still very much working his ranch, and—feeling healthier than he'd been in years—couldn't resist taking part in the annual roundup, which was always a wild day. He cut hay, fixed windmills, and rebuilt the garage using old ceiling panels he'd dragged home from the Skunk Works.

In August 1981, he bought Nancy a five-hundred-dollar riding horse named Dusty as an inevitable replacement for two of his rapidly aging favorites—My-O-My and Misty. "They are very old!" he wrote, then added: "Who isn't?"

By December, Kelly was pondering My-O-My's mortality, too. His old favorite had become unreliable and, at one point, "almost killed Nancy and I [*sic*]." Which left Kelly no choice but to have his old friend put down.

"My Gawd what a decision," he wrote. "I love our animals so much but *Nancy* more." Then, almost as an afterthought, after counting his calves and complimenting the beauty of his fields, Kelly raised an important question that he'd been unwilling to confront: "I am thinking of retiring for good! Will I?"

There was another big decision that summer regarding Kelly's legacy. He picked a writer named Maggie Smith to co-write his autobiography and got clearance from the CIA and the Air Force to tell at least the broad strokes of his story. By September, he'd sent a draft of his U-2 history to Jim Cunningham, one of his longtime CIA counterparts. He and Smith were off and running.

Life as a consultant, though, was often frustrating. The Skunk Works, under his handpicked successor, was flying again, but Kelly didn't feel essential in the way he had for so long. On the plus side, he was free of responsibility for the bad things. "Glad I'm not on the board now!" he wrote, as Lockheed struggled with the decision of whether or not to abandon the L-1011 TriStar commercial jet program, which would result in a staggering loss of $600 million.

It seemed almost as if Kelly was checking items off a list. Which, knowing him, is completely possible. That fall, he took Nancy east, for a talk about the U-2 at the U.S. Naval Academy, and then went north, via Detroit, for a final tour of his childhood stomping grounds in Ishpeming.

42

OUT OF THE SHADOWS

At 7:00 p.m. on Sunday, October 18, 1982, Kelly Johnson made the most public appearance of his life, when he sat down across from Morley Safer in the ranch house at Star Lane to be interviewed for *60 Minutes*, the most popular show in America.

This interview was a long time coming.

Forty distinct—and in most cases wildly innovative—airplanes could be directly attributed to this one man, Safer explained. "And he's about the last of the one-man designers."

What Safer meant is that it was no longer possible to have the kind of influence Kelly had had over Lockheed and the planes that emerged from his legendary Skunk Works. Airplane design had become a far more complicated process. It involved legions of engineers and designers who used computers instead of pencils. Hyperspecialization was now the norm.

An engineer who designed planes in 1982 came to the industry with, in most cases, an extreme focus on one particular facet of design—aerodynamics, say, or hydraulics. He was almost certain never to ride shotgun on a maiden flight or fly alongside a test pilot even one time in his own design.

The conversation that Safer and Kelly had was about what you'd expect if you knew the man, who rarely if ever bit his tongue, even when he should have. Kelly's comments were barbed, to say the least, and some of what he told Safer is just as true—and stinging—today.

"We're into an era where a committee designs the airplane," Kelly said, and the results, in his opinion, showed it. "You never do anything totally stupid, you never do anything totally bright. You get an average wrong answer. And very expensive."

It wasn't entirely a criticism. Kelly recognized that the planes of the 1980s were far more complicated than the ones he'd built. "The things you can do with an airplane has multiplied greatly. The electronics . . . and the power plants are very complicated."

And for that reason, at least in part, engineers had become specialists. "The average engineer does not get the kind of well-rounded experience that I got because I had stress analysis, weight and balance, flight testing, frontal testing, came through the whole thing and so I could supervise generally every aspect of the airplane and nowadays you have a specialist for everything."

Generalists weren't extinct; they were just rarer. Kelly pointed to Bill Lear, a self-taught electrical engineer and serial inventor who held 120 patents and created both the 8-track music cartridge and the Learjet.

"Most of the people that we have now coming out from engineering schools either specialize immediately" or they chase the money and "want to join the sales department with an unlimited expense account, become the vice president in charge of sales."

Kelly was, on the one hand, a bit of a curmudgeon. A Luddite, if you can say that about the man who built America's first jet, and the world's fastest plane, the SR-71 Blackbird. But he was also right that skipping straight to computers was preventing students from learning certain engineering fundamentals, like drafting.

The end result, he said, was a more expensive process. More men, more computers, more bureaucracy. If the cost of building a single plane were to increase at the same rate at which it was currently rising, from 1982 to the year 2000, "there would be enough money in the budget to buy one airplane for all the services."

Safer seemed eager to talk about the so-called stealth fighter, noting that published stories stated that the plane would be invisible to current radars, but was mostly stymied.

Anything he could freely discuss, Kelly told Safer, was obsolete.

Understandably, Safer was curious about speed. Did the man who created the fastest plane in history envision a Mach 4 plane? Mach 5? Maybe, someday, one that could exceed the speed of light?

Kelly didn't think so. Because it seemed unnecessary. Planes went fast to protect the pilots. To avoid being shot down. And now there were drones. Also, speed like that would cost money. Lots of money. "Too much money for what you get," Kelly said.

Even in his twilight, Kelly Johnson was ahead of his time. He thought that planes would become less important because of satellites. These orbiting communication relays would, in part, obviate our need to go anywhere.

In a sense, he was predicting Zoom calls.

"Have your conferences by TV. You don't have to do all this traveling. It's bad for the airline business, but you only travel when you're going to take a vacation."

Safer asked a question that revealed an assumption about Kelly—an assumption, I think, that we often make about genius types, especially in their twilights. He wanted to know if Kelly counted on his designs being "a hundred percent perfect."

The answer, Kelly said, was in his record. And that record suggested that he was "probably about 51 percent right." There were plenty of disappointments, too.

The XFV—"our vertical rising airplane that would sit on its tail and climb straight up"—came to mind. It worked great up in the air, when the pilot could fly around with no risk. When there was nothing to crash into. Unfortunately, the pilot had to land, too. "You couldn't look over your shoulder when you got close to the ground and guess how high you were," he said. As a result, "that last twenty feet was something."

Kelly was pleased with how he handled this failure. He recognized it early, and admitted defeat. He wrote a letter to the Navy.

Dear Navy, he said. *We're afraid to fly this thing.* "And this is the only time I ever had to eat that much crow on an airplane. And they said, *Well, we agree with you.*"

Kelly didn't put the experiment entirely out of his mind. Later, he realized that there was actually a simple analog fix for the landing problem. The plane could have had a rope that deployed when it was some safe distance off the ground, maybe a few dozen feet. At which point a ground crew could pull the plane down safely to the tarmac.

Safer suggested that there was something "quite insulting" about a plane this sophisticated "being pulled in by a rope." He expected, it seemed, for Kelly to agree.

"Yes, but if that's what you have to do, you do it," Kelly replied.

Kelly Johnson had also been thinking a lot about the Cold War in retirement. He'd made it clear, upon his exit, that America was losing the technology war. The United States may make better planes, in the end, but its programs take longer, and Russia never stops. Being autocratic, the Soviet Union didn't have layers of deciders. It was essentially flat management, which meant that it could decide faster, build faster, and iterate faster.

And you couldn't just focus on better planes, either.

Superior technology in the air did not necessarily advance American superiority because the enemy would just pour money into antiaircraft defenses at the same rate. "It's a losing battle," Kelly told

Safer, and by proxy the Pentagon. "It ends up and proves that the best airplane to be designed is going to be a crop duster."

He wasn't being flip. From where Kelly sat, as a man in his seventies staring at the planet in 1982, a plane with the sophistication of those he first saw on airfields in the 1920s was more valuable to America's health and safety than the SR-71. A crop duster "can feed us, it can keep us healthy, and save our country from starvation, and a lot of other countries. The crop duster is the plane of the future."

Near the end, Safer posed an interesting hypothetical. If Kelly Johnson were twenty and leaving college today, in 1982, would he still go into airplane design?

"I'd probably be inclined to do something else, go into space," Kelly answered. "I lived through a very interesting era—essentially when the Wright brothers were doing their early flying until fifty years later, and the whole aircraft industry has developed, blossomed, and gone through several wars."

The whole of modern aviation history to date, in one man.

On February 3, 1983, a few months after that appearance, Kelly wrote in his ranch log for the first time in a few months. As was often the case, the entry was backdated—here, to November 1, 1982—but was written in a less steady scrawl; Kelly's handwriting, never great, was noticeably shakier.

The date he'd chosen was to note an important event: his second triple bypass heart surgery. "A very rough operation," he wrote, and that wasn't the whole of it. "I had a mild stroke somewhere along the line that badly affected my vision"—the stroke being an unfortunate side effect of the bypass. "Note my poor writing! Nancy drives me to work and elsewhere." Kelly wasn't sure if he'd ever be able to drive again. (He did later pass a test and had his license renewed.)

Nancy was also, occasionally, guest-writing in the log. That

April, she appended an entry Kelly wrote about her roses, which had been planted in a box made of railroad ties, with a note about whom she tended them for. Namely, her predecessors: "Now they are blooming—In memory of *Althea mostly*, & also Maryellen," Nancy wrote. "And in thanks for the love of CLJ!!"

Kelly continued to go into Burbank three half days a week, though poor vision made it increasingly difficult for him to do much. He was given a secretary until the end, and that helped, but there wasn't a lot for her to do, either, other than answer Kelly's phone and respond to mail from fans.

On August 23, 1983, Lockheed held a special lunch on Kelly's behalf and gave him, he wrote, a "50-year pin celebrating my continuous service since 21 August 1933. It's been a glorious career for 50 years."

After a difficult stretch that almost ruined the company, Lockheed was finally on the rebound. The stock split three for one, the L-1011 program was finally, mercifully killed (after a loss of $2.25 billion), and the Skunk Works, Kelly reported, was profitable and "doing fine," with a $1.25 billion backlog of work and the next great military jet—the F-117 Nighthawk—preparing to step out of the shadows.

That September, President Ronald Reagan wrote to Kelly, care of the Skunk Works, with word of another great honor. Reagan planned to give him the National Security Medal, to recognize the ongoing importance of his work. Specifically, Kelly noted, "the intense and successful use of the U-2, TR-1 & SR-71," which had "given the US vital information on the Middle East, Russia, Chad and Central America." The U-2 alone was still making three flights a day over Chad.

It's hard to know how Kelly was spending his free time in these later years, beyond what he was still able to do on the ranch, which was less and less. He golfed, a little; rode, when able; read, a lot; and, to some extent, had begun to curate his legacy and personal collection.

He scrawled inside the cover of *The Ancient World from the Earliest Times to 800 A.D.*, one of Allyn and Bacon's series of school histories, that he'd owned this particular book, its pages long since yellowed, since he first read it at age thirteen. "I've read it 5 times," he wrote on March 2, 1983, and signed his name C. L. "Kelly" Johnson, which seems like the kind of thing a person only does for posterity.

The ranch was still a retreat, but it was also increasingly a burden. Running Star Lane cost the Johnsons $44,000 in 1982, and the ranch took in just $36,000. And with Kelly's health, especially his vision, failing, he could barely even enjoy it anymore.

So in the spring of 1984, Kelly made the very difficult decision to put his beloved Star Lane up for sale, at an asking price of $6 million.

It didn't sell right away. There were inquiries but no good offers. Kelly and Nancy continued to use the place sporadically well into 1985, even as the ranch managers went about the process of selling off the cattle. Not that any of it was easy for Kelly. "The ranch is beautiful—both Nancy and I regret selling it's too much of a thing for her to handle when I am gone," Johnson wrote in August.

In fact, he didn't write it. "I can neither see nor write well enough." That entry had been dictated to Maggie Smith, the co-writer of his autobiography, published that year.

Kelly: More Than My Share of It All is slim and sparsely told. It runs 205 pages in a not-small font and sprints across Kelly's career, omitting plenty of detail. It was, according to Smith, meant to give people who knew and cared about Kelly a look at who he'd been as a person.

"This book is not intended to be a history of aviation nor a documentation of specific aircraft development," Smith wrote in the book's intro. "It is the personal reflection of one man in his time."

43

COMING-OUT PARTY

With Have Blue and Senior Trend, the Skunk Works proved, again, that it could develop, test, and deliver an outrageous new design in complete secrecy, on time, and on budget, or close to it.

But eventually, secrecy became a burden. The Air Force needed to show its stealth fighter to the world so it could more easily implement the new weapon into its plans. It could, for instance, deploy the planes to bases overseas and do more testing in daylight.

The F-117 had been the subject of very public rumors for years—Morley Safer was asking about it on national TV in 1983—but the extreme secrecy under which it was flown and kept placed a great burden on the pilots and their families as well as on the airspace around Tonopah.

So on November 10, 1988, the most revolutionary airplane design since the SR-71 Blackbird was officially unveiled to the public. A Pentagon spokesman said at a news conference that the plane had first flown in 1981 and had been "combat ready" since 1983.

The Air Force shared just a single photograph, of a jet black Nighthawk in flight, banking left. The look was beyond radical, described thus in *The New York Times*: "The shadowy photograph showed it to be boxy and angular, unlike the smooth and sleek machine that previous speculation had suggested."

The so-called stealth fighter looked like nothing ever flown before. But that was it. No more details. No additional photos.

The F-117A (as it was officially designated by the Air Force) was a very special tool. It was a "silver bullet" with one very specific purpose, as Lockheed later explained in a promotional brochure: "a few aircraft penetrating hostile territory at night to destroy specifically designated, high-value targets."

The official Air Force fact sheet put it slightly differently, describing the Nighthawk as a "single-seat fighter designed to penetrate dense threat environments and attack high-value targets with pinpoint accuracy."

But the point in both was the same: This plane was designed to reach and destroy the kinds of targets that were heavily defended and basically impossible to reach by any previous aircraft. And missile guidance wasn't yet at a point where unmanned weapons could serve this important purpose—to cut the head off the enemy snake (by destroying command and control hubs) at the outset of a conflict.

The specificity of the F-117A's mission is why the Air Force ordered so few of them. But tests showed the plane to be more versatile than anyone thought, and the Pentagon began to broaden its potential, increasing its production order to fifty-nine planes.

On July 12, 1990—eight years after delivering the first operational Nighthawks—Lockheed held a ceremony in Burbank to mark the delivery of the fifty-ninth and final F-117A stealth fighter to the Air Force.

In a press release commemorating the event, the company pointed out that the program was ending "ahead of schedule and under budget" and quoted Ben Rich on "an odd looking flying machine" that was a "sterling example of what American ingenuity and hard work can create in response to a critical need." Rich also called the program "a model of the defense procurement process at its best," but recognized privately that the Nighthawk was probably the last plane the Skunk Works would ever make under a process—and conditions—that reflected Kelly Johnson's founding philosophy.

The average cost per unit for those fifty-nine planes was $42.6 million, which, Lockheed bragged, compared "favorably" with fighters that were produced "at rates and in quantities that exceed F-117A production by 10 times or more."

Ben Rich delivered remarks at the actual ceremony, too, standing at a podium on a sweltering afternoon in the San Fernando Valley and staring out at rows filled with Air Force officers, their chests of brass sparkling in the sun. "It's not often that one has the opportunity to develop and field an aircraft that represents a true technological breakthrough," he said, and then summarized this remarkable project as a triumph of the Skunk Works and its methods: A small group, operating with flat management, clear communication, and limited oversight resulted in one of the most revolutionary machines in history—in record time.

Just thirty-one months after Rich got the go-ahead, on June 18, 1981, Have Blue was flying. And twenty-eight months after that, the world's first stealth fighter was operational. Even better, Rich said, "it was done at low cost."

Kelly Johnson, who was too ill to make the event, would have been thrilled.

44

STEALTH

It took fifteen hours and seven aerial refuelings to get the first squadron of F-117A Nighthawks across the Atlantic to Khamis Mushait Air Base, in the mountains of Saudi Arabia. The location, some nine hundred miles from Baghdad, was chosen because it was outside the range of Iraq's Scud missiles.

Something funny happened there in Saudi Arabia, something that convinced Barry Horne, one of the pilots who'd soon be flying the Nighthawk in combat, of the plane's stealth abilities.

Each night, bats would come out and chase the insects that flew around inside the hangars. And the next morning, Horne would see dead bats lying alongside the planes. Bats navigate using sonar, and they just weren't picking up the plane's low RCS tail. So they kept flying smack into it.

Which was reassuring to Horne and his teammates because the stealth fighters, and their elite pilots, were about to get a real-life test.

Just before 3:00 a.m. local time on January 17, 1991, Major Greg Feest penetrated Iraqi airspace in an F-117A and kicked off Operation Desert Storm by destroying an installation that controlled air defense radars around Iraq's capital city, Baghdad, with a 2,000-pound laser-guided bomb.

The first wave of ten F-117As to hit Baghdad were, indeed, invisible to radar and also—because they flew only at night, with no lights—invisible to people on the ground, so the jets' arrival over the city that night was almost paranormal. (It was such a surprise, in fact, that the city's lights were still on when the first bombs fell. Had the planes been detected, all lights would have been extinguished to conceal targets.)

"Those early attacks, along with the next few waves, knocked the eyes and ears out of the Iraqis, so they were blind and deaf," said then Major Joe Salata, who flew one of the planes in that wave. It "really paralyzed them."

Which was a relief to the pilots, who'd been briefed ahead of time that they were flying into the densest assemblage of air defenses on the planet. Saddam Hussein, the men were told, had 16,000 missiles and 3,000 antiaircraft batteries surrounding Baghdad—which, if you want to put that in perspective, was more firepower than the Russians had around Moscow.

One source told *Aerospace Daily* that Iraqi radars, artillery, and surface-to-air missiles didn't even turn on until a plane was "leaving the target area. The only way you know it's been there is things start blowing up."

The Iraqis were completely blindsided; operators on the antiaircraft batteries had no idea what they were even shooting at. They just lit up the skies, firing willy-nilly in every direction with every gun and missile battery that was operational. The rounds went everywhere, and mostly fell back down on residential areas.

"It looked so dense I thought it would be impossible to fly through without at least getting a couple hits," Joe Salata said, and the pilots feared that maybe half the Nighthawks wouldn't make it home.

In fact, every plane survived. And the pilots themselves could hardly believe it. "The only analogy I could think of was being on a ramp above an exploding popcorn factory and not having one kernel hit you," Horne later said. "The law of averages alone would have made that impossible—and so I prayed."

But the Nighthawks weren't done. At 4:00 a.m., the second wave of F-117As swooped in over the city, followed by a third wave. That first night of the Persian Gulf War, twenty-eight of the forty-nine laser-guided bombs hit their targets, destroying Iraqi air defenses and eliminating risk for the non-stealth bombers that would follow.

Nighthawks continued to attack Iraqi targets for days and took out some of Saddam Hussein's most important installations, including the Ministry of Defense, the air force headquarters, the presidential palace and retreat, and, allegedly, a biological weapons plant that Iraqi officials later claimed was a "baby milk factory."

The fight was over almost before it started.

On January 28—eleven days after the first raids—Iraq surrendered, and the F-117A sorties ended. The Nighthawk was, unquestionably, a linchpin of the swift and decisive campaign. It was the only plane to fly over downtown Baghdad and finished the war with 1,669 direct hits and 418 misses over approximately 1,280 combat sorties. Nighthawks comprised just 2 percent of the total Allied sorties but did 40 percent of the damage to Iraqi installations on the ground. And their weapons, guided by lasers to targets marked by infrared sensors, were incredibly accurate. Briefings by Pentagon spokespeople often included videos taken by F-117s that showed the bombs hitting not just specific buildings but specific rooms within them.

"In World War II it could take nine thousand bombs to hit a target the size of an aircraft shelter," Air Force secretary Donald Rice said shortly after the war. "In Vietnam, three hundred. Today we can do it with one laser-guided munition from an F-117."

Not long after the Gulf War ended, Colonel Richard "Butch"

Sheffield, a former Blackbird pilot, took Ben Rich to Washington to meet Brigadier General Buster Glosson, who had helped the Air Force assign targets during the attacks on Baghdad.

Based on what he knew about the F-117s, Glosson gave those fighters the most difficult targets, with the most robust defenses, and he told Rich that he did this with a heavy heart, fully expecting at least a few of those planes to get shot down every night. But then, the first night, they all came back unharmed. And again the second night. And every other night, for forty-three nights.

As Sheffield tells it, Glosson got emotional explaining how this incredible plane, dispatched in waves to attack the most defended targets on Earth, somehow brought back its pilots every time. And that every night, those pilots gained confidence and got even more lethal. As one of them would later tell *60 Minutes*, it was the first time in the history of aerial warfare that pilots were able to focus completely on the task at hand—destroying targets—without being hugely distracted by trying to stay alive.

Rich got emotional hearing Glosson talk, too, and the two men hugged.

Any way you paint it, the F-117 program was an undeniable success. Lockheed got its contract for the fighter from the Air Force in December 1978 and flew the first prototype in June 1981, thirty-one months later. The world's first stealth fighter then went operational in October 1983, less than five years after contracts were signed. Lockheed built a total of fifty-nine F-117s for $2.5 billion, which works out to $42.6 million a plane. It was a bargain by any math, and especially so when you consider that Lockheed built only eight a year during the seven years the plane was under production and that it met every requirement the Air Force had asked for, including reliability and ease of maintenance in the field.

Ben Rich was a popular man in defense circles, and he told everyone who would listen how he did that: by adhering to the Skunk Works methodology, more or less the same one Kelly Johnson laid out when he first hung the circus tent—especially the part about keeping the program small, with minimal oversight outside the core teams at Lockheed and within the Air Force.

Problems are inevitable, especially with advanced technology; it's how you react to those problems that determines whether a program works or spirals out of control.

Bureaucracy, and especially paperwork, was burying innovation, Rich thought. If Kelly Johnson had to build the U-2 under 1990s DOD protocols and report requirements, "he would have dumped the program as not cost-effective."

Ben Rich played politics better than his predecessor, but as was clear from his many Nighthawk victory speeches, bureaucracy annoyed him, too. When the U-2 was built, Kelly Johnson's Skunk Works had fourteen departments. By 1990, coming out of the F-117A program, Rich was overseeing two hundred departments. And the acquisition process that drove Kelly so nuts had only snowballed.

The paperwork, Rich said in 1990, "comes out of your ears," and a favorite talking point in speeches was that the DEA had 1,500 agents—in total—working on drug enforcement in the United States, while the Pentagon employed 27,000 *auditors*. And those auditors were rewarded for identifying errors, even tiny and insignificant ones. "Now that's where we put our values," Rich said. "We're doing it to ourselves."

Kelly's beliefs were imprinted upon Rich. Memos and reports should be short and essential. Many things could be hashed out in a conversation instead of creating what he called "paper pollution." A favorite target of Rich's was the B-2 bomber. One Air Force general told him the program generated 33 million sheets of paper a month—more than a million sheets a day. "No wonder the danged airplane costs $1 billion," he told a reporter in 1994.

Frustrations with oversight also helped drive the Skunk Works out of Burbank. Rich struggled to meet OSHA requirements in fifty-year-old buildings that were thrown up, in large part, during wartime. He was fined for having equipment in the aisles, fined because there wasn't ample storage, and fined for not letting inspectors into certain buildings because they lacked a security clearance.

"Have you ever been to the Pentagon?" Rich cracked to a reporter. "If they sent OSHA there, they'd go to jail!"

OSHA had been a thorn in Rich's side for years by this point, and the lack of communication between federal agencies especially drove him crazy. His favorite example came from the F-117 days, when OSHA sent inspectors on a surprise visit and Rich turned them away because they didn't have top secret clearances.

"The plane was not supposed to exist!" Rich later said. "The program was classified at the absolute highest levels. So [I'd] had no choice but to deny them entry. But the story hit the media with a different spin. That being, 'Lockheed refuses entrance to OSHA inspectors.'"

By 1994, it was time to go, and there was no clearer symbol of the end of the Skunk Works as Kelly imagined it than this move from Burbank. Ben Rich's Skunk Works 2.0 and its 4,000 employees packed up and moved to their new, far larger home, on 542 acres right next to Edwards Air Force Base, in Palmdale, on the other side of the mountains. Directly abutting the land where so many of Kelly Johnson's designs were tested.

The Skunk Works was, by this point, a stand-alone company within Lockheed, with the same nickname and mascot but a new official title: the Advanced Development Company.

One of the largest structures on the new site was the 225,000-

square-foot Kelly Johnson Advanced Development Center, which would house "nearly 1,000 engineering and program management employees," according to press reports, and reporters who attended the building's dedication "were treated to flyovers by a U-2 and an F-117, one of which will also be put on display outside the Johnson building."

45

ON THE FUTURE

Ben Rich may have gotten along better with blue suiters than his mentor, but that was a low bar, and the longer he sat in Kelly Johnson's old chair, the more he, too, grew frustrated with the way military planes—especially fighters—were procured by the Pentagon.

That the Department of Defense had suffocated itself in bureaucracy when it came to fighter production was essentially canon in the aerospace world of the 1980s. As a result of inflation, production delays, and budget cuts, the United States was on the verge of pricing itself out of the combat plane market entirely.

The weight of U.S. designs kept climbing—thanks to an ever-growing list of requirements—and costs had soared, to an average of $40 million a fighter and $300 million per bomber. If that trend continued, Ben Rich told *Aerospace* magazine in 1985, "we're not going to be able to buy any airplanes by the turn of the century."

But Rich didn't just complain. He had one clever and radical solution to offer: The United States should build two types of planes. The first would be for peacetime, by which he meant planes that could fly for years on end as the country stood on a peace footing, training its pilots and providing a deterrent to America's enemies.

The other type of plane, purpose-built for war, would be designed for short-term use. These planes would be stored away only for combat and would make use of cheaper parts—like tires that were certified for one hundred landings, instead of one thousand.

It was a clever solution to a conundrum created by a system that drove costs to untenable levels, making America's fighters almost precious objects that the country literally couldn't afford to lose. Except that, in war, loss was inevitable. You expect it. Plan on it, even.

During the Vietnam War's Operation Linebacker bombing campaign, for instance, the United States lost ten B-52s in seventeen days. But America's latest bomber, the B-1, cost $300 million per plane. The U.S. Navy lost *five* fleet aircraft carriers during World War II. But a carrier in 1980 cost $2.5 billion—more than the GDP of many Third World nations.

Plus, the more complexity you add, the less reliable the weapons. Too much technology can make a plane, or a ship, unwieldy. You have to be a genius to diagnose the problem, let alone fix it. Rich estimated that the "availability of weapon systems" in 1985 was 30 percent. Which meant that you'd need three extremely expensive F-15s just to ensure one was fully capable at any given moment.

This fact, that the United States was continuing to price itself out of the innovation business, was troubling Kelly Johnson, too. Even in retirement, the father of the Skunk Works worried still—more, even—about Pentagon bloat than when he'd been running the place himself, perhaps because he had more time on his hands, or perhaps he just felt a duty to speak up, even if at heart he surely knew it didn't matter. The procurement world in which the Skunk Works of his imagination arose simply didn't exist anymore.

In a talk to Lockheed's marketing group in February 1981, Kelly told the crowd that he was breaking one of his own rules: "Don't try to predict anything in this silly business beyond five years in the future because you invariably will be totally wrong." He needed only the recent past to show how volatile and fast-moving aerospace was.

Even six years earlier, he wouldn't have believed that every new U-2R spy plane would require titanium landing gear, or that the resilient U-2's latest derivative, the TR-1, would cost $5 million more than the initial selling price of the L-1011 commercial passenger jet, or that the "electronic gear" going into some of the latest airframes of Lockheed's planes cost more "than the flying vehicle itself."

Skyrocketing costs had Kelly agreeing with Ben Rich's worry that America's golden age of aviation design was finished, at least in the defense sector. "I do not believe that our government will develop any new aircraft beyond those already on order," he said.

As he often did, Kelly pointed out how bloated bureaucracy played into this. He recalled an early case study of the recently canceled B-1 bomber program. That plane was going to cost a fortune and would take heavy casualties if called into action. The Skunk Works predicted a 37 percent loss rate per mission and shared that data with the Defense Department. This, he said, caused a "three-day argument involving thirty-one people on their side against two on ours." The result of this prolonged argument: "We arrived at a joint agreement that the loss rate would be only thirty-three percent per mission."

Both Kelly and Ben Rich were seeing a battlefield rapidly leveled by technology. Cheap weapons were a legitimate match for astronomically expensive ones. The Yom Kippur War—which saw hundreds of Israeli tanks and jets destroyed, often with light arms—made that abundantly clear.

In retrospect, this was prescient. That problem would worsen over time and would bedevil the Pentagon more and more over subsequent decades.

Kelly didn't want to be a scold. He hammered this point at any opportunity with the hope that decision-makers might listen. Progress had simply stalled, at least in the areas that most attracted Kelly, like structures and aerodynamics. The biggest leaps now were coming in electronics and propulsion. He felt fortunate to have been a part of a golden age in aviation, and the new world wasn't really for him. He'd aged out at the perfect time.

For the most part, Kelly believed, the planes the United States was flying were as good as they needed to be or as good as they were going to get. He thought the B-52 should probably be the last manned bomber ever built, because, in his opinion, the United States was better off using missiles with multiple warheads to reach long-distance targets. And you could just update the P-3, a surveillance workhorse, to combat submarines.

In Kelly's mind, we just weren't likely to see any more evolutionary leaps in airplane manufacture, at least not on the level of the Blackbird or the Nighthawk. Planes might get cheaper, more efficient, and certainly more automated, but we weren't going to reinvent our manned flying machines as Kelly had. Which is a bit of a shame if you consider how much we still had to learn. As an engineer I spoke to while working on this project said to me, when I asked if we were nearing the end of engineering in the skies: "Watch an airplane fly, then watch a bird, and tell me if you think we've reached the limits." In other words: not even close.

Which doesn't mean there's much incentive to get there, at least not with pilots in the seat. During a late-career speech to the Air Force, which had sought his counsel on the subject of what the branch might look like in the year 2000, Kelly said that investing in advanced technology fighters and bombers was a waste of money. A better option, in his opinion, was the unmanned plane—a drone operated by a pilot, from the safety and comfort of a building (or country) far away.

Kelly was an innovator of supersonic flight, the architect of the first and only plane to cruise at Mach 3. He and Lockheed had subsequently tried and failed to win a commercial SST project, and one of the biggest problems with building supersonic planes that carry passengers is the noise—specifically, sonic booms. You just can't have planes that take off every day from major cities and shatter windows around the airport. And you can't expect people to have to travel to remote airports, far from cities, just to use them. The whole point of a supersonic plane is to shorten trips.

Kelly didn't see this problem as impossible, though. He had a simple solution—aerial refueling. He'd helped to pioneer this process for the A-12, and Lockheed-built Blackbirds refueled in the air more than fifteen thousand times over that plane's run. Why not refuel the Concorde over water, far from populations, using retired passenger jets?

That way, the plane could take off light, under normal engine power, fly to some safe distance over the ocean, then gas up and hit the afterburners.

Problem solved.

To an engineer, it made perfect sense. To an untrained ear, it sounded nuts. Was Kelly really, honestly, proposing a *commercial jet* filled with tourists that refueled en route, in the air? "They think people would be afraid," he told one interviewer. "The people would not even know it was going on."

Lockheed had even looked at the possibility of a Mach 6 transport, but it just wasn't feasible. Such a plane was cool to imagine. It would fly at 4,200 mph but would need 30 percent of its fuel just to get to altitude and speed—leaving only about 30 percent for cruising at Mach 6, because you'd need to save the rest of the gas for deceleration and landing. Which meant that you'd build a plane that could use only 30 percent of its fuel for cruising at top speed. The math just didn't make sense.

Kelly was advising America until the end, whether or not anyone asked for it.

46

LAST DAYS

Sometime toward the end of 1986, Kelly had a nasty fall and broke his hip. That's a serious injury at any age. For a man of seventy-six, especially when that man weighs more than 200 pounds, it was devastating. Kelly was admitted to St. Joseph's—the same hospital where he'd had all his previous surgeries, just a few miles from the original Lockheed factory—and was never discharged.

Kelly Johnson spent the next four years at St. Joseph's, in a large suite, fading away, with Nancy at his side. The hip healed, but it didn't matter; the rest of him was failing. As Ben Rich would later say, there was "general physical deterioration," but the real problem was "advancing senility exacerbated by hardening of the arteries to his brain."

The Kelly Johnson who lived in that hospital suite over those final years was only occasionally the guy who had as much impact on aviation, defense, and innovation as anyone in America during his lifetime.

Ben Rich and his second wife, Hilda, were regular visitors, at least for the first couple of years. It wasn't easy for Rich to watch his mentor and idol deteriorate, but he kept coming for Nancy, who almost never left Kelly's side. And when she did, even for a moment, Kelly panicked. "Where's Nancy?" he'd yell. "I need her."

This was not a dignified decline.

"His skin became a white and dry parchment and he suffered from excruciating bedsores," Rich later wrote. "His eyes seemed unfocused and lifeless and he increasingly began to slip in and out of coherence."

During one visit, Kelly seemed to snap to and told Ben that he had an idea for an airplane. He wanted Rich to get Allen Dulles—who'd been dead for years—on the phone. When Rich pointed out that this was impossible, Kelly "became agitated" and ordered him, again, to "get Dulles on the line."

Kelly was in the third year of his stay at St. Joseph's when the SR-71 program officially ended, after twenty-one years and 3,551 operational sorties. No airplane had ever flown higher or faster, or evaded more missiles. "Like every good thing," Ben Rich said, "it must come to an end."

"When you look at it, it is a piece of artwork," says Steve Justice, a young engineer who later rose to run the Skunk Works. "I know its shape was derived by engineering principles, and aerodynamics, and propulsion needs, but the engineers that conceived the plane . . . created something visually that was unlike anything that had been seen before. It's been said that the SR looks part spaceship and part airplane and the pilots are part pilots and part spacemen. It's . . . a sinister black shape that looks like it's doing 1000 mph when it's sitting on the ground."

On January 26, 1990, Ben Rich gave a speech he didn't want to give at Beale Air Force Base, on the occasion of the SR-71's retirement from service, in front of a crowd that was missing one of his "favorite people"—"Kelly Johnson, father of the SR-71, seriously ill in the hospital."

Kelly, I think, would have approved of his protégé's message. Rich told the assembled, which included numerous representatives of the Pentagon, that he understood the budgetary reasons for killing the program, but he took exception to the idea that this plane was no longer needed. "I agree it is expensive, but so is the fire department, the police department, and life insurance. But that doesn't mean you get rid of it. What is freedom worth?!" It had been nearly a quarter century since the first flight of its sister, the A-12, and the SR-71 Blackbird became the first and only combat plane in Air Force history to retire without the loss of a single crew member. Blackbirds had flown "almost 65 million miles, half of them over Mach 3," Rich said, which was the "equivalent to 2,600 trips around the earth or 135 round trips to the moon or two round trips to Venus."

An SR-71 flew from San Diego to Savannah Beach, Georgia, in fifty-nine minutes and, on numerous occasions, Rich said, went "half way around the world and returned—from the US to the Middle East and back; from England to Lebanon and back; from Okinawa to the Persian Gulf and back."

The plane was immaculately engineered. So much so, Rich said, that Kelly Johnson had offered one hundred dollars to anyone at the Skunk Works who could save ten pounds from the final design. No one ever collected. Rich had suggested replacing the air in the tires with helium, then realized the gas would simply leak out through the rubber. His next idea, which Kelly didn't find amusing, was to give each pilot an enema before flying.

One of the last surviving Blackbirds was donated to the National Air and Space Museum of the Smithsonian Institution. This being the world's fastest plane, Lockheed didn't want to just deliver that plane on a truck.

So on March 9, 1990, two pilots took off from Palmdale before sunrise and broke the cross-country speed record, flying from L.A. to Washington, D.C., in 64 minutes, 2 seconds, at an average of 2,144.8 mph.

Rich regretted that the pilots hadn't been allowed to push the bird even more. "We could have done better if we had been allowed to fly different from the handbook," he said. "If it was my last flight, you know what I'd do? After all, it's going to a museum . . ."

But there was still one last farewell to go. When the final flight of the last operational Blackbird was scheduled—for May 8, 1990—Ben Rich called Nancy and asked if she thought it was possible to check Kelly out of the hospital, for even a few hours, to watch his greatest achievement fly for the last time. It just felt like the right thing to do.

All the surviving OXCART team members who were alive and able to get to the Mojave Desert were in Palmdale that day, standing outside the Skunk Works headquarters, and they cheered the arrival of the black limousine they knew held a man no one had seen in years. Rich understood that Kelly was in no condition to see any of his former teammates, or even move. Instead, he stayed in the limo, behind tinted windows.

Nancy and Ben were in there, too, and they let Kelly doze until they heard the roar of the jet's engines approaching the base. Nancy lowered a window and nudged Kelly awake as the SR-71 screamed in over the field, causing two sonic booms "as loud as thunderclaps." Rich looked at his friend and mentor and couldn't tell whether the moment registered or not.

"The pilot was saluting you," he told his friend and then, according to Rich's account, tears welled up in the corners of Kelly's eyes.

Kelly lived another seven months, finally succumbing to his various maladies around 6:30 a.m. on December 22, 1990, one day before Ben Rich retired and handed the keys to longtime engineer Sherm Mullin, who became just the third man ever to lead the Lockheed Skunk Works.

Nancy was there to the end, as was Rosemary Gomez, the nurse who'd helped her tend to Kelly for the last five years of his life. "He was sad every day near the end, but he always would smile or give my hand a little squeeze," Gomez told a reporter. "Those things mean so much to me."

The next day, Lockheed bought a full-page ad in the *Los Angeles Times* to honor Kelly. The ad was mostly blank space, bordered in black, with just a small graphic in the upper right-hand corner—the legendary Skunk Works skunk, with a single tear trickling down his cheek, along with the words "So long, Kelly."

AFTERWORD

LOOKING BACK

Kelly taught me that integrity was the most important thing in all my dealings with people; that none of us are smart enough to lie; don't do anything or build anything that you don't believe in; and communicate with everyone, from top to bottom.

—Ben Rich, in his retirement speech

Though he amassed enough awards—two Presidential Medals of Freedom among them—to fill a very large trophy case, the one Kelly Johnson was most proud of was the 1981 Daniel Guggenheim Medal because it was given for art and design. Kelly cared a lot about function, but he loved form more.

In the official citation for that award, the Guggenheim committee cited both Kelly's engineering record and his other, less-known but possibly more impactful legacy: "For his brilliant design of a wide range of pace-setting, commercial, combat and reconnaissance aircraft, and for his innovative management techniques which developed these aircraft in record time at minimum cost."

Kelly would say at the end of his career that he never worked on a plane he didn't believe in. To do that was to invite failure, because there's no way to do the kind of work he was doing—ambitious and fast—without believing in it.

And he loved all his designs, or at least the ones that worked out. "It is hard to say which design he was most proud of," his third and final wife, Nancy, wrote when asked about her husband's legacy. "He put himself into each one of them so completely. . . . But each design was like a child to him—it had its own uniqueness, its own personality, if you will."

A Lockheed press release states that the company registered Skunk Works as a trademark in 1973 and then added a trademark for the logo itself a few years later. The first official use of the term that anyone could find, according to the company's patent counsel, was on June 23, 1943, but by the 1960s, according to Ben Rich, the outfit's legacy was so widely recognized and admired "that many companies and people, including President Lyndon Johnson, used [the term Skunk Works] generically to connote well-managed programs accomplished in short spans of time and at low cost, following Kelly Johnson's motto: 'Be quick, be quiet, be on time.'"

The first reference, as a descriptor that isn't specific to Lockheed, seems to have appeared in 1976, in the *Harvard Business Review*. The next year, it appeared in *Newsweek* and, I guess, was formally adopted as an English-language phrase when it was added to the *Random House Unabridged Dictionary*. Its definition: "an often secret experimental division, laboratory, project or the like, for producing innovative designs or products, as in the computer or aerospace field."

And today, the term "skunk works" is such shorthand in the business and technology world that it borders on cliché. Seemingly every tech company in America claims to have a Skunk Works, where new ideas are spun up and tested. I kid you not that I once heard the term used unironically at a hearing aid company.

But the Skunk Works, in its original context, still stands for something very clear and bold. "Sometimes science takes place at its own pace," engineer–turned–Skunk Works boss Steve Justice said a few years back on a podcast. "But the development of the Blackbird wasn't like that. It was invention on a schedule. And that's an incredibly tough thing to go do. All too often we'll step back and say, 'Y'know, it's hard because this and this and this' and we talk ourselves out of it. What was magical about these Skunk Work engineers was they just assumed it could be done and determined how you do it. It fundamentally structured how I approach problems after working with the grandfathers of the Skunk Works. The impossible is something you've never seen before and when people look at the Blackbird they see the impossible as reality."

It's a powerful message for all of us, especially in the face of extremely difficult situations: Maybe nothing is impossible.

When Morley Safer sat across from Kelly Johnson for that *60 Minutes* appearance in 1982, even he was skeptical that the Skunk Works, as a place—let alone as an idea—could still exist in the modern world, where late-stage capitalism created behemoth corporations that slaked shareholder thirst for growth by buying more companies, to enter more markets. Safer wondered whether such a place would "ever be possible again with these enormous conglomerates where people are making . . . cupcakes and supersonic missiles . . . and book publishing, all under the same roof."

Could a Skunk Works, a real one, ever happen again?

"I think so," Kelly replied, predicting the future yet again, but only as a cynical ploy, to promote "new ideas." The world—especially in business and government—had just changed too much.

In 1956, when Kelly became chief engineer of Lockheed's California Division, he had 5,500 engineers working under him, and that number astounded him. *Just what the hell were all these people doing?* This wasn't a hypothetical. Kelly really wanted to know. So he set out to understand exactly how much bureaucracy he had accidentally enabled. Specifically, he wanted to know what percentage of that engineering staff "actually put a line on paper to make a drawing, or made a stress analysis, or an aerodynamic report that had anything to do with the hardware."

The answer: Six percent of the engineering man-hours of his employees were devoted to actual engineering. This horrified him. A Skunk Works, using his original model, could operate with 10 to 25 percent of that staff, provided that it was made up of the most adept engineers, because his guys would spend more like 60 percent of their hours on design.

Like so many of Kelly Johnson's ideas, this seems obvious. But it somehow wasn't the norm during much of his life, and still isn't today. Picking your brightest minds, and giving them the runway to stretch out and work—that's how great leaps get made.

Steve Justice came up under Ben Rich, on the F-117, and later ran the Skunk Works. He never overlapped with Kelly, but he worked with him a bit at the end, and the Skunk Works founder's fingerprints were still on so many things around him.

When Justice first saw the logs that Kelly and his engineers kept, he was amazed by the attention to detail, by the sheer precision of the work they were doing—with slide rules!

"The Skunk grandfathers wrote reports that just put us to shame," he says. "They had full text with embedded figures and they did complete analysis of what the data meant, drew conclusions, even under the huge schedule pressures that they had, wrote these incredible reports. Some of the reports on the Blackbird are so detailed about the operation of inlets and nozzles in the propulsion system and the aerodynamics of Mach 3 flight, things that worked as predicted, things that didn't work as predicted."

Justice is still inspired by those reports, and it's very hard to imagine the engineers of today being able to write them. "First of all, it's a tremendous lesson," he says. Also, when his version of the Skunk Works was asked to develop the RATTLRS Mach 3 cruise missile for the Navy (a possible successor to the famous Tomahawk), his engineers *actually used* those reports and that data.

Their design, in fact, was similar in size and shape to both the nacelle of the Blackbird and to Kelly's decades-old D-21 drone. Leveraging this work, Justice says, saved both testing and analysis, "because some of the answers were right there for us. And so that investment in the Blackbird in the late 1950s and early 1960s paid tremendous benefit for us in the early 2000s."

"His successes speak for themselves," Dick Heppe, one of Kelly's longtime engineers, said. "He left an absolutely indelible legend of the importance of first-rate engineering in our business. And I'd like to think that he set a standard for all of the engineering leaders we had over what they strove to live by. So I think one of his greatest contributions was his example that he set that he left behind for others to learn and live by in their times."

This effect of Kelly Johnson's legacy—his methods, more than his machines—was a thing that those closest to him talked about most, even when he was alive. Retired general Leo Geary worked with him for decades, over numerous programs. And when he spoke in front of Kelly, at the 1978 AIAA "Tribute to Kelly Johnson," this is what he spoke most emphatically about.

"The one thing that sticks in my craw—has and will—we just sort of ignore the real legacy that Kelly Johnson has given us," Geary said. "We concern ourselves with what he has done. Which is very proper in its own way. But the most important thing Kelly Johnson has done for any of us, and this country, is *how* he did it."

In other words, his process, not his products.

"How did the Skunk Works work? What made it work?" Geary

continued. "What committed him to do the kind of things he was able to do for the time and dollars? What I'm saying is he presented us with a system that was outside the system. To put it very bluntly, we don't seem to have the brains to take advantage of it. He has shown a method. The biggest problem we have today is people. The most expensive item in any budget is people. Kelly has shown us a way to get the most out of the fewest amount of people, and to get a better job done in a shorter amount of time. Now, when you're talking fewer people and less time, you're talking the ultimate bottom line of a weapon system, and that's dollars."

This is so sensible. Obvious. And to some degree it's the thinking that gave rise to Silicon Valley, when a whole new era of innovation was born in garages and basements, driven by tiny groups of people who didn't work in silos, or have piles of paperwork to file, and could wear all the hats. And even today, we hear echoes of the lessons Kelly Johnson taught us—like, *move fast and break stuff*, or *keep it simple, stupid*, or *remove walls between departments*—repeated ad nauseam by consultants and on tech bro podcasts. But it's not how most companies operate. Success leads to scale; requires it. And with scale comes structure, bureaucracy, and a need for certainty, which can also be thought of as a fear of risk.

That's not the path Kelly Johnson struck. And it's not the way you make giant leaps forward.

"Take the U-2, the SR—go back to the atom bomb, atomic submarine," Leo Geary said. "Every one of these systems has two things in common. All were eminently successful and all were done outside the system. Now, if that doesn't tell us something, I don't know what does. This is the legacy that Kelly has really left us."

Of course, here's the hard truth: This model really only works when you have a single person clearly and commandingly in charge. Someone smart, broadly capable (a human Swiss Army knife), and fearless enough to make hard decisions, without regret, so that people—and projects—can progress. Think Steve Jobs or Robert Oppenheimer. This type of person does not typically rise to the top of existing organizations because he or she threatens the less capable managers above or alongside them. So it sure helps, in Kelly's case, to have been there from the outset. Because while he clearly had the temperament to run through walls, those walls just got thicker and thicker over the years.

Plus, the temperament required—the one Kelly had—doesn't often endear a person to bosses. As Kelly wrote in a thank-you note to Leo Geary for those flattering comments: "I guess of all the things I'm proud of during my career, the thing I prize most is the friendship of you, Dick Bissell and my long suffering Lockheed friends in management who often by present standards should have fired me, but didn't!"

When I went to visit Kelly's grave a few years ago, in L.A.'s Forest Lawn Cemetery, it was Labor Day, and scorching hot. The cemetery sprawls up a hill from the Los Angeles Zoo until it hits Griffith Park and has lovely views of Burbank and the Valley.

It should come as no surprise that Kelly chose the plot himself—on a hill in one of the cemetery's highest points—probably because it would allow him to gaze down upon the valley where he made his name, in perpetuity.

According to Nancy's son, John, Kelly took his mom there on their first date and insisted they walk up the hill. Probably, John's wife said, to test her physical mettle. He was an engineer through and through.

When I made that same walk, a bagpiper was practicing nearby, and three deer hid out from the blast-furnace heat in the shade of a bush at the base of a scrubby mountain that looms over the site.

You can't quite see the runways of Burbank's busy airport from that spot, but there's a great view of the sprawling city, including the area where the Skunk Works was born in a parking lot and existed for decades, until packing up and moving to Palmdale, on the other side of the mountain range that looms across the valley from this spot.

When the Skunk Works began, when Kelly repurposed those boxes and a circus tent to build America's first jet fighter in 143 days, it would have looked far more suburb than city from here. Space was abundant, and there were still farms in the San Fernando Valley.

Now the Skunk Works site is a strip mall, the urban sprawl extends as far as you can see, and it's almost impossible to fathom that Lockheed once designed and built the U-2, Blackbird, and F-117 Nighthawk on the other side of the fence from an airport where thousands of people take off and land every day—on jets like the one I watched ascend out of the sprawl, climb out over the brown crags of the San Gabriel Mountains, and disappear into the smog.

ACKNOWLEDGMENTS

One of the hardest questions anyone can ask a writer is, "What's your next book?" Finding a subject that no other writer has cornered yet, and that can keep your interest piqued for the years it takes to report and write a book, well, that's a daunting task. Fortunately, in this case, someone picked for me. Sort of. I had just finished giving a talk about my last book—*The Taking of K-129*—at CIA headquarters, in Langley, Virginia, and was talking to the Agency's friendly in-house historian, David Robarge, about Kelly Johnson. Kelly plays a small but important role in that book, and the CIA-Lockheed partnership got a whole chapter, and I must have been singing Kelly's praises or something, because Dave just said (as best as I can remember it): *Why don't you write about the Skunk Works for your next book?*

It was, honestly, a great idea, and as soon as I got my phone back from CIA security (which doesn't happen until you leave the premises), I did a search to see what else was out there. Surprisingly little! So, a few days later, I started a proposal. Thanks, Dave! I owe you another talk, this time about spy planes.

No book can be sold without an agent, so fortunately my longtime literary agent, Daniel Greenberg, liked the proposal, too, and sent it to Dutton, which published my last book. My editor, Jess Renheim, was unfortunately on her way out the door to move to Sweden, but her boss, John Parsley, loved the pitch, and bought it. He passed me first to Brent Howard, but life intervened—there was a pandemic, and I also cofounded a podcast company, accidentally becoming an entrepreneur—and it took me so long to write this book that I had to move on to a third editor by the time I was ready. That was Nick Amphlett, who is great, and shaped the final product—after author, friend, and military history buff Kevin Maurer helped me wrestle a first draft to the ground.

Many books have a sort of angel/mensch lurking behind the scenes. Last time, that was David Grann. This time, it's Stephen Trimble, an aviation journalist who saw the announcement of my Kelly Johnson book online somewhere and reached out to say that he, too, was obsessed with this brilliant engineer and was planning a biography of his own. Most versions of this story would result in, at best, a friendly rivalry. Or at worst, some kind of arms race, where Steve and I raced to lock up sources while plotting ways to stymie each other. Steve, instead, wished me well. He was excited, in fact, to hear that I'd be writing the book he long planned and said that he wasn't going to get to it anytime soon himself, because his full-time job at *Aviation Week* was basically all-consuming. But that's not it, either. Steve shared all the research he'd assembled to date (print and digital), and offered to be both a cheerleader and a sounding board—a decision he may now regret, after five years of annoying emails from me asking insanely specific questions only he could understand. Steve also read the manuscript, fixing critical mistakes that surely saved me angry emails from airplane nerds, and introduced me to Kelly's only surviving relative, his stepson, John Horrigan.

John, in turn, welcomed me into his home outside L.A.; served me lunch; and shared files, photographs, and a mounted color print of the Merlin airplane design Kelly drew by hand, as a child, on the front of his sketchbook. It's now hanging on the wall of the game room in my house. Most of the materials his mom, Nancy, had collected from Kelly before her death had already been donated to the unbelievably beautiful Huntington Library, in Pasadena, where I spent a full month during the summer of 2019, digging through (not at all dusty) files and taking thousands of photos on my iPhone that became references for the book. I also used my time at the Huntington Library to have a series of lunches in the shade of the botanical garden's many exotic trees—including one with the author and journalist Annie Jacobsen, who knows as much about Kelly Johnson's secret test base, Area 51, as anyone.

There aren't many Skunks who knew Kelly left anymore, but I did get a tour of the Smithsonian's Udvar-Hazy collection, which includes a Blackbird, from museum docent Buz Carpenter, who flew sixty-five operational missions in an SR-71 and later trained pilots to fly that incredible machine. Former Skunk Works boss Steve Justice drove me around but not onto the premises of today's HQ, in Palmdale, sharing stories of working for Ben Rich and even Kelly when he was just a wee engineer. Keeper of the "Habubrats" Twitter account, Linda Miller (née Sheffield), sent me her father Butch's unpublished book about flying SR-71s for Kelly, as well as numerous anecdotes from Skunk Works program participants she'd met over the years. And for the second straight book, Thornton "TD" Barnes, who runs the Area 51 veterans collective Roadrunners Internationale, was a tremendous resource. That group's website is a little lo-fi, but it's chockablock with fascinating stories.

My uncle Larry, an engineer and pilot, helped me understand and then layman-ize some of the more confusing bits of airplane physics, then read the draft to make sure I wasn't embarrassing myself. He also flew his twin-engine down to Florida to pick up my dad, who's had a rough couple of years, and got him out of that swampy backwater and up to some cooler air in Michigan.

Speaking of my dad, I said at the end of the last book that he needed to stick around for a few more books, and he's lived up to the first part of that bargain, anyway. I didn't expect this book to take six years, either, but he's not off the hook yet. Gonna need your eyes on the next one too, Pops.

When I started this book, in 2019, my sons, Nicky and Charlie, were five and nine. They came to Pasadena with me for that fellowship, along with my wife, Gillian, and we had a fantastic but very hot month in a Mount Washington rental. They're now eleven and fifteen, which is ridiculous and means they're growing too fast, and I'm writing too slowly. Only one of those things seems fixable, unfortunately, but I'm lucky to have those two and their beautiful mom (also a journalist) around. I love you all. Maybe I can do research for the next one in New Zealand?

A NOTE ON SOURCES

Normally, you'd find my notes here, to prove that I diligently researched this book and didn't just make up a bunch of stories about an extraordinary man who, being very secretive for most of his life, wouldn't be hard to invent stuff about. I really put in the work, I swear. But if you read my last book, *The Taking of K-129*, what I'm about to say will sound familiar. I'm not doing notes, at least in the print edition. Kelly Johnson led a very full life, which resulted in a very full book, and to help keep this from becoming what editors call—in technical jargon—a "door stopper," we've decided to move the notes for *The Impossible Factory* to a dedicated website.

So, if you need to know what specific fact comes from which declassified CIA report, or interview with a retired SR-71 pilot, or from one of Kelly's scotch-influenced ranch log posts, please surf your way over to theimpossiblefactory.com, where I'll provide a complete list of notes, as well as additional photos, documents, and other miscellany I've collected along the six-year journey of writing this book. (As well as information on book tours, readings, public appearances, and encounters with retired Blackbirds spotted in the wild.)

Kelly Johnson didn't talk a lot to the press during his life, and when he did, he didn't say much, which is probably why there aren't a half dozen major biographies of the man in the world already. There's only one, his slim autobiography, *Kelly: More Than My Share of It All*, written with Maggie Smith, which helped me verify some things, but was a starting point for reporting journeys far more often than an end point. Any existing Kelly-head will have read it already, but if you're new to the gang, you should absolutely check it out, as well as Ben Rich's far better (and much more fun) *Skunk Works*, written with Leo Janos.

Kelly was, like many geniuses, not the most modest guy. He certainly understood his importance in the world—in aviation, engineering, and management philosophy—so he left a pretty decent archive behind. Most of that archive has been donated by Lockheed or Kelly's only surviving relative, his stepson John Horrigan, and now lives at the spectacular Huntington Library in Pasadena, alongside the papers of many other Skunk Works legends, including the Gross brothers, Ben Rich, and Willis Hawkins. I drew heavily from the detailed logbooks kept on all the Skunk Works programs, as well as Kelly's equally detailed and mostly dull—but occasionally quite revealing—ranch logs. You'll also find just about every memo or letter ever written to or by Kelly, as well as a comprehensive collection of news clippings, photographs, and internal Lockheed publications. Additional Lockheed papers can be found at the Smithsonian Air and Space Museum's Udvar-Hazy branch (just outside Dulles Airport), and at the current home of the Skunk Works, in Palmdale, California (where you can also visit the small but cool Blackbird Airpark, which is the only place in the world where you can see an A-12 and an SR-71 at the same place, as well as a D-21 drone and a U-2).

If you have a correction, fact, or additional story you'd like to share about Kelly or the Skunk Works for possible future editions, please drop me a line via an account I set up specifically for this: TheImpossibleFactory@gmail.com. You can also find me on my social media platforms, but not TikTok, because I'm too old. Just search for @JoshDean66. And if you've gotten this far but still can't get enough of me, why not check out some of my podcasts? More about those, and me, at campsidemedia.com.

SELECT BIBLIOGRAPHY

A full bibliography for *The Impossible Factory* would probably require fifty pages and thus kill way too many trees. It would also turn this already hefty book into a potential murder weapon. But, in short, much of my research relied on the piles of work material and correspondence that Kelly Johnson—thankfully and not surprisingly—left behind. I drew heavily on his letters, memos, and handwritten logs, which he kept quite religiously both at the Skunk Works and at home, on his ranches. I also read hundreds of letters to and from Kelly, as well as to and from his early bosses—the Gross brothers, especially—and piles of *The Lockheed Star*, a newspaper published by the company starting back in its early days. (Most of these Lockheed materials can be found at the aforementioned Huntington Library in Pasadena, California.) For nitty-gritty on the spy plane projects and Area 51, there's also the excellent digital archive kept by the Roadrunners Internationale, at roadrunnersinternationale.com. It is an incredible collection of declassified papers, personal stories, and lists of persons who worked on projects at Area 51. I've included some of the pieces that were most helpful to me, but there are hundreds of articles there that could easily swallow a full day of your time if you're not careful.

Anyway, here's a curated list of some of the most important materials to seek out should your interest be piqued by the story of this fascinating man and place.

SELECT BIBLIOGRAPHY

BOOKS

Allen, R. S. (1993). *Revolution in the Sky: The Lockheeds of Aviation's Golden Age*. Schiffer.

Anderson, J. D., Jr. (2018). *The Grand Designers: The Evolution of the Airplane in the 20th Century.* Cambridge University Press. https://doi.org/10.1017/9780511977565.

Barlett, D. L., & J. B. Steele. (2004). *Howard Hughes: His Life and Madness.* W. W. Norton.

Barnes, T. (2017). *The Secret Genesis of Area 51*. History Press Library Editions.

Beschloss, M. R. (1986). *MAYDAY: Eisenhower, Khrushchev, and the U-2 Affair.* Harper & Row.

Bissell, R. M. (1996). *Reflections of a Cold Warrior: From Yalta to the Bay of Pigs.* Yale University Press.

Boyne, W. J. (1998). *Beyond the Horizons: The Lockheed Story.* St. Martin's Press.

Brown, P. H., & P. H. Broeske. (2004). *Howard Hughes: The Untold Story.* Da Capo Press.

Goodall, J. C. (2021). *75 Years of the Lockheed Martin Skunk Works*. Bloomsbury Publishing.

Graham, R. H. (1996). *SR-71 Revealed: The Inside Story.* Zenith Press.

Graham, R. H. (2002). *SR-71 Blackbird: Stories, Tales, and Legends.* Zenith Press.

Graham, R. H. (2017). *SR-71: The Complete Illustrated History of the Blackbird, the World's Highest, Fastest Plane*. Zenith Press.

Hartung, W. D. (2012). *Prophets of War: Lockheed Martin and the Making of the Military-Industrial Complex*. Bold Type Books.

Jacobsen, A. (2012). *Area 51: An Uncensored History of America's Top Secret Military Base*. Back Bay Books.

Johnson, C. L. (1989). *Kelly: More Than My Share of It All*. Smithsonian Institution.

Killian, J. R. (1977). *Sputnik, Scientists, and Eisenhower: A Memoir of the First Special Assistant to the President for Science and Technology*. MIT Press.

Lovick, E., Jr., & E. Lovick. (2010). *Radar Man: A Personal History of Stealth.* iUniverse.

Merlin, P. W. (2008). *From Archangel to Senior Crown: Design and Development of the Blackbird.* AIAA (American Institute of Aeronautics and Astronautics).

Merlin, P. W. (2023). *Dreamland: The Secret History of Area 51*. Schiffer Military History.

Miller, J. (1995). *Lockheed Martin's Skunk Works: Updated Edition*. Midland Publishing.

Pocock, C. (2000). *The U-2 Spyplane: Toward the Unknown: A New History of the Early Years*. Schiffer Publishing.

Pocock, C. (2005). *50 Years of the U-2: The Complete Illustrated History of the "Dragon Lady."* Schiffer Publishing.

Powers, F. G., & C. Gentry. (2004). *Operation Overflight: A Memoir of the U-2 Incident*. Potomac Books.

Powers, F. G., & K. Dunnavant. (2019). *Spy Pilot: Francis Gary Powers, the U-2 Incident, and a Controversial Cold War Legacy*. Prometheus Books.

Rich, B. R. (1995). "Clarence Leonard (Kelly) Johnson." In *Biographical Memoirs*, vol. 67, p. 221. National Academies Press. https://doi.org/10.17226/4894.

Rich, B. R., & L. Janos. (1996). *Skunk Works: A Personal Memoir of My Years at Lockheed*. Back Bay Books.

Richelson, J. (2002). *The Wizards of Langley: Inside the CIA's Directorate of Science and Technology*. Westview Press.

Suhler, P. A. (2009). *From Rainbow to Gusto: Stealth and the Design of the Lockheed Blackbird*. Library of Flight.

ARTICLES

Anderson, J. D., Jr. (2015, April). Riding the Crest: A History of Michigan's Aerospace Engineering Department. *AIAA Journal* 53, no. 4, 805–16.

80 Years of Lockheed Martin Skunk Works® Innovation. (2023, June 26). Lockheed Martin. https://www.lockheedmartin.com/en-us/news/features/2023/80-years-of-skunk-works-innovation.html.

Head Man at the Skunk Works Talks Management. (1971, June). *Quality Management & Engineering Magazine*.

Johnson, C. L. (1940). Wing Loading, Icing, and Associated Aspects of Modern Transport Design. *Journal of Aeronautical Sciences* 8, no. 2, 43–54.

Kelly Johnson: Architect of Air. (2020, October 1). Lockheed Martin. https://www.lockheedmartin.com/en-us/news/features/history/johnson.html.

Kelly Johnson Looks at Deltas and Swept Designs and Finds . . . Thin Straight Wings Best Up to Mach 2. (1953, December 7). *Aviation Week*.

McIninch, T. P. (1971, Winter). The OXCART Story. *Studies in Intelligence* 15, no. 1, 1–34. https://www.cia.gov/readingroom/docs/CIA-RDP78T03194A000300010011-0.pdf.

Memories of a Lockheed Legend: A Candid Interview with Ben Rich. (2003, May). *Airpower.*

Slattery, C. (2014, August). Secrets of the Skunk Works. *Smithsonian.* https://www.smithsonianmag.com/air-space-magazine/secrets-skunk-works-180952122/.

Straight Wing or Delta for High Speed Flight? (1953, December). *Aviation Age.*

Ulsamer, E. A. (1968, January). "Kelly" Johnson: A Worried Planner. *Space Digest.*

Weinraub, B., & National Security Agency. (1968). Stop Firing You Bastards! The Seizure of the USS Pueblo—The NSA's Version (declassified and uncensored). National Security Agency. https://roadrunnersinternationale.com/AIJ_Stop%20firng%20you%20bastards.pdf.

SPEECHES

Anderson, R. (1990, December 27). Eulogy, Farewell to Clarence L. (Kelly) Johnson.

Johnson, C. L. (1971, April). Application of Aerospace Techniques to Solving Social Problems. To National Academy of Engineering, after receiving Sixth Annual Founders Medal.

Johnson, C. L. (1972, February). Prerequisites for a Successful Skunk Works. Seminar on Prototyping.

Rich, B. R. (1986, January 13). Remarks of Ben R. Rich. Asian Aerospace Technology Symposium Singapore.

WEB PAGES

Black Shield A-12 Missions. (n.d.). Roadrunners Internationale. https://www.roadrunnersinternationale.com/missions.html.

CIA OXCART Pilot Training. (n.d.). Roadrunners Internationale. https://www.roadrunnersinternationale.com/oxcart_pilot_training.html.

CIA U-2 Project Aquatone/Air Force U-2 Project Idealist. https://www.roadrunnersinternationale.com/u-2_cia.html.

Cuba—1962. Roadrunners Internationale. https://roadrunnersinternationale.com/u-2/cuba.html.

How They Got Here: The Challenge of Transporting the A-12s to Area 51. https://roadrunnersinternationale.com/transporting_the_a-12.html.

Managing Lockheed's Skunk Works—Good Science Project. (2024, May 11). https://goodscienceproject.org/articles/managing-lockheeds-skunk-works/.

Merlin, Peter W. *Groom Lake Timeline: The First Fifty Years.* (n.d.). Roadrunners Internationale. https://www.roadrunnersinternationale.com/area51_timeline.html.

Milgrom, Randy. *Kelly Johnson to the Rescue.* (n.d.). University of Michigan Heritage Project. https://heritage.umich.edu/stories/kelly-johnson-to-the-rescue/.

Murray, Frank. *"Suit Up"* (n.d.). Roadrunners Internationale. https://roadrunnersinternationale.com/suitup.html.

Pilots of the U-2 Project Aquatone, Idealists, and the Pilots Today. (n.d.). Roadrunners Internationale. https://m.roadrunnersinternationale.com/u-2/u2_pilots.html.

U-2 Dragonlady. (n.d.). Roadrunners Internationale. https://m.roadrunnersinternationale.com/u-2/u2.html.

U-2 During Cuban Missile Crisis. (n.d.). Roadrunners Internationale. https://www.roadrunnersinternationale.com/u-2_cia.html

VIDEOS

DroneScapes. (2022, September 7). *The Genius Behind Skunk Works: Kelly Johnson's Top Secret Airplane Designs That Will Blow Your Mind* [Video]. YouTube. https://www.youtube.com/watch?v=B1JHGNFU5cQ.

DroneScapes. (2024, January 7). *Genius of the Jet | The Invention of the Jet Engine: Frank Whittle | PART 1* [Video]. YouTube. https://www.youtube.com/watch?v=crRbwtWquvw.

Film Gate. (2019, October 4). *Kelly Johnson Talks About His Greatest Creation the SR-71, Uncut Interview. | Stock Footage* [Video]. YouTube. https://www.youtube.com/watch?v=n8kBiy6RkOs.

Film Gate. (2020, January 19). *Kelly Johnson and Lockheed Story* [Video]. YouTube. https://www.youtube.com/watch?v=bukQbeaP_Cw.

Lockheed Martin. (2023, July 19). *Skunk Works®: Pioneers of Stealth* [Video]. YouTube. https://www.youtube.com/watch?v=9DLL62rEjzM.

National Aviation Hall of Fame. (2019, December 17). *Benjamin Robert, Rich Bio* [Video]. YouTube. https://www.youtube.com/watch?v=50lO2mBnniE.

Vic. (2019, July 31). *60 Minutes—Skunk Works F117.* YouTube. https://www.youtube.com/watch?v=I9wAEjox2WI.

REPORTS

Hildebrant, D., SSgt, USAF, February 1967 through February 1974. (2002). *Time Line of the SR-71.* https://www.roadrunnersinternationale.com/sr-71timeline.pdf.

Johnson, C. L. (n.d.-a). *Prediction of Statistical Longitudinal Stability of Airplanes.* University of Michigan. Undergraduate paper, University of Michigan.

Johnson, C. L. (n.d.-b). *TRIP REPORT—C. L. "KELLY" JOHNSON 6 APRIL 1970 thru 15 APRIL 1970* (SP-5001).

Johnson, C. L. (1953). *Airplane Configurations for High Speed Flight, for Presentation at Society of Automotive Engineers (SAE) National Aeronautic Meeting.* Society of Automotive Engineers.

Johnson, C. L. (1954). *Sighting of a Flying Saucer by Certain Lockheed Corporation Aircraft Personnel on 16 December 1953* (to Commander, Air Technical Intelligence Center, Wright-Patterson Air Force Base, Ohio).

Johnson, C. L. (1960). *TRIP REPORT—ST. JOSEPH'S HOSPITAL Burbank, California Noon Tues. 7/19 to Noon Tues. 7/26/60 To: C. S. Gross.*

Johnson, C. L. (1963). *"Idiot Charts and Related Subjects."*

Johnson, C. L. (1981). *The Story of the Lockheed U-2 Airplane (for Jim Cunningham/CIA).*

Memo to Dr Murray Weiss on "Physical Examination of Clarence L Johnson." (1977).

Monesmith, B. C. (1955). *Memo to ALL Supervision Re: Mandatory 3 Week Leave for CLJ.* Lockheed Martin.

Pedlow, G. W., & D. E. Welzenbach. (1989). *The Central Intelligence Agency and Overhead Reconnaissance: The U-2 and Oxcart Programs, 1954–1974.* Central Intelligence Agency. https://www.archives.gov/files/declassification/iscap/pdf/2014-004-doc01.pdf.

Project Tagboard/Senior Bowl. (n.d.) https://roadrunnersinternationale.com/coldwarstories/Project_Tagboard_story_revised.pdf.

Udvar Hazy Center Training. (2010, June). *SR-71 Overview.* Udvar Hazy Center Training. https://roadrunnersinternationale.com/carpenter/uhc_sr-71_mstr_trn_c-Jun10.pdf.

PODCASTS

Inside Skunk Works. (n.d.). Lockheed Martin. https://podcasts.apple.com/us/podcast/inside-skunk-works/id1350627500.

TV SHOW EPISODE

Kelly Johnson interview on *60 Minutes.* (1982, October 18). CBS-TV.

ORAL HISTORIES/INTERVIEWS

Ed Baldwin on his career at Lockheed, interviewed in his home by R. E. Baldwin, December 6, 1992.

Alan Brown interview by Peter Westwick, November 15, 2010.

Frank Bullock interview with Volker Janssen on August 21, 2009.

John (Jack) C. Duffendack interview by Bill Deverell, Sherman Mullin, and Peter Westwick, June 12, 2008.

Oral History interview with Alan Brown.

Oral History Interview with Dick Heppe.

Lockheed Oral History Project Transcript of W. D. Perreault Interview with M. C. Haddon, February 5, 1982.

Lockheed Oral History Project Transcript of W. D. Perreault Interview with D. T. J. Haughton, January 1982.

Lockheed Oral History Project Transcript of W. D. Perreault Interview with W. M. Hawkins, February 1–2, 1982.

Lockheed Oral History Project Transcript of W. D. Perreault Interview with Hall Hibbard, February 1982.

Lockheed Oral History Project Transcript of W. D. Perreault Interview with V. A. Johnson, May 1982.

Lockheed Oral History Project Transcript of W. D. Perreault Interview with A. W. LeVier, January 26, 1982.

Lockheed Oral History Project Transcript of W. D. Perreault Interview with L. E. Root, February 3, 1982.

Memoir: Dr. Franklin Alanson Rodgers, University of Illinois, February 3, 1995.

Joseph Szep interview with Peter Westwick, November 23, 2010, at the Huntington Library, with Szep's daughter Mary Mingo present.

INDEX

INDEX